HYPERFINE INTERACTIONS (C)

CUSCO - PERÚ
SEPTEMBER 9 - 14

INTIHUATANA
Where the Sun stands still

LACAME '96

UNIVERSIDAD NACIONAL MAYOR
DE SAN MARCOS
1551 - 2001

PROCEEDINGS

OF THE

FIFTH LATIN AMERICAN CONFERENCE

ON

APPLICATIONS OF THE MÖSSBAUER EFFECT

Cusco, PERU
September 9 - 14, 1996

Editor:

V.A. Peña Rodríguez

J.C BALTZER AG, SCIENCE PUBLISHERS, AMSTERDAM, THE NETHERLANDS

PRINTED IN THE NETHERLANDS BY VE & ES DEVENTER BV

COMMITTEES

Chairpersons

Jorge Bravo Cabrejos
Víctor Peña Rodríguez

Latin American Executive Committee

Edilso Reguera Ruiz (CUBA)
Elisa Baggio-Saitovitch (BRAZIL)
Fernando Gonzáles Jimenez (VENEZUELA)
Germán Pérez Alcazar (COLOMBIA)
Roberto Mercader (ARGENTINA)
Víctor A. Peña Rodríguez (PERU)

Latin American Committee

Angel Bustamante Domínguez (PERU)
Bernardo Fernández (PANAMA)
Carlos Diaz Aguila (CUBA)
Celia Saragovi Badler (ARGENTINA)
Hercilio Rechenberg (BRAZIL)
Raúl Gomez (MEXICO)

Organizer

Facultad de Ciencias Físicas. Universidad Nacional Mayor de San Marcos

Local Organizing Committee

Honorio Barrientos
Jorge Bravo Cabrejos
Mateo Márquez Jácome
Máximo Poma Torres

PREFACE

The Fifth Latin American Conference on Applications of the Mössbauer Effect, LACAME'96, held in Cusco, Peru, between September 9 and 14, 1996, was co-hosted by the Physics Sciences Faculty of the San Marcos Major National University of Lima and the San Antonio Abad National University of Cusco. Contributions related to the traditional scientific subjects connected with applications of the Mössbauer Effect were presented in sixteen plenary talks, twenty one selected contributions for oral presentation and eighty-eight posters. The conference was attended by some eighty-five scientists from fourteen countries.

The Proceedings of LACAME '96 consist of two parts. The first one includes invited papers and selected contributed papers for oral presentations, which are published in the dedicated issue of the Journal on Hyperfine Interactions. The second one includes forty-eight selected papers for poster presentation, which are published in a special volume of Hyperfine Interactions C (Conferences). The editor is grateful to the authors and to the referees for their joint effort contributing to the scientific quality of both these proceedings.

We remember that, since 1988 several Latin American research groups, which work with Mössbauer spectroscopy, have met in Latin American Conferences in Rio de Janeiro (1988), Havana (1990), Buenos Aires (1992) and Santiago de Chile (1994). LACAME' 96 was the fifth of its kind in this series. The attendance to these meetings has been of about one hundred physicists coming mainly from Latin American countries. In addition, it has been usual to invite selected physicists from Europe and North America. Presently, in the Latin American scientific community there are about 20 groups working with Mössbauer spectroscopy. In the last 6 years the interactions among these different groups have increased due to these meetings. It has allowed the integration of this scientific community. However, we expect to strengthen these relations more through future new collaborations.

The Latin American scientists attending the Cusco meeting agreed that these conferences are extremely useful in the area and are committed to organize the sixth Conference LACAME '98, in Colombia.

March, 1997 **Víctor A. Peña Rodríguez** (editor, chairman)
Jorge Bravo Cabrejos (chairman)

Acknowledgements

This Conference has been sponsored and organized by The Physical Faculty of the San Marcos Major National University. We are grateful with the collaboration received from the San Antonio Abad University of Cusco.

Financial support has been provided by the following organizations:

Centro Latino Americano de Física (CLAF)
International Centre for Theoretical Physics (ICTP)
Organización Internacional para las Migraciones - Sede Perú (OIM)
Cinsejo Nacional de Ciencia y Tecnología - Perú (CONCYTEC)

CONTENTS

I. SOILS, MINERALOGY, GEOLOGY AND ARCHAEOMETRY

II. Amorphous Materials and Nanocrystalline Materials

III. Chemical Structure and Bonding

IV. CORROSION, CATALYSIS AND SURFACES PHENOMENA

V. MAGNETISM AND MAGNETIC MATERIALS

VI. Physical Metallurgy and Materials Sciences

VII. Solid State Reactions in Mechanical Milling and Ionic Implantation

VIII. SUPERCONDUCTING MATERIALS, THIN FILMS AND ARTIFICIALLY STRUCTURED MATERIALS

I. SOILS, MINERALOGY, GEOLOGY AND ARCHAEOMETRY

MÖSSBAUER STUDY OF THE MODIFICATIONS OF FE MINERALOGY INDUCED BY MANAGEMENT PRACTICES IN LATERITIC SOILS.

C. Saragovi(1), A. Mijovilovich(1) and H.Morrás(2).
(1) Depto de Fisica, Comisión Nacional de Energía Atómica, Argentina.
(2) INTA-CIRN, Instituto de Suelos, 1712, Castelar, Argentina.

ABSTRACT

Iron mineralogy in three different management conditions of an Ultisol soil from Misiones, Argentina, was studied by Mössbauer spectroscopy. Profile 1 corresponds to a virgin soil under the characteristic subtropical forest of the region; profile 2 pertains to a Maté plantation in which Elephant grass was implanted between rows and profile 3 corresponds to the traditional Maté cultivation in which weed control is performed with repeated passages of a disk harrow between rows.

In samples extracted from a depth of 100-150 cm (Bt_2 horizon) of profiles 1 and 3, goethite represents 50% of existing Fe oxides. At variance, a different behaviour is observed in the superficial horizon: the A1 horizon of profile 1 presents 56 % of goethite content, progressively decreasing to 50 % in profile 2 and to 44% in profile 3. These variations of the Fe states are related to modifications of the surface physico- chemical environments, in connection with cultivation practices.

INTRODUCTION

The high Fe content in red soils from Misiones, Argentina plays an important role on particles aggregation and consequently, on the physical behaviour and on the hydraulic dynamic of the soils. Despite of the good structure of these soils, recent studies [1] on an Ultisol showed degradation processes under Yerba Maté (*Ilex paraguariensis*) cultivation as a consequence of mechanical weed control. On the contrary, the Elephant grass (*Penissetum purpureum*) planted between Yerba Maté rows as a green manure and weed competitor, has shown an increase of soil organic matter, porosity and aggregate stability[2]., Modifications on several parameters (pH, cation exchange capacity, base saturation percentage) were also observed, indicating concomitant changes of the physico-chemical environment as a consequence of new physical and biological conditions. These changes suggest the interest on the study of the iron mineralogy of the red soils from Misiones under different management conditions.

EXPERIMENTAL

In this paper, we have studied the Fe mineralogy of an Ultisol (typic Kandihumult) from Campos Ramon, Misiones, Argentina. We have chosen samples from the A_1 and Bt_2 horizons corresponding to different soil profiles. a virgin one under forest (Profile 1), under a Maté plantation with Elephant grass between rows (Profile 2) and under a traditional Maté plantation where weed control is performed by mechanical methods (Profile 3).

We have characterized the samples by XRD and chemical analysis and used the Mössbauer spectroscopy to analyze the Fe mineralogy. The chemical analysis results are shown in Table 1.

The diffractograms were done in a Philips diffractometer with Cu - K_α radiation. Mössbauer spectra were collected in transmission geometry. A Co57 source in Rh matrix was used. The samples were prepared in powder form, mixing with sugar in order to avoid texture effects [3]. Analysis of the spectra was done using the Dist3e program [4] based on the Wivel - Mørup method.

	Profile 1 (virgin soil)		Profile 2 (Yerba Maté + Elephant grass)		Profile 3 (Yerba Maté)	
	A1	Bt_2	A1	Bt_2	A1	Bt_2
Organic matter %	6.7	1.07	4.2	0.9	3.76	0.76
Clays %	62.7	83.9	57.9	87.0	55.4	84.3
pH water (1:2.5)	6.0	5.5	5.0	5.0	5.2	4.9
CEC $Cmol_c\ kg^{-1}$	23.9	17.3	16	14.7	15.8	14.6
Base saturation %	76.5	30.6	43.1	30.6	58.2	25.3

Table 1. Chemical analysis.

RESULTS AND DISCUSSION

Kaolinite and quartz were found in all the soil samples analyzed by XRD. For the identification of the Fe oxides and oxi-hydroxides on the X-ray diffractograms, we have used as the characteristic line for hematite the [012] line, for goethite the [110] line and the [220] one for maghemite-magnetite pair. The other goethite lines were difficult to isolate due to the presence of a large amount of hematite and/or quartz. Our results showed hematite and the maghemite-magnetite pair in two horizons (A_1 and Bt_2) of profiles 1 and 2. Profile 2 Bt_2 horizon showed also some goethite. In both horizons of the profile 3 (degraded soil), we have identified the maghemite-magnetite pair and in the A_1 sample also some hematite. Since the highest intensity [313] line of

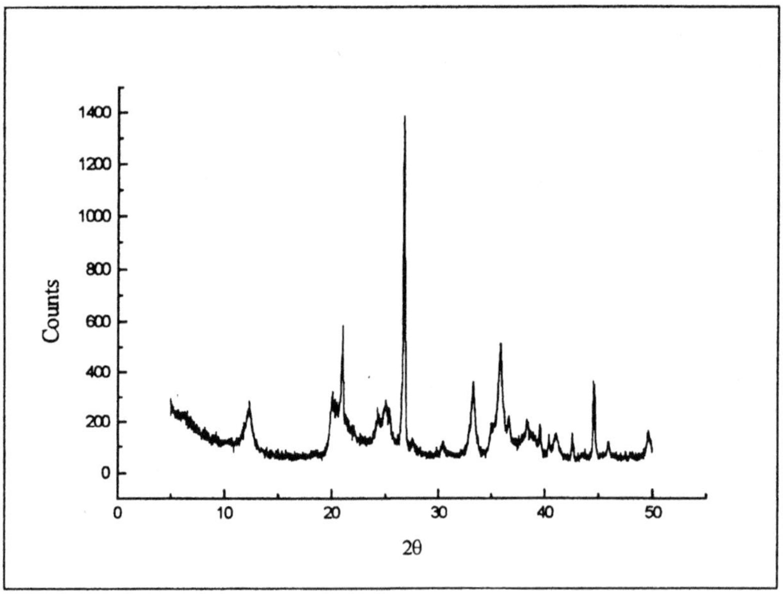

Figure 1. XRD diagram for the Profile 2, A1 horizon.

maghemite and the [311] line of magnetite overlap with the [110] one of hematite, and since the [220] line from maghemite-magnetite pair is a weak one, we have checked the presence of the maghemite-magnetite pair by making use of the magnetic character of these oxides. Using a hand magnet we have concentrated the magnetic compounds of the soils, looking at the produced variation of intensities on the diffractograms.We refer to the magnetite- maghemite pair as we were not able to resolve each one by XRD due to the similar crystal structures of both compounds. In Fig. 1 we show the whole soil X-Ray pattern.

Mössbauer spectra were performed at room temperature and at 15K. The best χ^2 value was obtained when fitting the spectra with three sextet distributions and one doublet distribution. The fitted parameters at 15K, are shown in Table 2 and some characteristic spectra are displayed in Figure 2. We assigned the doublet mainly to kaolinite. The broader sextet (S3) was assigned to goethite and maghemite-magnetite of low crystallinity. S1 and S2 were assigned to antiferromagnetic and weakly ferrimagnetic hematite, respectively. The presence of the weakly ferrimagnetic phase of hematite together with the presence of goethite hampered the identification of maghemite-magnetite even at low temperatures [5]. Also, the separation with hand magnet was unsuccessful in separating the maghemite-magnetite pair from the hematite and goethite.

	S1			S2			S3			D	
Horizon	H_M	Q_S	IS	H_M	Q_S	IS	H_M	Q_S	IS	Q_S	IS
Profile 1											
A_1	51.9	0.21	0.40	51.9	-0.27	0.42	49.5	-0.13	0.39	0.50	0.43
Bt_2	52.1	0.38	0.44	51.9	-0.21	0.42	49.5	-0.17	0.41	0.50	0.42
Profile 2											
A_1	52.1	0.29	0.43	51.9	-0.20	0.45	49.5	-0.18	0.43	0.50	0.44
Bt_2	52.1	0.17	0.38	52.0	-0.28	0.39	49.1	-0.14	0.36	0.50	0.40
Profile 3											
A_1	52.1	0.32	0.53	52.0	-0.21	0.52	49.0	-0.18	0.52	0.50	0.52
Bt_2	51.9	0.24	0.34	51.9	-0.26	0.38	49.6	-0.17	0.35	0.50	0.37

Table 2. Mössbauer parameters at 15K. H_M are in Tesla; QS and IS in mm/s.
IS is refered to Fe metal at RT.

We used the spectral areas at 15K as a measure of the Fe-bearing compounds concentration, mentioning that at low temperature the recoilless fractions *f*, values for all the different compounds become similar [6].

We have computed the relative population from each Mössbauer spectrum. The Bt_2 horizon subareas were normalized to the A_1 ones for each profile. We indicate the total magnetic fraction (Stotal) and the proportion of S3 to the Stotal in Table 3. In spite of there is some maghemite-magnetite involved, we have taken the value of S3 as a measure of the goethite content, and we have analyzed its variation with the different management conditions. For the A_1 horizons, the amount of goethite decreased from a value of about 56% to 50% and to 44% for the profiles 1, 2 and 3 respectively. This behaviour of Mössbauer areas may be related to a parallel decrease of organic matter in the same direction. This result suggests that the increase in organic matter and concomitant physico-chemical modifications produced by the Elephant grass in Profile 2, also produce a modification in the iron mineralogy. The goethite contents in Bt2 horizons of Profile 1 and 3 are similar, while the one corresponding to Profile 2 is lower. Up to the present, it is difficult to explain this value. It is worth while to observe that in the sample with the least goethite content as seen by Mössbauer spectroscopy (Profile 2, horizon Bt_2), we were able to identify directly the goethite by DRX , indicating a better crystallization than in the other samples.

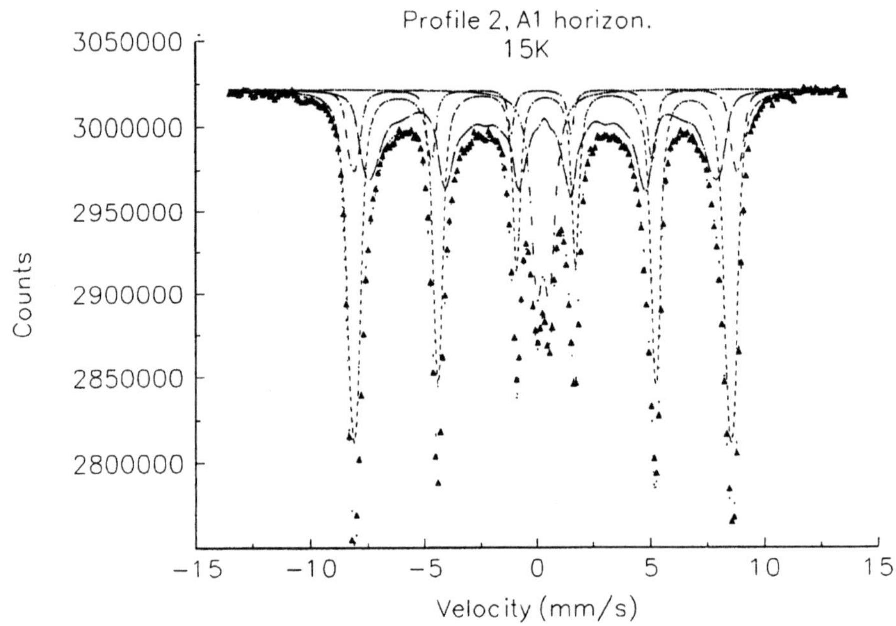

Figure 2. Mössbauer spectra of Profile 2, A1 horizon at 15 K.

Horizon.	S1	S2	S3	D	Stotal	S3/Stotal
			Profile 1			
A_1	5.7	31.5	48.5	14.3	85.7	56.5
Bt_2	3.4	32.6	37.4	10.8	73.4	50.9
			Profile 2			
A_1	4.9	39.2	44.2	11.6	88.4	50.0
Bt_2	8.9	34.7	28.2	10.7	71.8	39.2
			Profile 3			
A_1	5.8	42.5	38.3	13.4	86.6	44.2
Bt_2	5.7	33.9	49	18	88.6	55.3

Table 3. Subareas of the Mossbauer spectra at 15K related to the A_1 one for each profile.

The goethite content in Profile 1 has changed between the superficial and the deeper horizon, which indicates that the virgin forest soil has homogeneous conditions in depth. In Profile 3 there was an important difference between A1 and Bt_2 horizons, which can be explained by the higher exposure to the sun heating effect on bare surface: the dry periods accelerate the crystallization of anhydrous iron oxides [7,8], thus increasing the proportion of hematite in surface horizon. Profile 2 showed that the Elephant grass plantation has modified the goethite content in A1 horizon: this intermediate value may be the result of a slower rate of hematite crystallization as well as a new process of goethite formation , both processes are in agreement with the increase in organic matter and structural modifications which implies in a better retention of water and produces a decrease of soil pH. This interpretation may be related to Chauvel [9] findings, who also observed the transformation of hematite to goethite due to acidification and temporary hydromorphic conditions in ferrallitic soils.

CONCLUSION

Accordingly to preceding studies, an Ultisol from Misiones under different management conditions has shown important modifications on several physical and chemical characteristics of the superficial horizons. In this work, we show that the Fe also presents quali-quantitative differences in the three studied situations. The goethite is predominant on the superficial horizon of the virgin forest soil (Profile 1). On the contrary, on the more degraded soil (Profile 3) the relative proportion of hematite increased. This result, could be related to environmental conditions with periods of higher temperatures and less soil humidity which would promote the hematite crystallization. In the soil with Elephant grass (Profile 2), with an intermediate goethite content, the new environmental conditions produced by this cultivation (higher water retention, lower pH), would preclude the hematite to finish their crystallization and possibly would favor the transformation into goethite. The above mentioned results suggest the existence of very dynamic processes on the Fe mineralogy as a consequence of the ecosystem modifications and the generation of new physico-chemical micro-environments on the soil surfaces.

REFERENCES

[1] *a*. H. Morrás and G. Píccolo,:in XV Congr. Arg.Ciencia del Suelo, Santa Rosa, La Pampa, 1996, p 237; *b*. H. Morrás and G. Píccolo, in Abstracts, 9th Conference of the International Soil Conservation Organization, Bonn ,Germany, 1996, 266

[2] G. Píccolo, in: Resumenes XV Congr. Arg. Cien.del Suelo, Santa Rosa (L.P.) 1996 , 147.

[3] T. Ericsson and R. Wäppling, J. de Physique. Colloque C6, 37 (1976) 719.

[4] R.E. Vandenberghe, DIST3E computer program based on the Wivel- Morup method. Laboratory of Magnetism, University of Gent, Gent. Belgium.

[5] E. Murad, in: Advances in Soil Science, Vol. 12, de B.A. Stewart (Springer-Verlag, New York 1990) p 125.

[6] R.E. Vandenberghe, in: Mössbauer spectroscopy and applications to Geology (Geological Institute, Ghent, Belgium, 1991).

[7] U. Schwertmann, W. Fischer and R. Taylor, in: Trans.10th Int.Congr.Soil Sci 6 (Moscow, 1974) p. 237.

[8] B. Guillet and B. Souchier, in: Pedologie, de Bonneau et Souchier (Masson,Paris 1979) 16.

[9] A. Chauvel, Travaux et Documents de l'ORSTOM,62 (1977) 532 p.

Mössbauer Study of Ceramic Finds from the Galería de las Ofrendas, Chavín de Huántar, Perú

R. Gebhard[1], F. Kauffmann-Doig[2], L. G. Lumbreras[3], J. Riederer[4], F.E. Wagner[5], U. Wagner[5]

[1] Prähistorische Staatssammlung, 80535 München, Germany; [2] Universidad Peruana de Ciencias Aplicadas, UPC, Lima, Perú; [3] Institut für Völkerkunde, 53117 Bonn, Germany; [4] Rathgen-Forschungslabor, 14059 Berlin, Germany; [5] Technische Universität München, 85747 Garching, Germany.

Abstract: A classification of ceramic finds from the Galería de las Ofrendas in Chavín de Huántar was achieved by combining neutron activation analysis, thin section microscopy and Mössbauer spectroscopy. Sherds of different styles largely exhibit different types of Mössbauer patterns. The analyses indicate that the vessels were made from different raw materials and fired in different ways. This supports the archaeological notion that they were brought to Chavín, perhaps on the occasion of a festivity.

1. Introduction

We have studied 70 specimens of ceramics from the Galería de las Ofrendas, Chavín de Huántar, Perú, which was excavated in 1966/67 [1,2]. Although the Galería de las Ofrendas represents a closed context, the styles of the ceramics found there differ considerably from each other and from the local ware. The vessels are supposed to have been brought from other sites on the occasion of a festivity. Mössbauer spectroscopy, thin section microscopy and neutron activation analysis are combined in an attempt to classify the material. Data from a previous study of sherds from the Pyramid Tello and the site of Chavín [3] are included for comparison. A cluster analysis of the neutron activation analysis data yields several distinct groups. Here we report the results of Mössbauer spectroscopy.

2. Results and Discussion

The Mössbauer spectra of the ceramics of the Galería de las Ofrendas exhibit characteristic patterns which depend on the mineral content of the different raw materials and on the methods of firing. The spectra were taken at room temperature using material from the cores of the sherds. In this way effects of surface treatment do not interfere with the investigation. The spectra were sorted into groups according to the nature and fractional intensity of the predominant species. We have defined 8 different types of Mössbauer patterns. The grouping of the samples according to these types is in fair agreement with the groups obtained from the neutron activation analysis data. Mössbauer spectra of types 1 and 2 (Fig. 1) show two intense Fe^{3+} doublets and the magnetic patterns of magnetite and hematite, indicating a nonequilibrium state between oxidizing and reducing firing conditions. The minor contribution from Fe^{2+}-species is more pronounced in type 2. Type 3 contains a distinct third Fe^{3+} quadrupole doublet with a splitting of about 2 mm/s. An identification of the mineral causing this doublet on the basis of its Mössbauer spectrum is difficult. Samples with Mössbauer

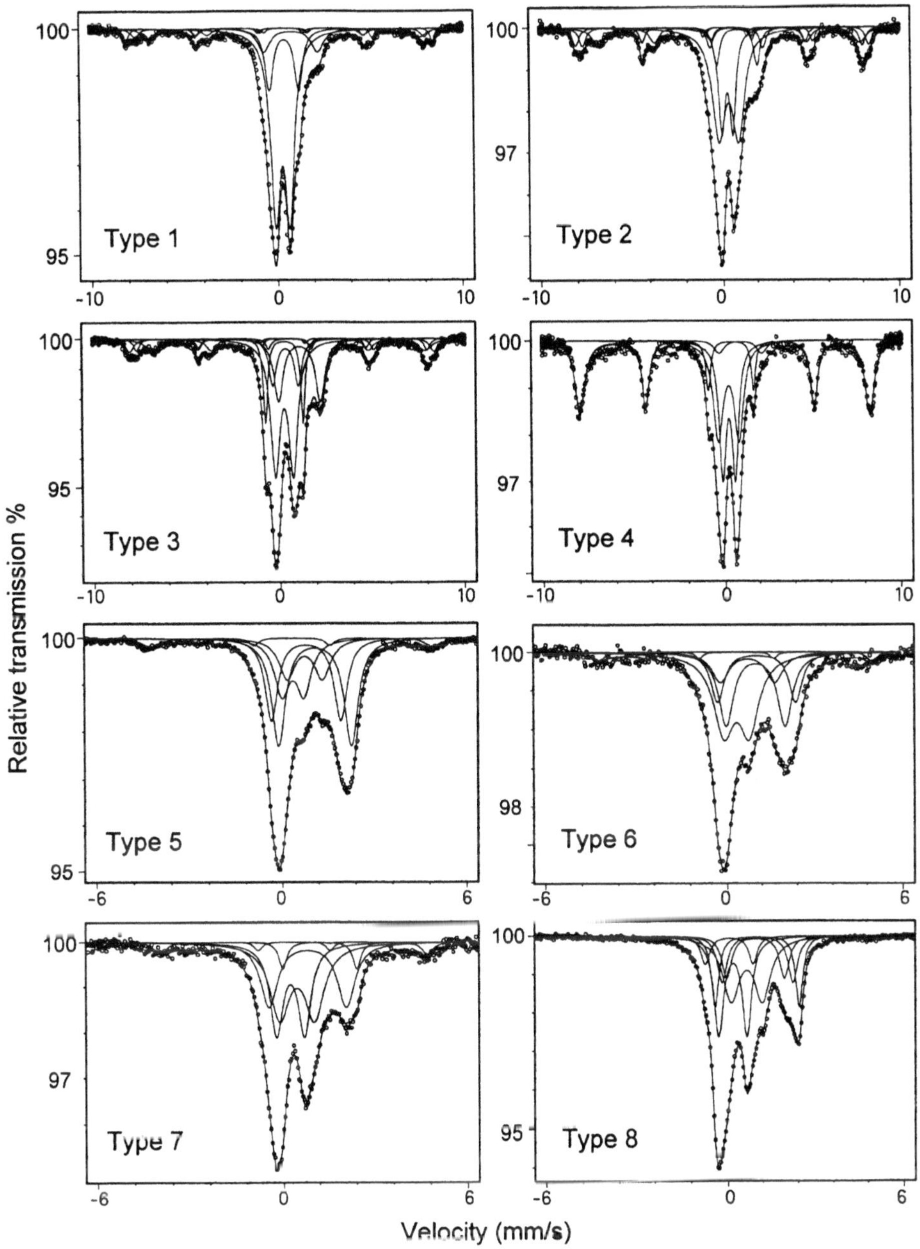

Fig. 1: The eight types of characteristic Mössbauer patterns as observed in material from the cores of sherds from the Galería de las Ofrendas and from the site of Chavín.

Tab. 1: Numbers of sherds of different styles from the Galería de las Ofrendas exhibiting Mössbauer spectra of the types defined in Fig. 1. Specimens having a red surface layer are marked with an *.

MOS-Type	1	2	3	4	5	6	7	8
Wacheqsa	6							
Raku					3	1	1	
Puca Orqo		1*		4*		1	2	
Ofrendas		3	1	1	1*	1	6,8*	
Floral	1		1					1
Dragoniano		1					1	
Qotopukyo			1				3	
Mosna				1			6	
Puksha		4						

spectra of type 4 were fired in an oxidizing environment, as is shown by the presence of differing amounts of hematite and the absence of magnetite. Mössbauer spectra of types 5, 6 and 7 are characteristic for firing in a reducing atmosphere with a brief oxidizing step at the end of the firing cycle [3,4]. The latter causes the surface of the sherds to become red, while the cores remain gray and hardly oxidized. Types 5, 6 and 7 differ in the amounts of Fe^{2+} species and all contain little magnetic material. Type 8, a very characteristic pattern with no magnetic contribution at all, is observed in one specimen from the Galería de las Ofrendas only. However, 11 sherds from the site of Chavín were found to have type 8 Mössbauer patterns. Fig. 2 demonstrates the similarity of the spectra of these 11 sherds and the one from the Galería de las Ofrendas.

The finds from the Galería de las Ofrendas were grouped according to stylistic criteria [1,2]. The names of the different styles and the number of studied sherds exhibiting Mössbauer patterns of certain types are given in Table 1. Wacheqsa, Mosna and Puksha sherds exhibit Mössbauer spectra of types 1, 2 and 7, respectively. Only one specimen of Mosna ware is strongly oxidized and therefore of type 4. Sherds of Wacheqsa, Mosna and Puksha style also form three distinct groups according to the results from neutron activation analysis. Each of these groups is quite homogeneous in its trace element content. Mössbauer spectroscopy shows additionally that the sherds of each group were not only made from the same raw materials, but also fired by the same procedure. This supports the notion that the vessels of each group were made at the same time and location, perhaps for a special occasion.

Sherds of the Ofrendas gray and red ware are highly uniform in appearance, but show Mössbauer spectra of all types except type 8. This is in agreement with the results of neutron activation analysis. Ofrendas ware appears in 7 of the eight groups resulting from a cluster analysis of the element concentrations determined in sherds of all styles. This indicates that Ofrendas style represents a plain ware which was produced from different raw materials, presumably at different sites, and that it was fired

Fig. 2: Three-dimensional plot of the type 8 Mössbauer spectra of 11 sherds found at the site of Chavín (no.1-11) and one sherd from the Galería de las Ofrendas (Of) showing the high degree of similarity of the latter with those from the site of Chavín.

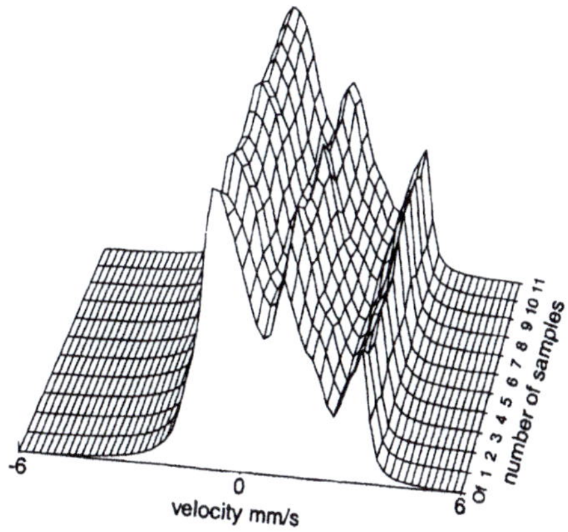

in different ways. Six pieces of Ofrendas ware, all with an oxidized surface and having gray cores of Mössbauer type 7, form a group in the cluster analysis of the element concentrations and therefore seem to be from the same production site. Regardless of surface colour, the majority of sherds of Ofrendas ware was fired in an oxygen-deficient atmosphere, as can be seen from the strongly reduced state of the core. Some pieces were oxidized on the surface during the end of the firing.

According to neutron activation and Mössbauer analysis, 61 finds from the Pyramid Tello [5] form a separate group [6, 7]. By comparison with a recent local clay it could be shown that the finds from the Pyramid Tello are of local production. Most of them were fired in oxidizing conditions. The results reported here, however, support the notion that the vessels deposited in the Galería de las Ofrendas came from other locations. On the other hand, sherds of all types, except Mosna and Ofrendas red ware, were also found at the site of Chavín, indicating that the vessels were not exclusively used as offerings in the Galería des las Ofrendas.

3. Conclusions

This study is being continued. It seems to hold the key to the understanding of Formative ceramics as soon as enough samples have been studied and the different material groups can be attributed to the sites of origin of the vessels.

[1] Lumbreras, G.L. - Chavín de Huantar, Excavaciones en la Galería de las Ofrendas. Mat. zur Allgemeinen und Vergleichenden Archäologie 31, 1993.

[2] Burger, R. - In: Chavín and the Origins of Andean Civilization, Thames Hudson, London, 1992.

[3] Wagner, U., F.E. Wagner, A. Stockklauser, R. Salazar, J. Riederer, F. Kauffmann-Doig, Hyperfine Interactions, 29 (1986) 113-1116.

[4] Wagner, U., R. Gebhard, E. Murad, J. Riederer, I. Shimada, F.E. Wagner. - In: Archaeometry of Pre-Columbian Sites and Artefacts. D.A. Scott and P. Meyers, edts., Getty Institution Press, 1994, p. 67-84.

[5] Kauffmann-Doig, F., M. Gonzales. - In: 24 planos arquitectónicos de Chavín de Huántar, Arqueológicas, Lima, 22 (1993) 10-74.

[6] Wagner, U., R. Gebhard, E. Murad, J. Riederer, I. Shimada, C. Ulbert, F.E. Wagner, A.M. Wippern. - In: Tecnología y Organización de la Producción Cerámica Prehispánica en los Andes. I. Shimada, edt., Pontífica Universidad Católica del Perú, 1994 p.121-156.

[7] Salazar, R., U. Wagner, F.E. Wagner, W. Korschinsky, M. Zahn, J. Riederer, F. Kauffmann-Doig. - In: Proc. of Archaeometry-84, edts., J.S. Olin and M.J. Blackmann, Smithsonian Institution Press, 1986, p. 143-151.

MÖSSBAUER SPECTROSCOPY : ONE CLUE FOR THE DECIPHERING OF OBSIDIAN ARCHAEOLOGICAL ARTEFACTS PROVENIENCE?

S. Petrick, R. B. Scorzelli and A. M. Rossi
Centro Brasileiro de Pesquisas Fisicas
Rua Xavier Sigaud 150, 22290-180 Rio de Janeiro, Brazil.
e-mail : scorza@novell.cat.cbpf.br

G. Poupeau
UPRES-A n°5025 of CNRS
Institut Dolomieu, Université Joseph Fourier, 38031 Grenoble, France.
e-mail : poupeau@ujf-grenoble.fr

A. Seelenfreund
Roberto Peragallo 6644, Las Condes, Santiago, Chile.

We analysed by Mössbauer and EPR spectroscopies obsidians from various Andean and Mediterranean sources. Typical spectra, from three chilean obsidians, are commented. EPR spectra seem to be composed of at least three lines due to Fe^{3+} sites, the relative intensity of these lines varying from sample to sample. Mössbauer spectra may present two different patterns, one due to at least one Fe^{2+} and one Fe^{3+} sites, also with relative intensities varying from sample to sample, the other presenting in addition a magnetic spectra. The use of Mössbauer and EPR techniques in provenience studies of archaeological obsidian artefacts is discussed.

I. INTRODUCTION

With the introduction in the decade of the sixties of physico-chemical methods of characterization of archaeomaterials, a new field opened in archaeology, that of *provenience studies*. Concerning the lithic industries, these studies may, under favorable conditions, show over which geographical extension were distributed, at certain times in the past, the artefacts made out from a raw material taken from some well geographically located geological *source*. This information is then used by the archaeologists to better understand raw materials procurement strategies, ancient trading routes and inter-cultural exchanges.

Obsidian appeared from the beginning of provenience studies as one favorite material for the analyst. First of all, because, due to its mechanical properties, obsidian, a glassy volcanic rock, was widely used during the Neolithic to produce artefacts, from arrow points to various kinds of tools and decorative objects. Second, as a glassy material, obsidian presents a rather homogeneous chemical composition, which makes it easy to characterize chemically, eventually even on a non-destructive basis. Since the variations of chemical composition for the obsidians from a given source (lava flow or volcanic tuff and their erosional remnants) are generally smaller than between obsidians from different sources, it might be possible, at least in regions of easy access, to identify (if it still outcrops) the volcanic mother rock of obsidian artefacts [1].

However, it may happen that the usual geochemical methods of analysis cannot reveal differences in the composition of two obsidian sources [2, 3]. It may also happen that some obsidians present variable compositions inside a given source, because of uneven mixing of two magmas before eruption [4], evolutionary paths of magma chambers [5] or/and eruption dynamics [6]. In such cases, provenience studies must call for other types of characterization, as fission track dating [7] or Mössbauer spectroscopy.

The use of Mössbauer spectroscopy for obsidian sources characterization was first suggested by Herzenberg [8]. The conclusions of a pioneer work by Longworth and Warren [9] on obsidians from Western Mediterranean islands was not very optimistic. More recently, Schmidbauer *et al.* [10], by combining Mössbauer spectroscopy with magnetization studies, showed that obsidian from Eastern Mediterranean islands and that from the Western Mediterranean Lipari island could be distinguished from each other.

During the course of a project on obsidian provenience studies in the andine area, it appeared that geochemical characterization alone could not characterize univoqually the source of artefacts and that some other kind(s) of approach(es), as fission track dating, could be necessary [11, 12].

In the frame of this obsidian sourcing project, and considering the limitations of the purely geochemical approach, we started one program of obsidian characterization relying on Mössbauer and EPR spectroscopies and magnetization studies. We present below some preliminary results.

II. RESULTS AND DISCUSSION

II.1 Mössbauer Spectroscopy

Mössbauer absorption spectra on ^{57}Fe were taken at room temperature in a transmission geometry, using a $^{57}Co/Rh$ source.

The spectra of the analyzed samples show two basically different patterns. The most common one, illustrated in figure 1 by two obsidians from Chili, from respectively Radal (7 Tazas, VII region) and Pretil (Laguna del Maule, VII region) presents a broad asymmetrical doublet, due to the superposition of at least two quadrupole doublets corresponding to Fe^{2+} and Fe^{3+} sites respectively, as observed by other authors [9, 10]. The second pattern, again illustrated by one chilean obsidian, from Estero La Plata (VII region), shows an important magnetic contribution in addition to the previously mentioned Fe^{2+} and Fe^{3+} quadrupole doublets.

The hyperfine parameters corresponding to the Fe^{2+} doublets present small variations between samples and their large linewidths $\geq$0.60 mm/s can be interpreted in terms of a distribution of non-identical sites. The Fe^{3+}/Fe^{2+} ratio shows considerable differences from one sample to another. In the context of obsidian provenience studies, these variations might contribute to the characterization of obsidian sources.

II.2 EPR spectroscopy

EPR studies were carried out at room temperature in a Bruker ESP300 spectrometer using microwave frequency of 9.5 GHz, a microwave power of 1 mW and a magnetic field modulation amplitude of 0.35 mT.

EPR spectra of samples from various obsidian sources from the Mediterranean area and the Andes were obtained. In general, they are composed of lines belonging to at least three paramagnetic Fe^{3+} species. The EPR spectra presented in figure 2 correspond to the three obsidians whose Mössbauer spectra are given in figure 1. The species at g=4.2, with an 12 mT linewidth, is attributed to a structural iron in rhombically distorted sites in the silicate structure. Broad lines (>200 mT linewidth) at g=2.7 and 2.00 can be associated to iron oxides, antiferromagnetically coupled Fe^{3+} and Fe^{2+} and/or to precipitate ferrimagnetic iron phases. The relative concentrations of these species varies from sample to sample.

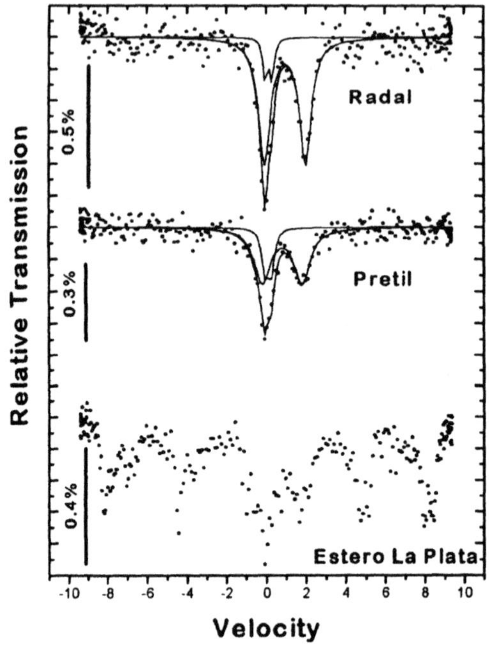

Figure 1.- Mössbauer spectra at room temperature of obsidians from Radal, Pretil and Estero La Plata (Chile).

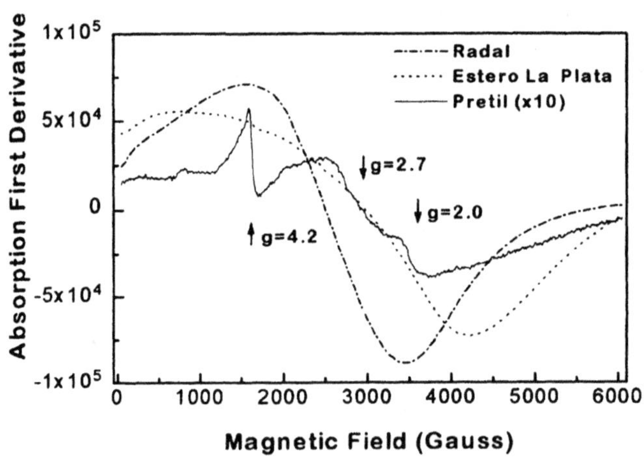

Figure 2.- EPR spectra at room temperature in X-band of obsidians from Radal, Pretil and Estero La Plata.

III. CONCLUSIONS

The Mössbauer and EPR spectra of obsidians from various sources, among which andine ones, exhibit complex structures which reveal that iron atoms may have different coordination geometries with their surroundings ions in the silicate structure. They also inform us about the presence of crystalline inclusions or precipitates. The observed variability between obsidians from different sources of the Fe^{3+}/Fe^{2+} ratio seen by Mössbauer spectroscopy and the relative abundance of Fe^{3+} in three paramagnetic sites obtained by EPR suggest a possible use of these observations in provenience studies.

Further characterization of structural Fe sites in obsidians will be pursued using EPR in X- and Q-bands, ENDOR and Mössbauer spectroscopies, with and without sample thermal treatment, in association with magnetization studies.

The suitability of this overall approach for obsidian provenience studies, combined with the back scattering electrons petrography of obsidians, will be tested by studying the intra-source versus inter-sources variability of the parameters involved in selected volcanic areas of interest.

REFERENCES

[1] J. R. Cann, *in* Petrology of Archaeological Artefacts, D. R. C. Kempe and A. P. Harvey Eds., Clarendon Press, Oxford (1983) 227.

[2] B. Gratuze, J.-N. Barrandon, K. Al Isa and M.-C. Cauvin, Archaeometry 35 (1993) 11.

[3] V. Francaviglia, Paleorient, in press.

[4] H. R. Bowman, F. Asaro and I. Perlman J. Geol., 81 (1973) 312.

[5] D. I. Godfrey-Smith, J. Kronfeld, A. Strull and J. M. d'Auria, Geoarchaeology 8 (1993) 385.

[6] R. E. Hughes and R. L. Smith, Geol. Soc. Amer. Special Paper 283 (1993) 79.

[7] G. Bigazzi, Z. Yegingil, T. Ercan, M. Oddone and M. Ozdogan, Miner. Petrogr. Acta 37 (1992) 17.

[8] C. L. Herzenberg, Mossbauer Effect Methodology 5 (1970) 209.

[9] G. Longworth and S. E. Warren, J. Archaeol. Sci. 6 (1979) 179.

[10] E. Schmidbauer, E. Mosheim and N. Semioschina, Phys. Chem. Minerals 13 (1986) 256.

[11] L. Bellot-Gurlet, Th. Calligaro, O. Dorighel, J.-C. Dran, G. Poupeau and J. Salomon, Intern. Assoc. Obsidian Studies Bull. 16 (1996) 3.

[12] G. Poupeau, L. Bellot-Gurlet, O. Dorighel, Th. Calligaro, J.-C. Dran and J. Salomon, C. R. Acad. Sci. Paris, 322, série IIa (1996) 443.

IDENTIFICATION OF THE IRON COMPOUNDS FROM THE NIGER (CM2) METEORITE

T. V. V. Costa[1], V. W. Vieira[1], M. A. B. de Araújo[1], I. Souza Azevedo[2] and R. B. Scorzelli[2]
1 Depto. de Física, Universidade Federal do Ceará, 60451-970 Fortaleza, Brazil
2 Centro Brasileiro de Pesquisas Físicas, 22290-970 Rio de Janeiro, Brazil

Abstract

The Niger meteorite is classified as a CM2 carbonaceous chondrite and its composition is very similar to that of the Murchison (CM2) meteorite. ^{57}Fe Mössbauer spectroscopy, nevertheless reveals an intriguing difference between these two meteorites: the Mössbauer spectrum of Murchison at room temperature hardly reveals the presence of magnetite (Fe_3O_4), whereas in the Niger spectrum ca. 50% molar of the iron can be attributted to non-stoichiometric magnetite. This large amount of magnetite has two interesting features: i) it permits a more precise determination, by Mössbauer spectroscopy, of the characteristics of this mineral in the CM meteorites; ii) it points out to the question of the relation (if there is any) between the amount of magnetite and the different degrees of hydrothermal alteration shown by the meteorites. A detailed investigation of Niger by means of X-ray diffraction and Mössbauer spectroscopy at low temperatures is being carried out in order to identify the paramagnetic and superparamagnetic phases (probably cronstedtite and tochilinite, respectively), to characterize the magnetite and to verify the eventual presence of maghemite associated to magnetite.

Introduction

Several authors [1, 2, 3, 4, 5] have investigated the carbonaceous chondrites by means of the Mössbauer spectroscopy in order to determine the mineralogy, the oxidation states and the crystal chemistry of iron in these meteorites. In the present work we intend to better understand the mineralogy and the oxidation states of the iron-bearing assemblages in the Niger meteorite. Niger is a typical CM2 carbonaceous chondrite and its chemical composition is very similar to that of the Murchison (CM2) meteorite. Despite this similarity the iron mineralogy of Niger differs from that of Murchison, specially by the presence of appreciably larger amounts of magnetite.

Previous Mössbauer measurements of CM carbonaceous chondrites [2, 3, 4, 6] confirmed the presence of phylosilicates and magnetite. Most of the Fe-rich phylosilicate in Murchison was identified as cronstedtite with the ideal formula $[Mg,Si,Fe^{3+}]_2[Fe^{2+},Fe^{3+}]_3O_5(OH)_4$, by means of transmission electron microscopy and electron microprobe [6].

The Murchison meteorite also contains a non-stoichiometric iron sulfide phase, initially called FeSON, with the approximated formula $Fe_{1\ 3}SO_{1\ 4}Ni_{0\ 1}$. This phase has a structure similar to that of iron-rich tochilinite $\{6Fe_{0\ 9}S_5(Fe, Mg)(OH)_2\}$ [7].

Mössbauer spectra of CM2 meteorites revealed the presence of tochilinite [5]. The Mössbauer spectrum of Niger measured at 30K suggests that the most abundant iron minerals in this meteorite are magnetite, cronstedtite and tochilinite.

Discussion

a) paramagnetic compounds

Figure 1 shows the room temperature Mössbauer spectrum of the Niger meteorite. As previously reported by Scorzelli et al [4], the spectrum indicates the presence of two magnetically splitted components due to iron in tetrahedral (A) and octahedral (B) sites of magnetite, and two quadrupole doublets due to paramagnetic Fe^{2+} and Fe^{3+} compounds.
These paramagnetic phases represent about 52% total area under the spectrum.

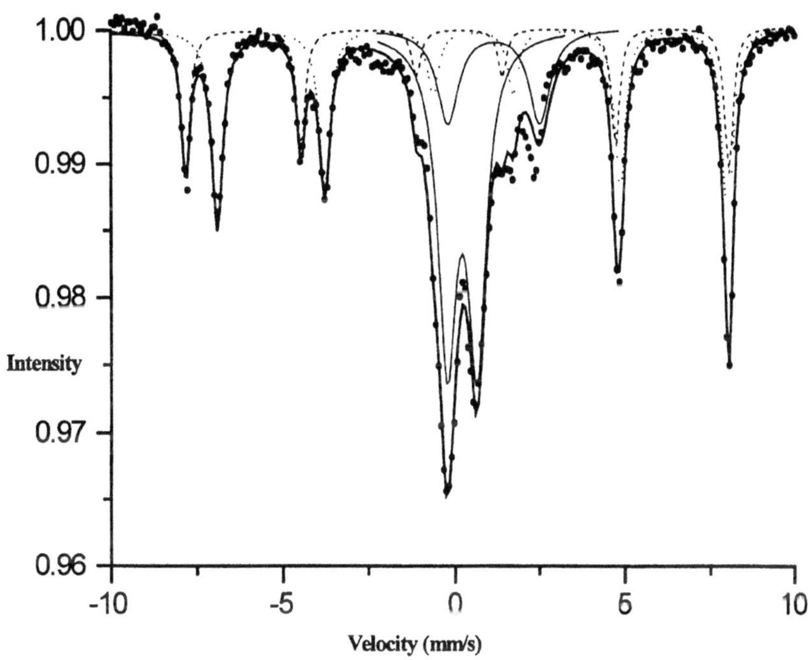

Fig. 1 - Room temperature Mössbauer spectrum of the Niger meteorite.

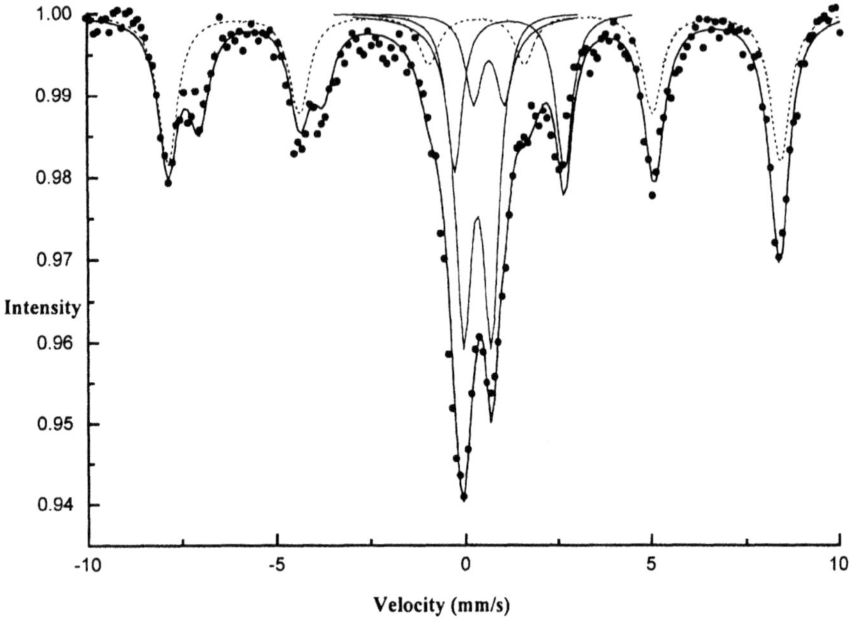

Fig. 2 - Mössbauer spectrum of the Niger meteorite measured at 30K.

Figure 2 shows the Mössbauer spectrum of Niger, measured at 30K. Scorzelli et al [4] investigated the meteorite through the temperature range 4.2K≤T≤300K and suggested that about 15% molar of the iron in the meteorite is in the form of a superparamagnetic compound.

Superparamagnetic components were observed in the Mössbauer spectra of Murchison (CM2) and Orgueil (CI) [2,3]. It was suggested that the component responsible for the superparamagnetic behaviour is the FeSON phase.

We measured the Mössbauer spectrum of Niger at 30K looking for the presence of tochilinite in this meteorite. According to Burns [5], at that temperature it is possible to identify coexisting cronstedtite and tochilinite. The close proximity of Fe^{2+} and Fe^{3+} ions in the octahedral sites of cronstedtite favours thermally induced electron-hopping in cronstedtite so that Fe^{2+} and Fe^{3+} ions are irresolvable in the room temperature spectrum [8]. At 30K the electron-hopping should be over and we do not have yet the magnetic ordering of cronstedtite (<10K) which produces a very complicated Mössbauer spectrum.

The paramagnetic part of the spectrum measured at 30K was fitted with three doublets (Table I) which consist of one Fe^{3+} and one Fe^{2+}, assigned to iron cations in cronstedtite and a third doublet which is assigneable to a iron sulphur phase, tochilinite. The fitted parameters for this phase are very close those reported values for other CM chondrites [5], δ=0.67mm/s-0.72mm/s and Δ=0.76mm/s-0.88mm/s.

Table I - Paramagnetic Mössbauer paramenters of Niger iron compounds, measured at 30K

Doublet	δ_{Fe}(mm/s)	QS (mm/s)	Mineral	A/A_{total}
1	0.33	0.75	cronstedtite	0.27
2	1.21	2.97	cronstedtite	0.13
3	0.67	0.84	tochilinite	0.07

The peak areas from the Mössbauer spectrum suggest that the (Fe,S)-rich phase represents about 7.4% molar of the total iron and ≈15.6% molar of the paramagnetic iron. In the meteorite Murchison, the (Fe,S)-rich phase represents ≈14.4% molar of the total iron. If this iron-sulphur phase is responsible for the observed superparamagnetic effects in the Niger meteorite, as related by Scorzelli et al [4]., the relative proportion of this phase to the total iron is being subestimated here. Nevertheless, our experiments can not testify for the existence of superparamagnetic effects (figure 2). There seems to be a slight increase in the area of the magnetic spectrum at low temperatures but the difference is within the experimental error.

b) magnetic compounds

Stoichiometric magnetite is usually represented under the formula $[Fe^{3+}]_{tet}[[Fe^{2+},Fe^{3+}]_{oct}O_4$, where tet stands for sites with tetrahehedral coordination and oct represents sites with octahedral coordination. At temperatures above 120K there is rapid exchange of electrons between Fe^{3+} and Fe^{2+} in octahedral sites. Thus the Mössbauer spectrum of stoichiometric magnetite shows two magnetic sextets, one corresponding to Fe^{3+} in the tetrahedral sites, and the other being assigned to iron in the octahedral sites with a intermediary valence, $Fe^{2.5+}$. The area ratio Fe(oct)/Fe(tet) in stoichiometric magnetite is close to 2.

In non-stoichiometric magnetite the introduction of vacancies or substitutional cations seems to favour a pair localized electron exchange process in which the non-paired ions do not participate in the electron-hopping [9].

The magnetic lines present in the room temperature Mössbauer spectrum of Niger (figure 1) can easely be identified as the two sextet pattern of the magnetite (Table II). The sextet (I) with a larger hyperfine field can be assigned to Fe^{3+} in the tetrahedral sites and unpaired Fe^{3+} in the octahedral sites which do not participate in the electron exchange. The sextet (II) with a smaller hyperfine field corresponds to $Fe^{2.5+}$ in the octahedral sites.

Table II - Mössbauer parameters for Niger magnetite at 300K

Sextet	δ_{Fe}(mm/s)	QS(mm/s)	B(Tesla)	A/A_{total}
1	0.125	-0.021	49.22	0.17
2	0.547	0.009	46.27	0.30

The area ratio of these two sextets is Fe(II)/Fe(I) ≈ 1.75, which indicates that magnetite in Niger is closer to stoichiometric than the magnetite from other carbonaceous chondrites. In magnetite from Orgueil, for example, that ratio is about 1.35 [10]. A detailed investigation of the magnetite at low temperature is being carried on.

References

[1] T. J. Wdowiak and D. G. Agresti, Nature **311**, 140(1984).

[2] M. B. Madsen S. Mørup, T. V. V. Costa, J. M. Knudsen and M. Olsen, Nature **321**, 501(1986).

[3] M. B. Madsen S. Mørup, T. V. V. Costa, J. M. Knudsen and M. Olsen, Hyp. Inter. **41**, 827(1988).

[4] R. B. Scorzelli, I. S. Azevedo and J. Danon, Proceed. LACAME-88, pp. 373, World Scientific, Singapore (1990).
[5] R. G. Burns and D. S. Fisher, Hyp. Inter. **91**, 571(1994).
[6] K. Tomeoka and P. R. Buseck, Geochim. Cosmochem. Acta **49**, 2149(1985).
[7] I. D. R. Mackinnon and M. Zolensky, Nature **309**, 240(1984).
[8] J. M. D. Coey, A. Moukarika and C. M. McDonagh, Sol. State Commun. **41**, 797(1982).
[9] J. M. Daniels and A. Rosencwaig, J. Phys. Chem. Solids **30**, 1561(1969).
[10] M. B. Madsen S. Mørup and J. M. Knudsen, Hyp. Inter. **50**, 659(1989).

MINERALOGICAL ANALYSIS OF BLACK SANDS FROM ZARZAL FORMATION IN COLOMBIA

M. F. Ruiz, G. A. Pérez Alcázar, A. Bohórquez and J. D. Fabris*

Departamento de Física, Universidad del Valle, A. A. 25360, Cali, Colombia

*CNPMS / EMBRAPA, Sete Lagoas, MG, Brazil

Eleven samples of black sands were collected from the Zarzal Formation in Colombia and studied by Mössbabuer spectroscopy and x-ray diffraction at room temperature. By x-ray diffraction it was possible to detect the structures corresponding to hypersthene, magnetite and albite. These structures are in agreement with EDAX analysis which detected the majority presence of Fe, O, Si, Mg and Al. Mössbabuer studies of in situ samples showed that the major phase (near 70% of spectral area) is hypersthene which is present as two doublets one with a quadrupole splitting of 2.21 mm/s and isomer shift 1.23 mm/s and other with a quadrupole splitting of 2.10 mm/s and isomer shift of1.12 mm/s. Previous works showed that these two doublets correspond to the two octahedral sites of the Fe^{2+} atoms in its orthorhombic structure. Also, the presence of a little quantity (near 5% of the spectral area) of a magnetite phase was detected in magnetically separated samples with hyperfine fields near to 457 and 499 kOe, respectively. An additional doublet was obtained with a quadrupole splitting near 0.62 mm/s and isomer shift near 0.43 mm/s, which can be attribute to the albite phase.

1. INTRODUCTION

In South America black sands are very common in some beaches, like those of Espirito Santo State (Brazil) or the Pacific Coast of Colombia and Ecuador. Black sands from Brazil were earlier studied [1] by Mössbauer spectroscopy, x-ray diffraction and x-ray fluorescence. Those studies showed that a major quantity of hypersthene that appears as an orthorhombic cell in x-ray diffraction and two doublets in Mössbauer spectra. It was also detected impurities of ilmenite.

Orthopyroxenes are single chain silicate mineral with orthorhombic crystal structu [2,3] and chemical composition $(Fe_xMg_{1-x})_2Si_2O_6$ [2]. The hypersthene, $(Mg,Fe)SiO_3$ is one of these orthopyroxenes consisting of chains of SiO_3 groups linked by bands of cations extending along C axis and two cations sites, M_1 and M_2, of six-fold coordination with oxygen ions over which the available Fe^{2+} ions are distributed [3,4]. The M_1 site has an octhahedral symmetry with the cation lying almost at the center of symmetry of the site, but the M_2 site is considerably distorted from octahedral symmetry and is non centrosymmetric. Also the M_2 site is more enriched in Fe^{2+} atoms that the M_1 sites and metal oxygen distances increase with rising Fe^{2+} contents at both sites.

In this paper we report the study by Mössbauer spectroscopy, x-ray diffraction and EDAX at room temperature of black sands from the Zarzal Formation (75° 54.1' W and 4° 44.5' N), in Colombia, which consists of sedimentary sequence composed by diatomites, clays and tobasic sands. This sedimentary sequence is presented in horizons of a black and fine sand in a cross stratification nearly 15 cm wide and white clays interstratifications. In total there exist nine layers of black sands and two upper layers of volcanic ashes and from each a sample was collected (samples 1, 2,...,11, respectively).

2. EXPERIMENTAL

The samples were ground in order to obtain powders for Mössbauer spectroscopy, x-ray diffraction and EDAX. Some part of these powders were magnetically separated and the magnetic fractions were also analyzed by Mössbauer spectroscopy and x - ray driffaction. The magnetic separation were performed using a permanent magnet. This method was not very efficient but permit us to detect the magnetic phase as it will be showed in the results. Mössbauer spectra were recorded using a conventional spectrometer, a Co^{57}/Rh source and the velocity calibration was done using a thin α iron foil. (95.2 % Fe^{57}).

3. RESULTS AND DISCUSSION

Figs. 1a and 1b show the Mössbauer spectra at room temperature of the non-magnetically separated samples and those magnetic samples obtained after magnetic separation. The nine spectra in Fig 1a obtained from the whole black sands were fitted with three doublets. The corresponding fitted parameters (isomer shift and quadrupole splitting) are showed in the table 1.

Table 1.Hiperfine parameters of black sands

SITES	δ (Fe) (mm/s)	Δ (mm/s)
Hyperstene	1.12 - 1.33 ± 0.01	2.12 - 2.34 ± 0.01
	1.03 - 1.19 ± 0.01	2.09 - 2.14 ± 0.01
Albita	0.38 - 0.49 ± 0.01	0.45 - 0.80 ± 0.01

The Mössbauer spectra of the magnetic part of the separated samples are showed in Fig. 1b. In this case, the spectra of these nine samples were individually fitted with two additional sextets.

Table 2 Hiperfine parameters of black sands magnetically separated

SITES	δ (Fe) (mm/s)	Δ (mm/s)	H (kOe)
Hyperstene	1.18 - 1.29 ± 0.01	2.06 - 2.35 ± 0.01	----
	1.12 - 1.20 ± 0.01	2.05 - 2.11 ± 0.01	----
Albita	0.37 - 0.43 ± 0.01	0.43 - 0.76 ± 0.01	----
Magnetite	0.44 - 0.58 ± 0.01	----	446.26 - 467.97 ± 3.00
	0.33 - 0.41 ± 0.01	----	494.82 - 504.30 ± 3.00

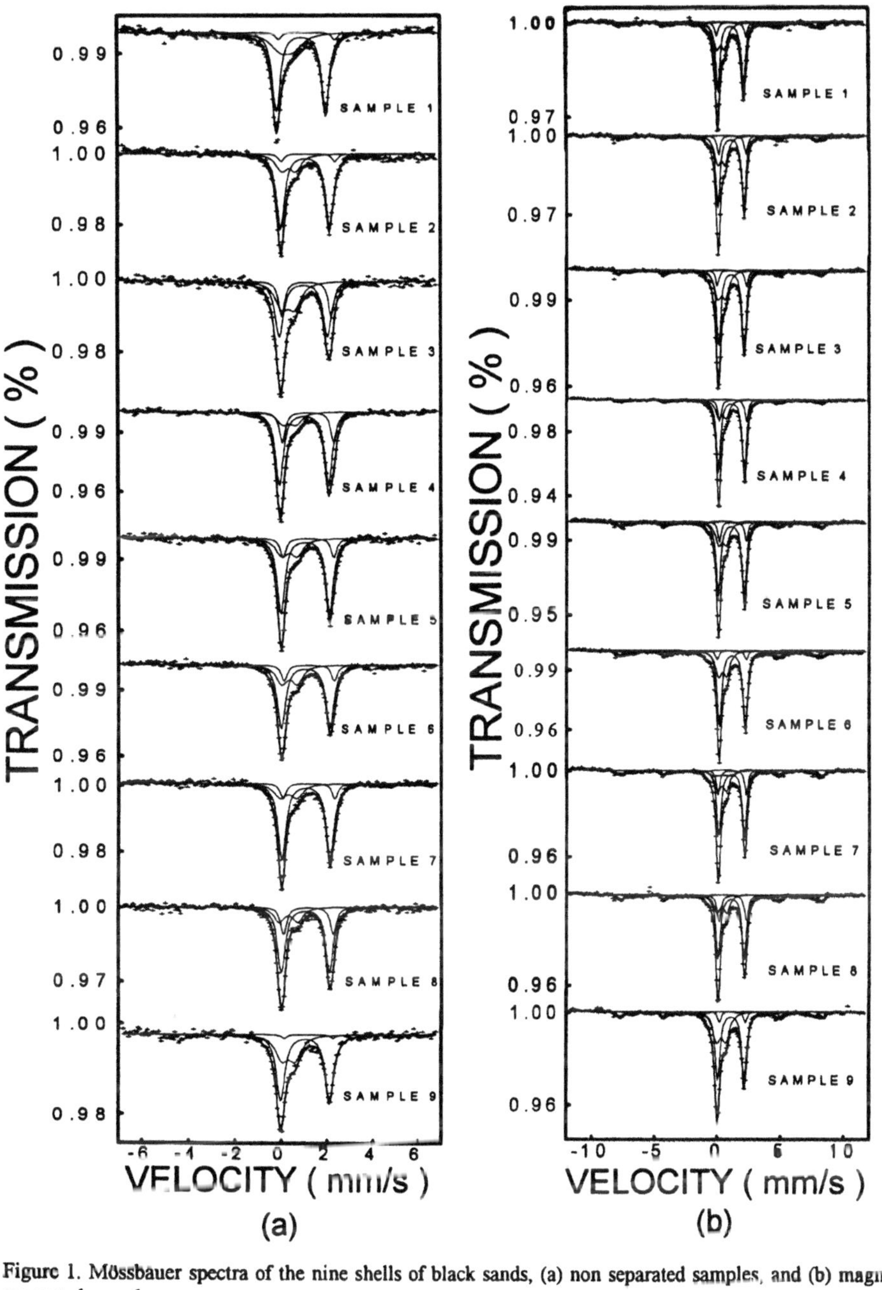

Figure 1. Mössbauer spectra of the nine shells of black sands, (a) non separated samples, and (b) magnetic part of separated samples.

The Mössbauer spectra of the last two samples, the volcanic ashes, are showed in Fig. 2. They were fitted with only one doublet with only one doublet with hyperfine parameters given by $\delta = 0.37 - 0.42 \pm 0.01$ mm/s and Q.S. = $0.43 - 0.46 \pm 0.01$ mm/s.

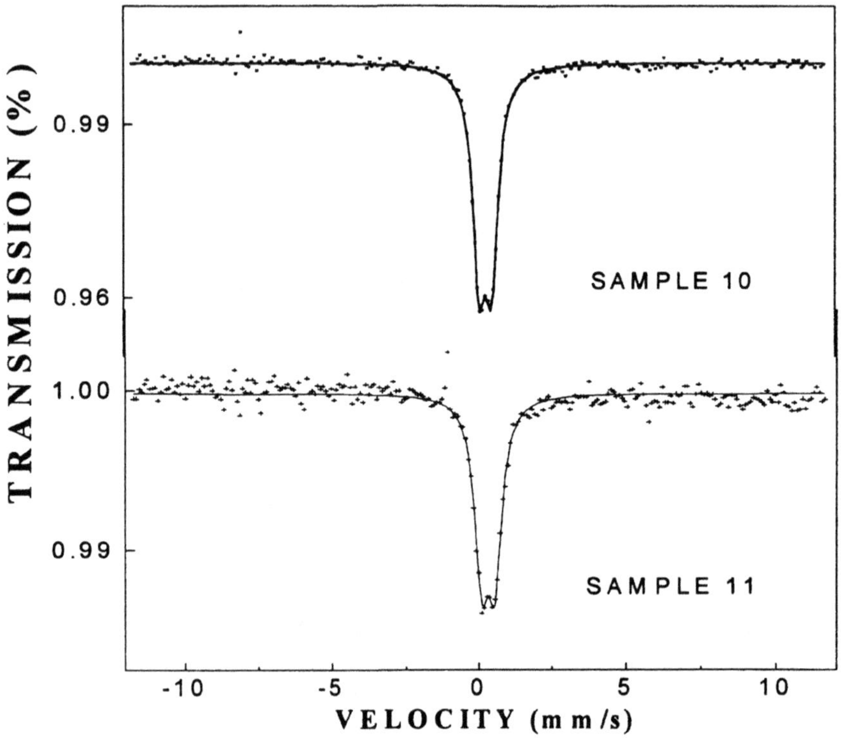

Figure 2. Mössbauer spectra of the volcanic ashes shells

By x-ray diffraction the peaks correponding to hypersthene and albite and some least intensive peaks of magnetite were detected for the first nine non - separated samples. The last two samples showed only the albite peaks. The x-ray diffraction studies of the magnetic part of nine samples detected the presence of hypersthene, magnetite and albite. Fig. 3 shows three typicaldiffractograms for samples 1, 6 and 11. In the first one we only mark the peaks of hypersthene, in the second one the peaks of the magnetite and in the third one only the albite peaks. In coincidence with x-ray studies we can attribute the first two doublets obtained from the Mössbauer spectra of the nine black sands, corresponding to sites M_2 and M_1 of hypersthene and the third doublet to the albite. The same is valid for the magnetically separated samples but in this case the fraction of magnetite increases and appears as two sextets corresponding to the octahedral and tetrahedral sites of Fe in the cubic shell of the magnetite.

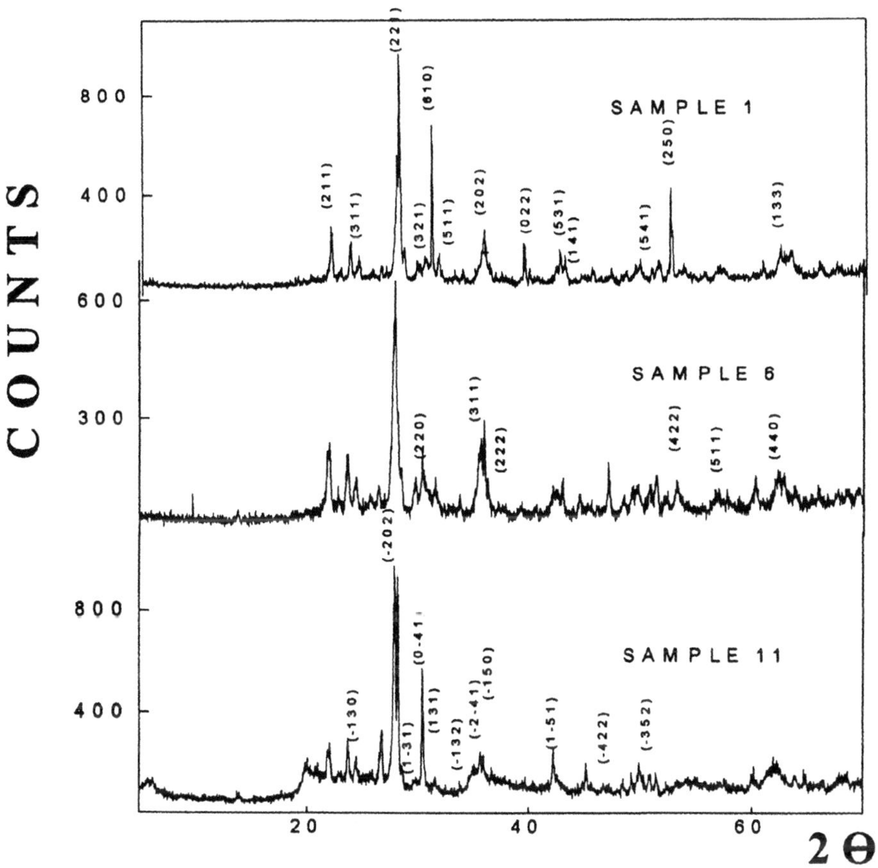

Figure 3. X- ray diffractograms of the two black sands 1 and 6 and the volcanic ashes 11. In sample 1 the peaks of the hypersthene are marked, in sample 6 those of magnetite are marked and the peaks of sample 11 correspond to those of the albite

In Table 3 is showed the chemical analysis obtained by EDAX of the samples 6 and 11, respectively. It can be noted the presence of some oxides as Na_2O, Fe_2O_3, and MgO. Also the elements Si, Al, Na, Mg, Ca and Fe were detected and these oxides and elements are those necessaries to form the compounds detected by Mössbauer and x - ray diffraction such as hypersthene, $(Mg,Fe)SiO_6$, albite, $(AlSi_3O_8)$ and the magnetite, Fe_3O_4. From the table we can note two things: the first one is that Na content is bigger to Ca content and this indicate that the volcanic ashes present more albite than anorthite and the second one is that the Fe_2O_3 content is bigger in the sand than in volcanic ashe and this explain the little presence of magnetite detected in sample 6 and its absent in the 11 sample as showed by Mössbauer and x - ray diffraction.

Table 3. Chemical analysis by EDAX of samples 6 and 11

CHEMICAL COMPOUND	PERCENTAGE OF COMPOUND SAMPLE 6	PERCENTAGE OF COMPOUND SAMPLE 11	CHEMICAL ELEMENT	ATOMIC CONCENTRATION SAMPLE 6	ATOMIC CONCENTRATION SAMPLE 11
Na_2O	5.07	1.62	Na	3.42	1.07
MgO	2.66	1.58	Mg	1.38	0.81
Al_2O	16.73	17.36	Al	6.87	6.99
SiO_2	63.91	73.38	Si	22.28	25.08
K_2O	1.93	0.57	K	0.86	0.25
CaO	3.23	0.96	Ca	1.21	0.35
TiO_2	1.00	0.57	Ti	0.26	0.15
Fe_2O_3	5.48	3.94	Fe	1.44	1.01
			O	62..28	64.28

According with the fitting of the Mössbauer spectra, for the hypersthene phase we have a strong inner doublet (Q.S. = 2.05 - 2.11 mm/s) and a weak outer doublet (Q.S. = 2.06 - 2.25 mm/s) that we attribute to the M_2 and M_1 sites, respectively, in agreement with several reports (see Refs. 1, 2, 3, 4 and more recently Ref. 5). The M_1 site corresponds to the Fe^{2+} cation in symmetric octahedral sites with a weak absortion spectra and the M_2 site correspond to the Fe^{2+} cation in a non-symmetric octahedral sites with a strong absortion spectra. These results indicate that the Fe in the M_2 sites are more abundant that in the M_1 site.

Finally, it can be noted that the parameters assigned to the albite in the sands are nearly the same obtained for the volcanic ashes, but with a Δ very different to that reported in the literature [6] for which Δ = 1.22 mm/s. More work is now in progress in order to (i) to explain this diference and (ii) to a better characterization of the iron spinel phase by selectively removing silicates with chemical attack.

ACKNOWLEGMENTS

The authors would like to thank COLCIENCIAS, Colombia Agency, and to the ALAMED program of the Community Economy European for their financial suppor.

REFERENCES

[1] O. D. Rodrigues, R. Garg, E. Galvâo Da Silva, J. C. Machado and V. K. Garg, Hyperfine Interactions **35**, 913 (1987).

[2] R. G. Burns, Mineralogical Applications of Crystal Field Theory (London : CUP) p 90 (1970).

[3] K. K. P. Srivastava, J. Phys. C : Solid State Phys. **19**, 6407 (1986)

[4] T. P. Sinha, D. Das and M. Bhattacharya, Hyperfine Interactions **77**, 29 (1993)

[5] A. Van Alboom, E. De Grave and R. E. Vandenbergue, Phys. Chem. Minerals **20**, 263 (1993).

[6] Mössbauer Handbook, Mineral Data, September 1983, pag. 49.

AUSTEMPERING KINETICS OF COMPACTED GRAPHITE CAST IRON

K. Laneri, J. Desimoni and R. C. Mercader
Departamento de Física, Facultad de Ciencias Exactas, UNLP
C.C.N°67, 1900 La Plata, Argentina.

R. W. Gregorutti and J. L. Sarutti
Laboratorio de Entrenamiento Multidisciplinario para la Investigación Tecnológica
52 e/121 y 122, 1900 La Plata, Argentina.

Abstract

Mössbauer spectroscopy, standard metallographic techniques and hardness measurements were used to study the kinetics of phase development in the first stage of high Mn alloyed compacted graphite cast iron austempered at 640 K. The starting point of the processing window has been determined using the microstructural criterion. Experimental results are interpreted in the frame of nucleation and growth models.

Introduction

Austempering is the name given to a treatment carried out on cast iron when the sample is heated at T > 1120 K ("austenitized"), followed by a quenching in a salt bath held at 540 K < T < 720 K. The transformation suffered by cast irons can be divided in two stages. The first stage begins with the austenite transformation into bainitic ferrite and high carbon austenite. The second stage begins when the metastable high carbon austenite breaks down into the thermodynamically more stable ferrite and carbides. The amount of retained austenite increases in stage I up to a "plateau", called "processing window", after which the austenite fraction decreases. During the processing window, optimum mechanical properties are obtained.

The behavior of cast irons to austempering heat treatment as well as their final matrix characteristics, mechanical and duty features are directly influenced by several factors like chemical composition and grade of allowing, as cast matrix [1-3]. Several studies have revealed that the transformation kinetics during austempering heat treatments depends on the morphology of the segregated graphite (nodular, vermicular or laminar)[1].

Taking advantage of the sensitivity of Mössbauer spectroscopy to monitor metallic phase changes, we have started a systematic study to verify the correlation between kinetics of transformation and graphite morphology and determine the processing window. In this work, we have investigated the behavior of vermicular graphite cast iron under austenitizing treatment at 1173 K followed by subsequent austempering at 640 K at times ranging from 1 to 90 minutes. Standard metallographic analyses and hardness measurements have also been performed.

Experimental

Ingots of vermicular cast iron were prepared by melting in electric furnace with the following composition : C (3.34 wt.%), Mn (0.57 wt.%), S (0.02 wt.%), P (0.03 wt.%), Si (2.32 wt.%). Samples of 20 mm diameter and 3 mm thickness were austenitized at 1173 K during 30 min, and quenched in a salt bath held at 640 K during different times in the range t = 1-90 min. The samples were specially prepared for Mössbauer spectroscopy by grinding and polishing down to $\approx 70\mu m$ thickness.

Mössbauer spectra were taken in transmission geometry using a ^{57}Co in Rh matrix source of ≈ 5 mCi and recorded in a standard 512 channels conventional constant acceleration spectrometer. Velocity calibration was performed against a 12 μm thick α-Fe foil. All isomer shifts are referred to this standard at 298 K.

All spectra were fitted to lorentzian line shapes with a non-linear least-squares program with constraints.

Standard metallographic analysis were performed with a Reichert microscope. Nital 2 vol.% solution of nitric acid in alcohol was used to enhance the phase contrast. Hardness measurements were made using a standard Vickers pyramid hardness tester with a load of 30 kg.

Results and discussion

Typical spectra are displayed in Fig.1 for three austempering times, showing the characteristic six lines belonging to the different iron-carbon configurations in the martensite/ferrite phase, and a central peak, indicative of the existence of austenite in the sample.

Austenite subspectra were fitted with the parameters suggested by Uwakweh [4]. Three different hyperfine interactions were used in the case of the magnetic subspectra (I_1 : H_1 = 332 kOe, δ_1 = 0.018 mm/s; I_2 : H_2 = 311 kOe, δ_2 = 0.044 mm/s, and I_3 : H_3 = 278 kOe, δ_3 = 0.10 mm/s, and null quadrupole splitting for all interactions).

The evolution with austempering time at 640 K of the relative fraction associated to austenitic phase is shown in Fig. 2 (assuming the same recoilless fractions for both phases). The change in the austenite amount follows a sigmoidal curve. In the first stage (up to t $\approx$ 10 min), an increase of the retained austenite fraction is noticeable and a decrease of hardness is observed (Fig. 3) as a consequence of the diminution of the martensitic fraction. For longer times, a "plateau" is manifested in both figures. This "plateau" indicates that the "processing window" has been reached. In concordance with the microstructural criterion, it starts at $t_1 \approx 9$ min, however, its end has not been achieved even after 90 min. This is in agreement with metallographical observations showed in Fig.4.

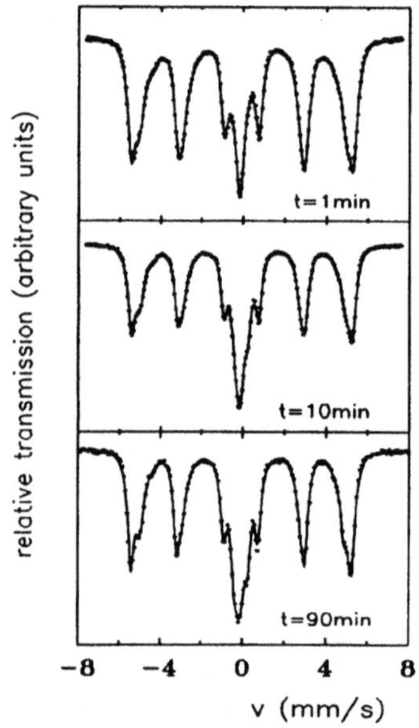

Fig. 1. Mössbauer spectra of austempered samples at different times.

Austempered structure is composed of bainitic ferrite and high carbon austenite in a metastable equilibrium. Fig. 4a shows the bainitic ferrite nucleation at graphite/matrix interface after 1 min of treatment. The precipitated ferrite rejects carbon atoms to the surrounding austenite which raises its carbon concentration and becomes thermodynamically stable. At that time, the matrix is predominantly martensitic due to the fact that the austenite has not been enriched enough in carbon to be stabilized. Fig.4b illustrates the microstructure after 10 min of transformation (when the "processing window" begins) composed of bainitic ferrite and high carbon austenite. Fig. 4c exhibits the microstructure after 90 min of treatment in the "processing window".

Theoretical calculations of the transformation kinetics considering nucleation and growth events indicates that it is an Avrhami-type law, $f = A+B\ (1\text{-}\exp(\text{-}C\ t^n)$ with n=4, suggesting a random volume nucleation with constant rate [5]. In spite of the studies related to the determination

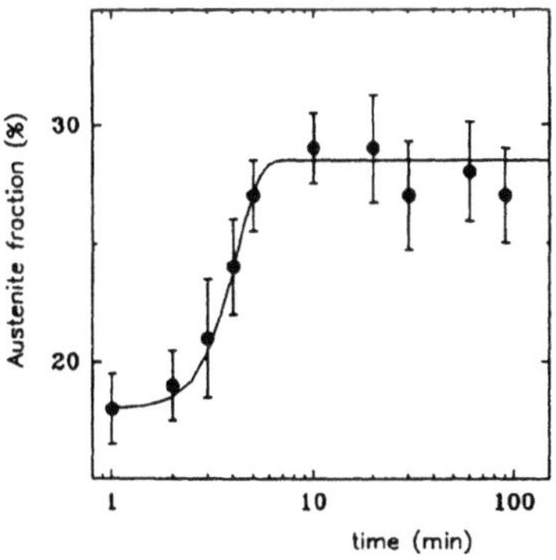

Fig. 2. Austenite fraction evolution with austempering time.

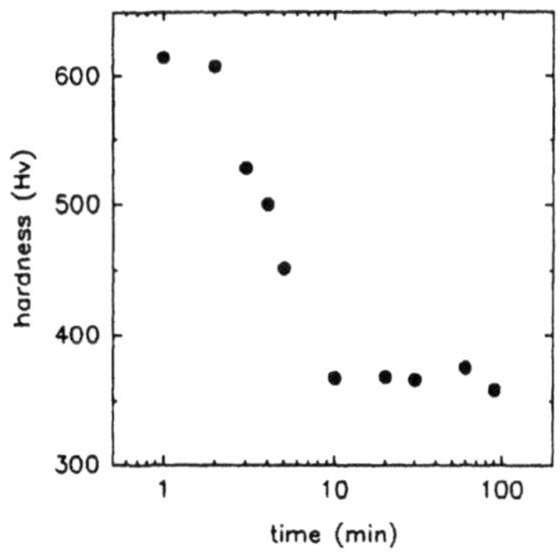

Fig. 3. Vickers hardness progression as a function of austempering time.

of the austempering kinetics and processing window in cast irons (see for example ref.2 and 3, and references therein), we have not found in the literature interpretations of the type of law followed by the transformation kinetics.

Due to a non-uniform definition or designation for the types of iron environments used by diverse investigators [6], it is not possible to perform a unique assignment of the hyperfine interaction to the different Fe-C configurations in the magnetic phase. Nevertheless, the analysis of the evolution of the relative normalized intensities of the magnetic subspectra (F_i^M) suggests that mean carbon concentration does not vary (see Fig.5) in the studied time interval.

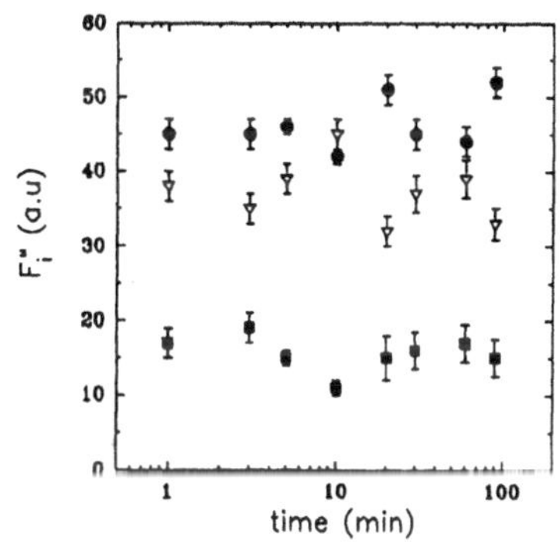

Fig. 5. Evolution of the relative normalized fractions of magnetic subspectra with austempering time.

Conclusions

Mössbauer spectroscopy indicates that transformation in austempered compacted graphite cast iron follows an Avrhami-type kinetics associated with a random volume nucleation with constant rate.

The retained austenite fraction reaches a "plateau" at 9 min of transformation. This indicates the start of the "processing window", in agreement with hardness measurements. Further experiments are in progress to determine the end of this "processing window" and the evolution of C concentration in the austenite.

Acknowledgemets

Authors are gratefully indebted to Consejo Nacional de Investigaciones Científicas y Técnicas (CONICET) and to Comisión de Investigaciones Científicas de la Provincia de Buenos Aires (CICPBA), Argentina, for the financial support. J. D and R. C. M are members of CONICET and J. L. S. is member of CICPBA.

References

[1] I.Riposan and M.Chisamera, Conference Proceeding , Cast Iron IV (1990) 195.

[2] H.Bayati and R.Elliot, Mat. Sci and Tech. 11 (1995) 118.
[3] N.Darwish and R.Elliot, Mat. Sci. and Tech. 9 (1993) 572, and 586.
[4] O. N. C. Uwakweh, J. P. Bauer, and J. M. R. Genin, Metallurgical Transactions A **21** (1990) 589.
[5] J.W.Christian, in "Theory of transformations in metals and alloys", Pergamon Press, Oxford, 1981.
[6] M. Ron, in "Applications of the Mössbauer Spectroscopy", Vol. 2, p.329, ed. by R.L.Cohen, Academic Press, New York, 1976.

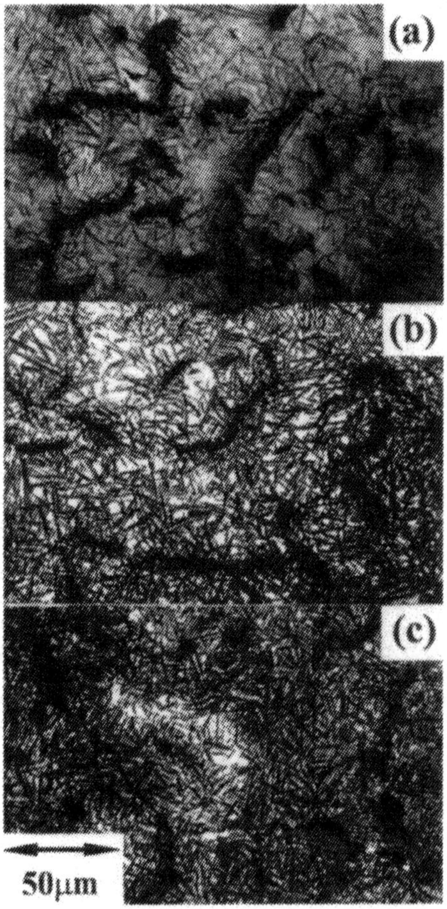

Fig. 4. Microstructures after a) t=1min, b) t=10min, and c) t=90min of austempering.

ON THE ANALYSIS OF THE MÖSSBAUER SPECTRA OF TOURMALINES

G.M. da Costa[1], C. Castañeda[2], N.S. Gomes[2], A.C. P. Soares[3], and G.P. Santana[4]

1- Departamento de Química, Univ. Fed. de Ouro Preto, 35400, Ouro Preto, MG, Brazil.

2- Departamento de Geologia, Univ. Fed. de Ouro Preto, 35400, Ouro Preto, MG, Brazil.

3- Instituto de Geociências, Univ. Fed. de Minas Gerais, 31270, Belo Horizonte, MG, Brazil

4- Departamento de Química, Univ. Fed. de Minas Gerais, 31270 , Belo Horizonte, MG, Brazil.

Abstract- Several samples of tourmalines that originated from pegmatitic deposits in the Minas Gerais State, Brazil, have been investigated by Mössbauer spectroscopy. Two ferrous contributions could be recognized for the two green samples and for a blue one. These components are ascribed to Fe^{2+} in Y and Z crystallographic sites. The black samples show the presence of an intermediate iron-valence component, in addition to the Fe^{2+} (Y) and Fe^{2+} (Z) doublets. The contribution of this component is reduced significantly at 85 K. For some spectra, the fitting routine was found to converge to slightly different solutions, depending on the initial values of the adjustable parameters, and hence different Mössbauer parameters, especially the relative area-ratios . This finding implies that the distribution of Fe^{2+} among the two available crystallographic sites cannot be unambiguously determined from the Mössbauer spectrum.

The tourmaline mineral is a borosilicate with general formula $XY_3Z_6(BO_3)_3(Si_6O_{18})(O,OH,F)_4$. The crystalline structure is very complex, and a large number of elements can occupy the crystallographic sites. There are two types of octahedral sites, hereafter called Y and Z. The Z site is smaller and more distorted than the Y site, and is usually occupied by Al^{3+}, although significant amounts of Fe^{2+} and Fe^{3+} can replace the Al ions [1,2,3]. The Y site can contain a variety of elements such as Fe^{2+}, Mg^{2+}, Fe^{3+}, Mn^{2+}, Cr^{3+}, Ti^{4+}, Li^{+}, and Al^{3+} [1]. The X site ideally contains Na ions, but K^{+}, Ca^{2+} or vacancies reportedly can replace these sodium ions [1]. The end members in which the Y site contains Mg, Fe, and Li/Al are named dravite, schorl and elbaite, respectively.

The Mössbauer spectra of these minerals essentially consists of a single asymmetric doublet. In some cases minor additional contributions are clearly recognized. Although it appears to be a simple task to fit these spectra, some inconsistencies regarding the derived parameters have been reported in the literature. Based on the expectation that the Y and Z sites can contain both ferrous and ferric species, it can be anticipated that a superposition of four components might be necessary to interpret the spectra. However, it has been proposed that up to six components are required to describe just the occupation of the Y octahedra [5,6]. As an example, Korovushkin et al. [5] claimed that eight subspectra are necessary to fully reproduce their experimental data. They employed several constraints in fitting their spectra, including a fixed line width, and only one isomer-shift value for all nonequivalent Fe sites at each position.

In this work the Mössbauer spectra of some tourmalines belonging to the elbaite-schorl series are investigated. The samples were collected from a pegmatite deposit in the Northeast of the Minas Gerais State, Brazil, and were chosen on the basis of their iron contents and macroscopic colors.

Mössbauer spectra (MS) were collected at room temperature (RT) and 85 K using a conventional spectrometer. Powder samples, as well as single-crystal slices, were used. The isomer shifts are quoted relative to α-Fe. Microprobe analyses were done for all major elements, and for the zoned samples the analyses were done in different regions of the crystal. The X-ray diffractograms were obtained in the region of 5-70° (2θ) using Cu Kα radiation.

The unit-cell parameters, macroscopic colors, and partial composition of the six powder samples and three single crystals used in this study are displayed in Table 1.

Table 1- Sample characteristics, cell parameters and microprobe analysis of Fe, Mg and Mn.

Sample	color	a (nm)	c (nm)	Fe (wt %)	MgO (wt %)	MnO (wt %)
PA2SN	black	1.600	7.146	14.3	0.3	0.3
PI2SN	black	1.567	7.276	9.1	0.3	0.2
LA5SA	blue	1.605	7.155	2.2	0.0	2.0
SC2A	blue	1.594	7.128	8.7	0.1	0.3
CA1SV	green	1.609	7.144	5.5	0.3	1.2
L2QV	green	1.588	7.094	5.6	0.1	1.6
PA3A	blue	1.600	7.122	9.0	0.1	0.6
SC4N	black	1.602	7.153	12.6	0.4	0.1
SC4V	green	1.620	7.098	3.6	0.2	1.0

Samples PA2SN and PI2SN are predominantly black, occurring as trigonal prisms with up to 15 cm in the base and a few centimeters in length. Homogeneously colored samples, such as LA5SA and SC2A (blue) and CA1SV and L2QV (green), also appear as prisms, but with less fractures than the black samples. Sample PA3A is blue, with an external ring of green color perpendicular to the c axis. Finally, samples SC4N and SC4V refer to a unique single crystal, but in which the colors are black and deep green respectively. From the unit-cell parameters given in Table 1 it is concluded that all samples are closer to schorl than to elbaite. From this result, and from the chemical analyses, it is expected that the majority of the iron atoms are located on the Y sites.

Representative spectra at RT and 85 K are shown in Figure 1 for sample LA5SA. The spectra of samples SC2A, CA1SV and L2QV are similar to those shown in Figure 1. A first trial fit was done with two symmetrical doublets, with independent line widths, isomer shifts and quadrupole splittings. These two doublets can be associated with Fe^{2+} atoms on the Y and Z sites [7]. As seen in Figures 1a and 1b the fitted spectra are in good agreement with the experimental data, as far as reproducing the spectra is concerned. However, depending on the initial values of the adjustable parameters, the fitting routine was found to converge to different solutions, all having approximately the same χ^2. Although both the quadrupole splittings and isomer shifts show little variation, the line widths, and hence the fractional areas, were found to be affected by a large variation. For instance, the quadrupole splitting of Fe^{2+} (Y) was found to be in the range 2.44-2.48 mm/s, whereas the area is in the range 53-75%. This finding implies that the relative occupation of the Y and Z sites cannot be unambiguously determined from the Mössbauer spectrum. The spectrum collected in the magic angle ($\theta \approx 55^{o}$) is very similar to that shown in Figure 1, and did not show any improvement with respect to a better separation of the two subspectra. The results listed in Table 2 refer to one of the several fits, and in view of the above discussion, they must be taken with some caution.

Figure 1. Mössbauer spectra at room temperature (a) and at 85 K (b) of sample LA5SA. Dots represent the experimental data, and full lines represent the calculated subspectra and their sum.

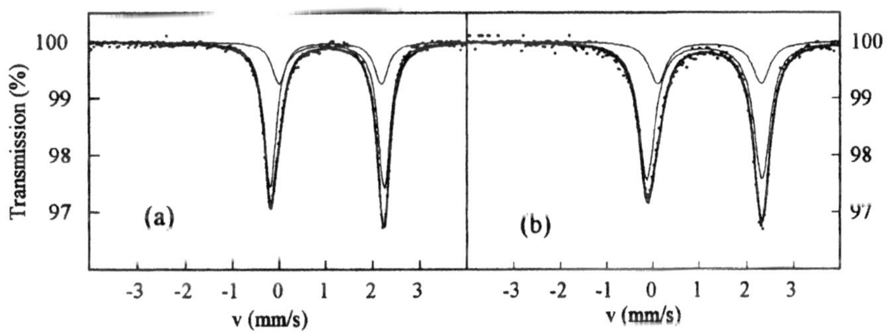

The RT spectra of samples PA2SN and PI2SN are quite different from the previous ones, as shown in Figures 2a and 2c. A minimum of four (PI2SN) or five (PA2SN) doublets had to be included. No restrictions were imposed to fit these components, and the derived parameters are given in Table 2. Assignment of the two main doublets is made according to literature, i.e., they are related to Fe^{2+} atoms in Y and Z sites [7]. The component with the smallest isomer shift is very ill defined, but probably it can be associated to Fe^{3+}, either in Y or Z. Finally, the subspectra with intermediate isomer shifts and quadrupole splittings are usually associated with iron atoms in a process of electron exchange ($Fe^{2.n+}$) [7]. This electron hoping occurs between the two octahedral sites, and can in principle involve three types of interactions: Fe^{2+}(Y)-Fe^{3+}(Y); Fe^{2+}(Y)-Fe^{3+}(Z); and Fe^{2+}(Z)-Fe^{3+}(Z) [8]. At 85 K there is a significant decrease in the charge-transfer process, as seen in Figures 2b and 2d. In fact, the spectrum of sample PA2SN required only two doublets (compared to five at RT), but the line width of the second doublet suggests that additional components might be present. The spectra shown in Figure 2 are very similar to those reported by Scorzelli et al. [9] for a black tourmaline from the same region. Nevertheless, these authors fitted the RT spectrum with only two components, one of which had ΔE_Q = 1.45 mm/s, δ = 1.15 mm/s and a line width of approximately 0.65 mm/s. This component was interpreted as being due to an electron-charge transfer between Fe^{2+} and vacancies.

Figure 2. Mössbauer spectra at room temperature and 85 K of samples PA2SN (2a and 2b) and PI2SN (2c and dd).

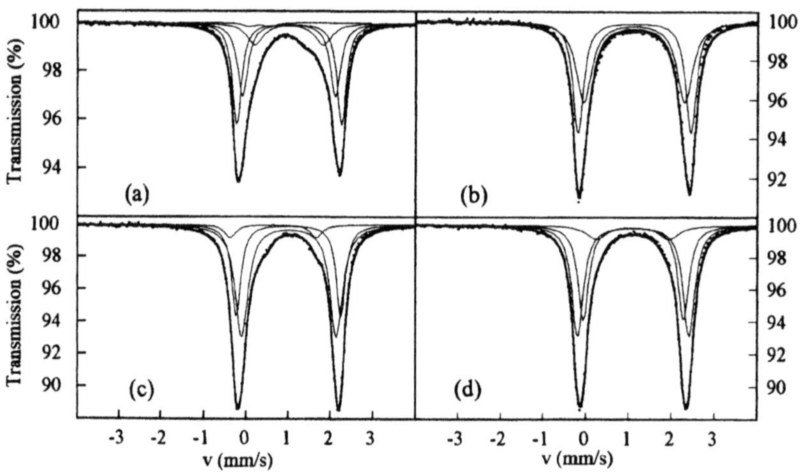

The spectrum of PA3A (single crystal) at room temperature is shown in Figure 3a. Only one ferrous component could be included in the fit, together with one ferric doublet. The derived parameters are also included in Table 2.

The spectra of both black and green regions of SC4 are shown in Figures 3b and 3c (SC4N and SC4V) respectively. The spectrum of SC4V could be well fitted with three doublets, and the numerical parameters are included in Table 2. It is to be noted that the mixed valence state accounts for approximately 4% of the total spectral area, whereas the ferric doublet has 15% of the total area. On the other hand, the spectrum of SC4N required 4 doublets, two for Fe^{2+} in the Y and Z sites, one for Fe^{3+}, and one for the mixed valence state (Table 2). Moreover, the total contribution of this latter component is 12%, in contrast with the 4% found for the green part of this sample.

Figure 3. Mössbauer spectra at room temperature of samples PA3A (a), SC4N (b) and SC4V (c).

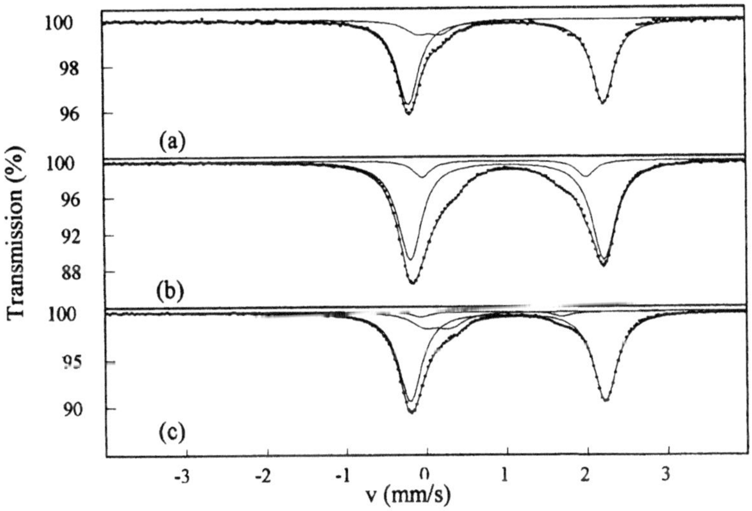

Table 2- Mössbauer parameters derived from the room temperature spectra. ΔE_Q and δ are given in mm/s. S is the relative area ratio in %.

sample	ΔE_{Q1}	δ_1	S_1	ΔE_{Q2}	δ_2	S_2	ΔE_{Q3}	δ_3	S_3	ΔE_{Q5}	δ_5	S_5
SC2A	2.38	1.10	78	2.32	1.22	12	1.76	0.87	6	0.36	0.48	4
LA5SA	2.44	1.13	75	2.19	1.20	25	-	-	-	-	-	-
CA1SV	2.45	1.13	82	2.16	1.24	18	-	-	-	-	-	-
L2QV	2.39	1.11	83	2.14	1.21	17	-	-	-	-	-	-
PA2SN	2.50	1.12	40	2.21	1.11	33	1.63	1.11	16	0.50	0.41	2
PI2SN	2.49	1.11	33	2.26	1.12	56	2.04	0.75	6	0.24	0.48	5
PA3A	2.42	1.13	86	-	-	-	-	-	-	0.30	0.19	14
SC4V	2.42	1.12	81	-	-	-	1.75	0.93	4	0.32	0.25	15
SC4N	2.40	1.13	67	2.03	1.09	8	1.73	0.95	12	0.23	0.36	12

At this point it is tempting to establish a qualitative correlation between the presence of chromophore atoms (iron and manganese in the present case), the Mössbauer spectrum and the color of tourmalines. The two black samples (PA2SN and PI2SN) have almost no manganese, and show a considerable presence of a mixed-valence state. This component is also present in sample SC4V (deep green) but in an amount 4 times smaller than in the black part of the same crystal (SC4N). Thus, it seems that in the absence of other chromophores the charge transfer causes the appearance of darker colors. The Mössbauer spectra of the green samples (CA1SV and L2QV) do not show any sign of such mixed valence state, but these samples contain an appreciable amount of Mn. Thus, in this case it seems that the Mn ions are the main responsible for their colors. The blue color of samples SC2A, L5SA and PA3A is more difficult to explain. In SC2A there is a very ill-defined component in a mixed-valence state, but its low intensity prevents the retrieval of reliable parameter values. Both SC2A and L5SA contain Fe^{2+} in the Y and Z sites, in contrast with PA3A in which only the Y site is occupied by Fe^{2+}. In addition, this latter sample seems to contain approximately 14% of the total iron in a ferric state.

It is clear from the preceeding discussion that the interpretation of the MS of tourmalines in terms of discrete doublets can lead to incorrect values for some parameters. The fact that the quadrupole splitting and isomer shift are similar for Fe^{2+}(Y) and Fe^{2+}(Z) suggests that a better aproach to analyse these spectra might be the use of a distribution of quadrupole splittings. In addition, complementary measurements at both low and high temperatures are mandatory to fully characterize these samples. It was also proposed that a relation might exist between the shape of the Mössbauer spectrum and the color. However, it is necessary to investigate more samples in order to firmly establish such correlation. At the moment work is in progress to contemplate these aspects.

References

[1] Barton R. Jr., Acta Crystallogr. B25 (1969) 1524.

[2] F.F. Jr Foit., and P.E. Rosenberg, Am. Mineral. 64 (1979) 788.

[3] R.G. Burns, Trans. Am. Geophys. Union 64 (1982) 1142.

[4] F.F. Jr Foit., and P.E. Rosenberg, Contrib. Mineral. Petrol. 62 (1977) 109.

[5] V. Korovushkin , V.I. Kuzmin, and V.F. Belov, Phys. Chem. Minerals 4 (1979) 209.

[6] J. Kraczka, A. Pieczka, A.Z. Hrynkiewicz, and W. Zabinski, Hyp. Int. 29 (1986) 1121.

[7] E.A. Ferrow, H. Annersten, and R.P. Gunawardane, Mineral. Mag. 52 (1988) 221.

[8] E.A.Ferrow, Hyp. Int. 91 (1994) 689.

[9] R.B. Scorzelli, E. Baggio-Saitovitch, and J. Danon, J. Physique Colloq. 37:C6 (1976) 801.

SCHORL AND SCHORL-OXIDIZED: A COMPARATIVE ANALYSIS BY MÖSSBAUER SPECTROSCOPY AND X-RAY DIFFRACTION

H. S. de Amorim[1], *M. R. do Amaral Jr., E. C. Torres Tapia, Ligia F. Moreira, Marcelo A. Neves and Paulo H. P. Domingues.*
Instituto de Física, Universidade Federal do Rio de Janeiro, C. P. 68.528
21.945-970, Rio de Janeiro/RJ, Brazil

A Fe^{+2} rich tourmaline was heated in air up to 800°C. The before and after heating samples was analyzed by Mössbauer spectroscopy and powder x-ray diffraction. All Fe^{+2} content was oxidized and the original crystal structure preserved. The results indicate a possibility of a continuous thermal transformation Schorl → Buergerite with implications to studies relative to paragenesis of these minerals.

1. Introduction

Tourmalines are generally described by an ideal chemical formula of relative complexity, $XY_3Z_6B_3Si_6O_{27}(O,OH,F)_4$ where X, Y and Z represent different cations. The end-members of tourmalines group are indicated in Table 1 [1]. Several isomorphic series among the different end-members are known, one of the most frequent occurring, is the Dravite-Schorl series [2].

Table 1
Tourmaline end-members.

End-Member	X	Y	Z	(O, OH, F)
Dravite	Na^{+1}	Mg^{+2}	Al^{+3}	$(OH)_4$
Schorl	Na^{+1}	Fe^{+2}	Al^{+3}	$(OH)_4$
Ferridravite	Na^{+1}	Mg^{+2}	Fe^{+3}	$(OH)_4$
Chromodravite	Na^{+1}	Mg^{+2}	Cr^{+3}	$(OH)_4$
Uvite	Ca^{+2}	Mg^{+2}	$(Mg^{+2})_{1/6}(Al^{+3})_{5/6}$	$(OH)_4$
Elbaite	Na^{+1}	$(Li^{+1})_{1/2}(Al^{+3})_{1/2}$	Al^{+3}	$(OH)_4$
Liddicoatite	Ca^{+2}	$(Li^{+1})_{2/3}(Al^{+3})_{1/3}$	Al^{+3}	$(OH)_4$
Buergerite	Na^{+1}	Fe^{+3}	Al^{+3}	O_3F
Olenite	Na^{+1}	Al^{+3}	Al^{+3}	$O_3(OH)$

The currently accepted crystalline structure of the tourmalines was originally described by Buerger and Donnay [3]. The crystal lattice belongs to the rhombohedral system (R3m), which in a hexagonal basis has cell parameters typically of a ≅ 16Å and c ≅ 7Å. One of its main structural characteristic is the presence of silicon tetrahedra SiO_4 sharing two oxygen in order to form a ring with six tetrahedra of composition Si_6O_{18}: an almost perfect hexagonal ring. The cations Y and Z are situated in the center of relatively distorted octahedra, sites y (point symmetry *m*) and z (point symmetry 1), respectively, to which are linked O, OH and F. Three octahedra related to the sites-y are linked by a common edge. The common vertex is on a three-fold rotation axis of symmetry and contains a hydroxyl group (OH(1)). A second hydroxyl group (OH(3)) is situated on the opposite vertex in each one of the octahedra. The octahedra related to the sites-z are coordinated to the ones of the sites-y by a common edge and in addition, they are sharing a hydroxyl group at a common

[1] Corresponding author: Fax: +55-21-280-7693; e-mail: hsalim@if.ufrj.br

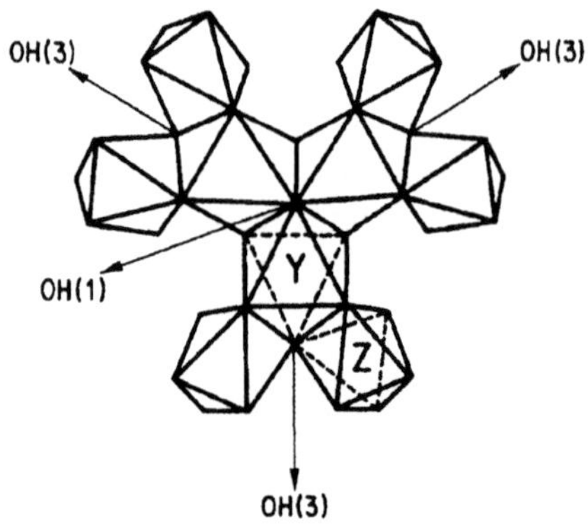

Fig. 1. A fragment of tourmaline structure showing sites-y, -z and the hydroxyl groups. The center is on a three-fold rotation axis of symmetry.

vertex, with two sites-z for each sites-y (fig. 1). The sites-z are connected to one hydroxyl group and sites-y to two, thus being different regarding the compositions of the first sphere of coordination.

Table 2
The Mössbauer parameters at room temperature.

		Tourmaline (natural)	Oxidized Tourmaline (800°C - 24h)
	AREA(%)	12.1	
Fe^{+2}-y	QSPL(mm/s)	2.561(8)	
	ISSH(mm/s)	1.160(3)	
	AREA(%)	22.9	
Fe^{+2}-y	QSPL(mm/s)	2.473(6)	
	ISSH(mm/s)	1.067(2)	
	AREA(%)	30.4	
Fe^{+2}-y	QSPL(mm/s)	2.26(1)	
	ISSH(mm/s)	1.063(1)	
	AREA(%)	31.5	
IVCT	QSPL(mm/s)	1.72(1)	
	ISSH(mm/s)	1.047(2)	
	AREA(%)	3.1	32.4
Fe^{+3}-y	QSPL(mm/s)	0.265(8)	1.037(7)
	ISSH(mm/s)	0.248(6)	0.251(2)
	AREA(%)		29.3
Fe^{+3}-y	QSPL(mm/s)		0.979(8)
	ISSH(mm/s)		0.365(5)
	AREA(%)		38.3
Fe^{+3}-y	QSPL(mm/s)		1.052(10)
	ISSH(mm/s)		0.493(2)

QSPL - Quadrupole splitting
ISSH - Isomer shift with respect to Fe-metal.

A sample of Fe^{+2} rich tourmaline of the Dravite-Schorl series, coming from Araçuai, Minas Gerais (Brazil), an important area of tourmaline production, was used to study the oxidation process of these iron cations by heating in the atmospheric air. Previous studies made by Gorelikova et al. [4] indicated the possibility of a complete oxidation of Fe^{+2} without altering the basic integrity of the crystal lattice of tourmaline. The oxidation reaction seems to be associated to a partial loss of hydroxyls according to the equation:

$$Fe^{+2} + (OH)^{-1} \rightarrow Fe^{+3} + O^{-2}.$$

2. Experimental techniques

The original sample was characterized by powder x-ray diffraction (XRD) and Mössbauer spectroscopy (MS). Its chemical composition was determined by x-ray photoelectron spectroscopy

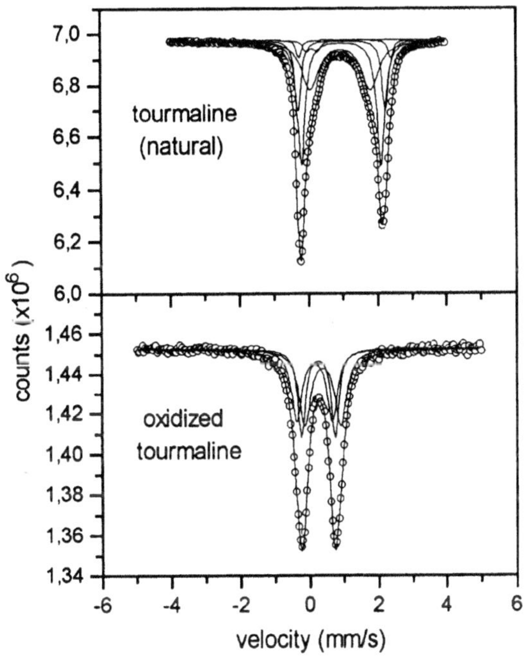

Fig. 2. Mössbauer spectra at room temperature for natural tourmaline sample (upper) and after heating sample (lower).

(XPS) with a spectrometer Perkin-Elmer, mod. 1257, using the method described by Wagner et al. [5] and based on the model of 31 oxygen by chemical formula. The determined composition led to the formula,

$$(Na)^{+1}\,(Mg_{1.6}Fe_{1.4})^{+2}\,(Al^{+3})_6\,(BO_3)_3Si_6O_{18}(OH)_4\,,$$

with a small Fe^{+3} fraction that is not indicated explicitly above. The structure refinement revealed that all iron cations are situated in sites-y and the cell parameters refined with 169 independent reflections was a = 15.9528(6)Å and c = 7.1467(1)Å.

One part of the original sample was finely ground and heated in a tubular oven in atmospheric air at 800°C for 24h and then analyzed by XRD (fig. 2) and MS (fig. 3). Another part of the sample was analyzed by thermogravimetric analysis (TGA) with an equipment Shimadzu TGA-50H and by differential thermal analysis (DTA) with the help of an DTA 1600 Cell of TA Instruments. The XRD analysis was made in an automated powder x-ray diffractometer Siemens type-F, in a scanning mode θ-2θ with steps of 0.025°. A graphite monochromator was used to select Cobalt Kα doublet. The samples were prepared using the procedure described by Klug and Alexander [6] and alignment and calibration evaluated with a sample of α-Al_2O_3. The results were analyzed by Rietveld method using the program DBWS-9411 by Young et al. [7]. The analysis by MS was made with a spectrometer CMTE in a standard transmission geometry controlled by a microcomputer IBM/PC-AT with an emulator card EG&G Ortec ACE-MCS with 512 channels. The spectrometer was calibrated with sodium nitroprusside and the isomer shift was measured relative to iron-foil of 25μm. The program NORMOS [8] was used in the analyses of the spectra.

3. Results and discussion

In Table 2 we present the results of MS analysis. The spectrum of original unoxidized sample was properly described by five doublets, one of them is characteristic of intervalence charge transfer (IVCT), another one of Fe^{+3}, and the three remaining of Fe^{+2}. The spectrum of the oxidized sample reveals only the presence of Fe^{+3} and was properly described by three doublets. The values of the quadrupole splitting are relatively large and indicate a strong distortion of the octahedra sites. The differences in the sites-y are attributed to the second sphere of coordination according to the ideas proposed by Korovushkin et al. [9]. The analysis of x-ray diffractogram (fig. 2) reveals the dominant presence of tourmaline with the cell volume slightly smaller than the one of the original sample and a small amount of another phase of low crystallinity. The presence of this extra phase is seen more clearly through a side line to the tourmaline reflection (012). It is also noticed other side lines to the most intense peaks of the oxidized tourmaline which bring us to admit that this extra phase has the tourmaline structure with different cell parameters, compatible with Dravite. This fact seems to indicate that during the oxidation process, one part of the sample rich in magnesium, can segregate itself giving rise to a Dravite type phase.

If we observe the thermogravimetric curve we see some structure in steps (particularly at 276 °C and 722°C are found sensible loss of mass) that seem to indicate a selective deprotonization of the sample. This property seems to be associated to the fact that the natural analyzed sample is not an "ideal" Schorl with hydroxyl groups associated to different compositions of the sites-y. One of these groups is found in the common vertex to three sites-y centered octahedra (Wyckoff notation 3a, point symmetry *3m*). These sites-y must have their composition with combinations of the form Fe-Fe-Fe, Fe-Fe-Mg, Fe-Mg-Mg and Mg-Mg-Mg implying different energies (temperatures) to the deprotonization reaction and consequent oxidation of the cations Fe^{+2}. The same three combinations with Fe^{+2} are associated to the three characteristic doublets used to describe the observed Mössbauer spectrum.

The structure refinement presents a sensitive improvement if we introduce a phase type Dravite what reinforce the above hypothesis. To this phase we adopted the atomic coordinates obtained from the natural sample and cell parameters of a pure Dravite as deduced by Epprecht [13]. The mass concentration for this magnesium-rich phase was found to be 8%. The obtained R-factors [12] was R-WP = 12.9%, R-P = 10.1% and S = 2.0 and atomic coordinate are in good agreements with those values obtained by Barton [10] for the Buergerite. In the Table 3 we present a comparison of cell parameters of the involved samples.

The DTA indicates an endothermic reaction with onset at ~900°C followed by a exothermic reaction with onset at 955°C and going on to about ~1100°C. We can notice in the thermogravimetric curve a significative loss of mass from ~900°C. Up to this temperature the mass losses observed are not sufficient to completely deprotonizate the sample. So, only above 900°C such a process is completed and a severe destruction of the original structure follows.

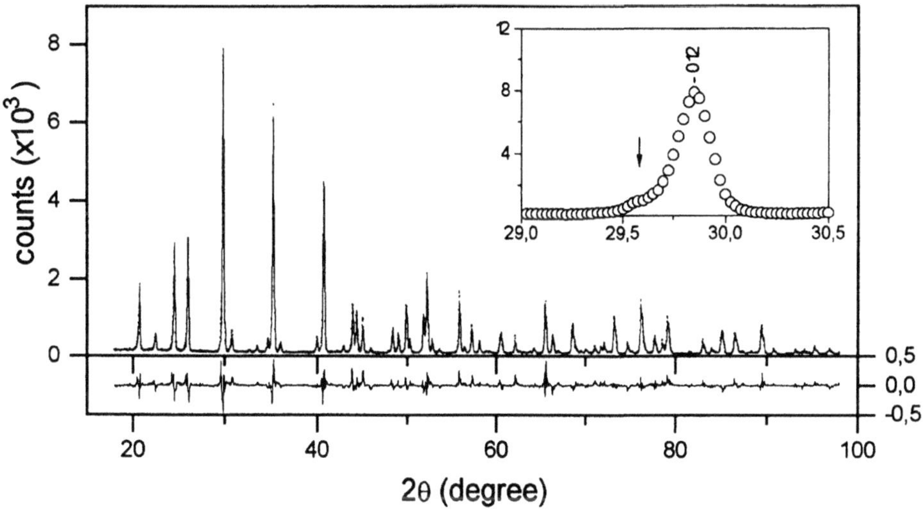

Fig. 3. *(Top)*: Observed (open circle) and calculated (solid line) diffractogram of oxidized tourmaline collected with Cobalt Kα radiation. *(Bottom)*: The graphic of difference between observed and calculated data. In the upper-right side, a detail of diffractogram where we can see the presence of an extra phase.

4. Conclusion

Our results have shown that a hydroxi-tourmaline, member of Dravite-Schorl series, had its

Table 3. Cell parameters for different samples involved.

	Schorl (*) "ideal"	Tourmaline (natural)	Tourmaline (oxidized)	Buergerite(**)
a(Å)	16.000	15.9528(6)	15.8638(3)	15.869(2)
b(Å)	16.000	15.9528(6)	15.8638(3)	15.869(2)
c(Å)	7.135	7.1467(1)	7.1668(1)	7.188(1)
V(Å^3)	1581.9	1575.1(1)	1562.0(1)	1567.6

(*) Values supplied by Epprecht [13].
(**) Values supplied by Barton [10].

original content of Fe^{+2} completely oxidized on a thermal treatment, in the atmospheric air at 800°C for 24h, without the destruction of the characteristic crystal lattice. The evolution of the cell parameter is similar to the variation observed from Schorl to Buergerite (diminishing of a-parameter and increasing of c-parameter - see Table 3). In the oxidized sample there are indications of a phase segregation of a low cristallinity magnesium-rich tourmaline phase. Further studies by electronic microscopy are necessary to clarify this question.

Acknowledgments

The authors would like to thank Mr. Rubio Brandão Larrubia for the supply of crystal used in this work, Mr. Newton Terjlen for the cuts in the crystal made in his gems lapidation workshop, and

Dr. R. H. Pratt (University of Pittsburgh - USA) for his valuable observations about the use of the anomalous scattering factors in ions case.

We also thanks to FINEP, FAPERJ, FUJB and CNPq for the financial support given.

References

[1] D. M. Burt, American Mineralogist, 74 (1989) 826.
[2] R. V. Dietrich, in: The Tourmaline Group (Van Nostrand Reinhold Company, New York, 1985).
[3] G. Donnay and M. J. Buerger, Acta Cryst. 3 (1950) 379.
[4] N. V. Gorelikova, Y. D. Perfil'yev and A. M. Bubeshkin, Int. Geol. Review 20 (1976) 982.
[5] C. O. Wagner, W. M. Riggs, L. E. Davis, J. F. Moulder and G. E. Muilenberg (Editor), in: Handbook of X-ray Photoelectron Spectroscopy (Published by Palmer-Elmer Corporation, Minnesota, USA, 1990).
[6] H. P. Klug and L. E. Alexander, in: X-Ray Diffraction Procedure for Polycrystalline and Amorphous Materials, (John Wiley & Sons, 1972).
[7] R. A. Young, A. Sakthivel , T. S. Moss and C. O. Paiva-Santos, J. Appl. Cryst. 28 (1995) 366.
[8] R. A. Brand, Nuclear Instruments and Methods in Physics Research 328 (1987) 398.
[9] V. V. Korovushkin, V. I. Kuzmin and V. F. Belov, Phys. Chem. Minerals 4 (1979) 209.
[10] R. Barton Jr., Acta Cryst. B25 (1969) 1524.
[11] F. F. Foit Jr., American Mineralogist 74 (1989) 422 .
[12] R. J. Hill and R. X. Fischer, J. Appl. Cryst. 23 (1990) 462.
[13] W. Epprecht, Schweiz. Min. Petrogr. Mitt. 33 (1953) 481.

MINERALOGICAL ANALYSIS OF A PEDOSYSTEM DEVELOPING ON STEATITE

G. P. Santana[a], *M.F. de Jesus Filho*[a,†], *J.D. Fabris*[b], *A.T. Goulart*[c] *and G.M. da Costa*[d]

[a]Departamento de Química, Universidade Federal de Minas Gerais, 31270-901, Belo Horizonte, MG, Brazil
[b]Visiting Researcher, Departamento de Química, UFMG, 31270-901, Belo Horizonte, MG, Brazil. Permanent address: EMBRAPA/CNPMS, Caixa Postal 151, 35701-970, Sete Lagoas, MG, Brazil
[c]Departamento de Química, Universidade Federal de Viçosa, 36571-000, Viçosa, MG, Brazil
[d]Departamento de Química, Universidade Federal de Ouro Preto, 35400-000, Ouro Preto, MG, Brazil

Non-magnetic portions from a pedon developing on steatite, including its fresh and altered parent materials, were studied by chemical analysis, Mössbauer spectroscopy and powder X-ray diffraction. The fresh rock sample was found to contain talc, chlorite, dolomite, tremolite and ilmenite. Hematite and goethite were additionally identified in the altered rock and soil samples. As a general trend, weathering processes in steatite occur firstly by removal of dolomite, followed by that of chlorite. Talc is relatively stable during weathering, in this lithology. Hematite formed through the alteration of magnetite is well crystallized and has the hyperfine parameters of the bulk oxide. This soil forming on steatite shows at least two types of hematite, with distint physical properties and origins.

1. Introduction

Tropical soils typically contain high amounts of quartz, kaolinite and cation (e.g., Fe^{3+}, Al^{3+}, Ti^{4+}) oxides. As far as structural, physical (e.g., magnetic) and chemical properties and mineralogenesis pathways are concerned, iron oxides (including hydroxides and oxyhydroxides) are relatively less well understood. Hematite (α-Fe_2O_3) and goethite (α-FeOOH) are the most common iron oxides in tropical soils. Magnetic pedosystems also contain magnetite (Fe_3O_4) and/or maghemite (γ-Fe_2O_3). All those oxides commonly occur as very finely divided particles, in variable degree of crystallinity and amounts of isomorphical substituents. These factors often make them less readily detectable by conventional powder X-ray diffraction (XRD). Therefore, Fe^{57} Mössbauer spectroscopy is a specific technique for iron and would be, in principle, useful in characterizing this group of minerals from tropical pedons.

In recent work, Santana et al. [1] have shown that a nearly nostoichiometric magnetite is the only iron-rich spinel occurring in both rock and soil samples, from a magnetic pedosystem developing on steatite, located in Minas Gerais, Brazil. The purpose of the present study was to identify the iron-bearing non-magnetic phases of this same soil profile, in an attempt to establish the main mineralogenic pathways by which the primary rock-minerals are transformed during pedogenesis.

[†]Professor Milton Francisco de Jesus Filho deceased on 02 January 1996.

2. Experimental

Samples of steatite and from the B horizon of its overlying magnetic soil were collected from a pedosystem nearby the city of Congonhas do Campo (20° 30' 05'' S, 43° 51' 39'' W), state of Minas Gerais, Brazil. The soil samples were firstly broken up by hand, left to dry in air and sieved, to obtain the fine earth (mean diameter of particles, ϕ < 2 mm, saturation magnetization 1.9 J T^{-1} kg^{-1}), which was then dispersed with 50 mL of NH_4OH 1:1 taken into 500 mL of the soil suspension under stirring in a mixer. The suspension was then sieved, to separate the sand fraction (ϕ = 0.05 mm - 2 mm; 17 % weight). The silt (ϕ = 0.002 mm - 0.05 mm; 15 % weight) and the clay (ϕ < 2 μm, 68% weight) fractions were separated by centrifuging the remainder, according to Jackson [2]. The altered rock material was separated from about 3-4 cm nearly circular brownish zones clearly identified thorough the cut surface of the greenish matrix of the rock samples. The non-magnetic separates from the fresh groundly, altered rock samples, from soil-sand and soil-silt were obtained from a dilute slurry in water while stirring with a magnetic rod stirrer. The magnetic particles were then separated with a small hand magnet, and non-magnetic samples were collected for this study. The whole altered rock sample was treated with a mixture of dithionite - citrate - bicarbonate (DCB) in an attempt to selectivelly remove some crystalline iron oxides [3], particularlly hematite and goethite. The four non-magnetic extracts (altered rock and soil sand, silt and clay) were treated with 5 mol L^{-1} NaOH [5]. XRD were obtained with a Rigaku Geigerflex diffractometer with a graphite diffracted beam monochromator using Cu(Kα) radiation and NaCl as internal standard. Mössbauer spectra were collected at room temperature (RT), and at 85 K, in a constant acceleration transmission set up with a ^{57}Co/Rh source. Isomer shifts are quoted relative to metallic iron.

3. Results and discussion

XRD pattern of fresh rock showed the existence of tremolite, chlorite, talc and dolomite, in agreement with the reported mineralogy of steatite [6]. The altered rock was found to contain less dolomite as compared to the fresh rock, whereas a decrease in the crystalinity, evidenced by the line broadening of the diagnostic reflection of chlorite, was observed in the soil sample. Neither hematite nor goethite could be identified in any of these soil samples

The Mössbauer spectra of the untreated samples (fig. 1, lefthand side) indicated a partial alteration of some steatite mineral, on going from the fresh to the altered rock and from the sand to the clay soil samples. This is seen as an increase of the intensity of the central doublet, which has characteristic parameters of high-spin Fe^{3+}. An incipient magnetic splitting is visible at RT both in the silt and clay fractions, but it is clearer in the 85 K pattern. The contribution of this magnetic fraction increased with decreasing temperature, indicating that part of the central doublet is due to supeparamagnetic particles. The NaOH treatment of the altered rock, silt and clay samples (Fig. 1, righhand side) leads to a large contribuition of the magnetic component, probably due to the removal of silicates and eventually of the smaller iron oxides particles [5].

The hyperfine parameters derived from the 85 K spectrum of the fresh rock (Table 1) are in agreement with those reported for chlorite, talc and dolomite [6]. Additionally, Mössbauer analysis of the altered rock sample treated with DCB revealed the presence of ilmenite, with an isomer shift relative to αFe, δ = 1.07 mm s^{-1}, quadrupole splitting, Δ = 0.64 mm s^{-1} and relative spectral area, RA = 26 % (spectrum not shown), which are consistent with values reported for the Fe^{2+} site of ferric ilmenite from mafiic systems, by Goulart et al. [7], Doriguetto et al. [8] and Pinto et al. [9]. The spectra in which the magnetic splitting is present were fitted with a model-independent hyperfine field distribution (B_{Hf}). The B_{Hf} values ranged from 20.0 to 54.0 T at steps of 1 T, and derived parameters are also included in Table 1. The maximum-probability hyperfine field ($P(B_{Hf})_{max}$), isomer shift and quadrupole shift of the altered rock suggest that goethite is the main iron oxide in this sample. The shape of the hyperfine field distribution probability curves (Figs. 2a and 2a') suggests that an uniform particle size distribution exists in this sample [3]. Goethite and hematite are the main iron-bearing phases in the soil-silt and -clay fractions that account for about 70 % of the whole Mössbauer spectral area (Table 1).

Table 1
85 K Mössbauer parameters of untreated samples, obtained from hyperfine field distribution.

Sample	Mineral	$P(B_{hf})_{max}$/T	B_{hf}/T	δ/mm s^{-1}	$2\varepsilon_Q$/mm s^{-1}	RA/%
fresh rock	Chlorite			0.50	0.93	23
	Talc			1.24	3.00	64
	Ilmenite			1.19	1.05	3
	Dolomite			1.39	1.93	10
altered rock	Chlorite			0.34	0.93	14
	Talc			1.24	3.00	22
	Ilmenite			1.06	0.68	6
	Goethite	46.7		0.44	-0.25	58
silt fraction	Chlorite			0.35	0.70	8
	Talc			1.26	3.07	17
	Ilmenite			1.03	1.16	2
	Goethite	46.0		0.48	-0.25	42
	Hematite		52.0	0.46	-0.18	31
clay fraction	Chlorite			0.35	0.65	12
	Talc			1.23	3.06	17
	Goethite	43.7		0.48	-0.25	40
	Hematite		51.8	0.46	-0.18	31

The Mössbauer spectrum at 85 K of the sample after NaOH treatment (Table 2 and Fig.s 2b and 2b') indicates that hematite is the main iron oxide in the silt fraction. The quadrupole splitting ($2\varepsilon_Q$ = -0.18 mm s^{-1}) is characteristic of hematite above the Morin transition. The suppression of the Morin transion down to this temperature is probably due to the effect of isomorphic substitution, most likely by Al, since chlorite is relatively rich in this element.

Table 2
85 K Mössbauer parameter of NaOH-treated samples, obtained from hiperfine field distribution.

Sample	Mineral	$P(B_{hf})_{max}$/T	B_{hf}/T	δ/mm s^{-1}	$2\varepsilon_Q$/mm s^{-1}	RA/%
altered rock	Chlorite			0.35	0.70	21
	Talc			1.26	2.96	39
	Ilmenite			1.06	0.68	7
	Goethite	47.1		0.43	-0.25	33
silt fraction	Chlorite			0.35	0.70	18
	Talc			1.24	3.06	19
	Ilmenite			1.06	0.68	2
	Hematite	51.8		0.43	-0.18	61
clay fraction	Chlorite			0.35	0.70	6
	Talc			1.26	3.12	3
	Goethite	45.2		0.44	-0.25	50
	Hematite		51.4	0.46	-0.18	41

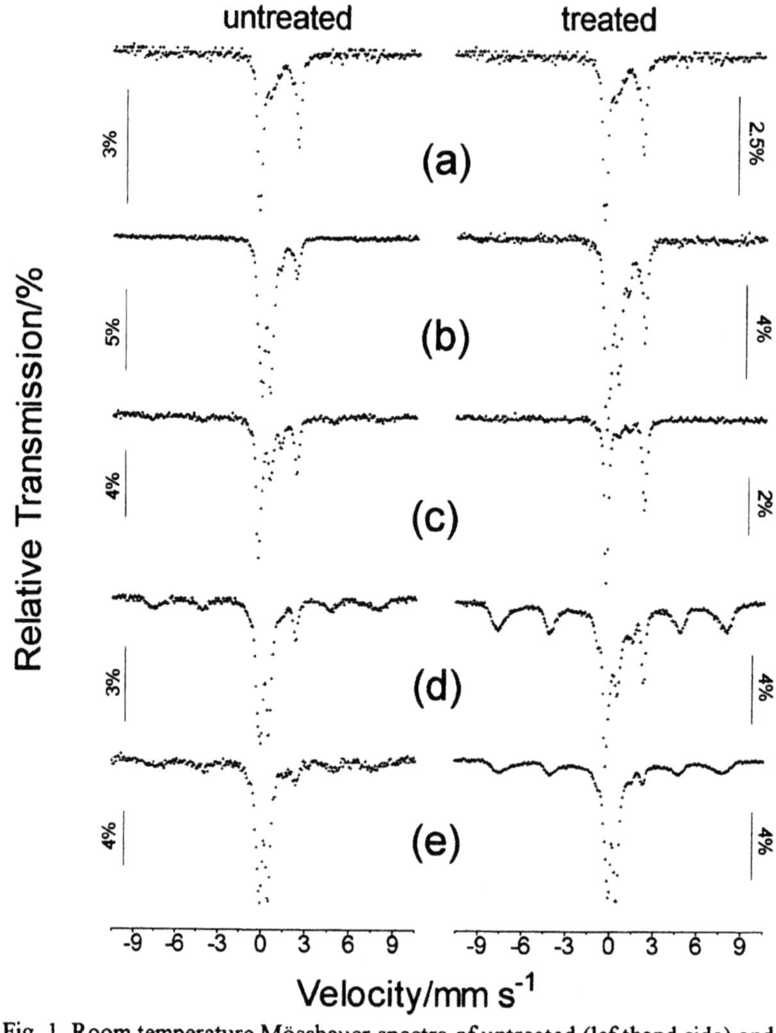

Fig. 1. Room temperature Mössbauer spectra of untreated (lef thand side) and NaOH-treated (right hand side) samples of (a) fresh rock; (b) altered rock; (c) soil-sand; (d) soil-silt and (e) soil-clay.

A typical 85 K Mössbauer spectrum of the clay fraction and respective hyperfine field distribution profile are presented in Fig.s 2c and 2c'.

From previous data reported data by Santana et al. [1], hematite is somehow formed from magnetite in this pedosystem. The present results indicate that a less crystallized and smaller hematite and goethite are formed from the decomposition of silicates during pedogenesis.

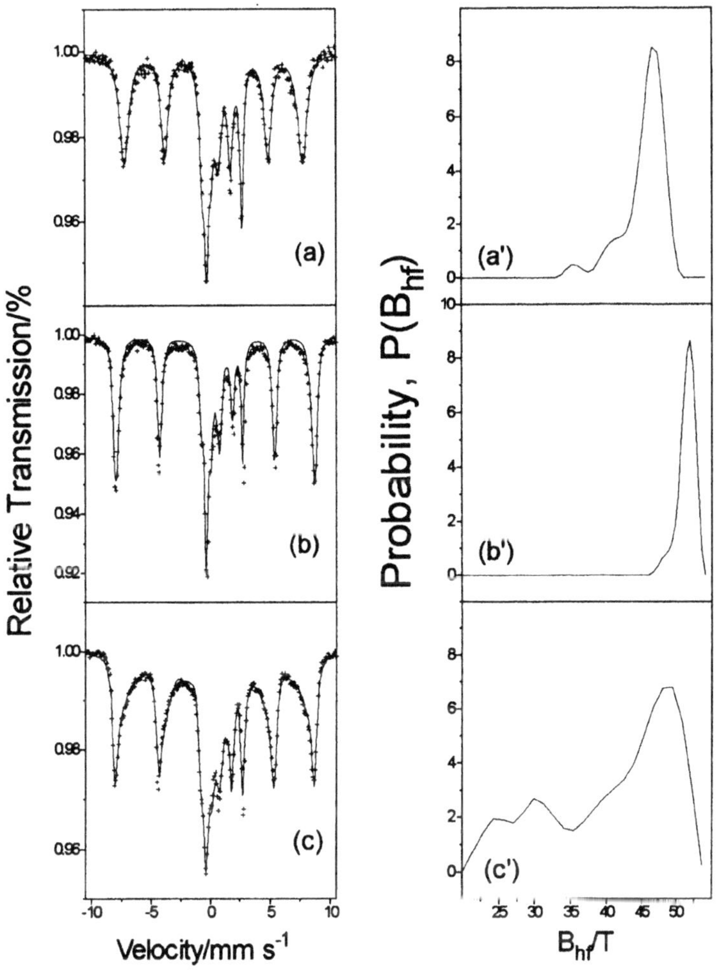

Fig. 2. 85 K Mössbauer spectra and respective hyperfine field distributions relative to the goethite subspectra of (a) - (a') altered rock; (b) - (b') NaOH-treated soil-silt and (c) - (c') untreated soil-clay samples.

4. Conclusions

As a general trend, weathering processes in steatite occur firstly by removal of dolomite, followed by that of chlorite. Talc is relatively stable during weathering, in this lithology. The decomposition of silicates leads to the formation of finely divided hematite and goethite. Differently from this pathway, hematite formed through the alteration of magnetite is well crystallized and has hyperfine parameters of the bulk oxide. As a consequence, this soil forming on steatite shows at least two type of hematite, with different physical properties and origins. The other soil-occuring minerals are goethite, also formed during decomposition of rock-silicate, and residual primary minerals, *viz* chlorite and tremolite, but mainly talc, in all fractions. Ilmenite was detected in all but the soil-clay sample

References

[1] G. P. Santana, M. F. de Jesus Filho, J. D. Fabris. and A. T. Goulart, to be published (1997).
[2] M. L. Jackson, in: Soil chemical analysis: advanced course, 3rd, (Madison, Wisconsin, 1969) p. 894.
[3] E. Murad, Am. Min. 67(1982) 1007.
[4] O. P. Mehra. and M. L. Jackson, in: Proc. 7th Nat. (U. S.) Conf. Clays and Clay Minerals. (Pergamon Press, London, 1958).
[5] N. Kämpf and U. Schwertmann, Clays and Clay Min. 30 (1982) 401.
[6] M. F. de Jesus Filho, M. A. Gonçalves, J. C. Bosh Neto and V. K. Garg, (1992). Hyp. Int. 70 (1992) 961.
[7] A. T. Goulart, M. F. de Jesus Filho, J. D. Fabris and J. M. D. Coey, Hyp. Int. 91 (1994) 771.
[8] A. C. Doriguetto, A. T. Goulart, M. F. de Jesus Filho, J. D. Fabris and G. P. Santana, (1996) Europ. J. Soil Sci. (1996). Submited.
[9] M. C. F. Pinto, M. F. de Jesus Filho, A. T. Goulart, J. D. Fabris and G. P. Santana, Hyp. inter. (1997). These Proceedings.

MINERALOGICAL ANALYSIS OF A MANGANESE ORE

Paulo A. de Souza Jr.[(1)], *J. D. Fabris*[(2)], *Wagner da N. Mussel*[(3)] *and Vijayendra K. Garg*[(4)]

[(1)]Departamento de Física, UFES, 29060-900 Vitória, ES, Brazil
[(2)]Visiting Researcher, Departamento de Química, Universidade Federal de Minas Gerais, 31270-901 Belo Horizonte, MG, Brazil. Permanent address: EMBRAPA/CNPMS, Caixa Postal 151, 35701-970 Sete Lagoas, MG, Brazil
[(3)]Departamento de Química, UFMG, 31270-901 Belo Horizonte, MG, Brazil
[(4)]Departamento de Física, UnB, 70910-900, DF, Brazil

A carbonate-rich sample from manganese deposits of Conselheiro Lafaiete, MG, Brazil, has been characterized by Mössbauer spectroscopy (MS), chemical analysis, X-ray diffraction, magnetization and electron microprobe (EMS) analysis. It is found to contain mainly manganese, iron, nickel and chromium. Room temperature (RT) MS showed a two-blocks quadrupole splitting distribution. The RT-MS of the sample treated with 5 mol L^{-1} NaOH shows additionally a broad hyperfine field. The EMS data of the treated sample reveal a highly heterogeneous powder, but some probed Fe-Ni-rich grains have a mass ratio Fe:Ni of 8.8. The presence of kamacite may explain the relatively strong saturation magnetization (σ_{sat} ~53 J T^{-1} kg^{-1}) for the chemically treated sample.

1 Introduction

Carbonate-rich rocks are an important constituent of the earth's crust. Since their deposition as sediments, the history of carbonate rocks overlaps the history of dissolution, cementation, recrystallization and dolomitization. Rhodochrosite ($MnCO_3$) occurs in manganese ores as anhydrous normal carbonate. It is hexagonal (a:c 1:0.8183) in structure [1]. Distinct crystals are not common, but rhombohedral rounded and composite thick tabular prismatic crystals are found. Parallel growths have been observed of rhodochrosite with dolomite. Mn is found to be substituted by Fe, Ca or Zn or by Fe and Ca, forming a complete series ranging from rhodochrosite to siderite ($FeCO_2$) to calcite ($CaCO_3$), respectively. It occurs as a primary gangue mineral in moderate to low temperature hydrothermal veins, especially of silver, lead, zinc, and copper ores, where it is associated with calcite ($CaCO_3$), siderite ($FeCO_2$), dolomite ($MgCa(CO_3)_2$), fluorite (CaF_2), barite ($BaSO_4$), quartz (SiO_2), manganese, alabandite (MnS), tetrahedrite ($Cu_{12}Sb_4S_{13}$), sphalerite (ZnS), etc and in high-temperature metasomatic (or metamorphic) deposits associated with rhodonite ($MnSiO_3$), garnet ($A_3B_2(SiO_4)_3$, A = Ca^{2+}, Mg^{2+}, Fe^{2+} or Mn^{2+} and B = Al^{3+}, Fe^{3+} or Cr^{3+}) , braunite ($Mn^{II}Mn^{IV}O_3$), and hausmannite ($MnMn_2O_4$). It also occurs as a secondary mineral in residual deposits of iron or manganese oxides.

In the present study a rhodochrosite-rich ore from Conselheiro Lafaiete (MG, Brazil, 20° 36' S, 45° 46' W) has been analyzed by powder X-ray diffratometry, Mössbauer spectroscopy, magnetization measurement and electron microprobe technique in order to identify and characterize its principal mineralogical phases.

2 Experimental

The ore sample was crushed and powdered to 74 μm. The electron microprobe analysis was performed with a Leica Cambridge Steroscan 360 scanning electron microscope equipped with an energy-dispersive X-ray spectrometer, that indicated the presence of Mn, Ca, and Fe as the predominant elements in the whole sample. The conventional chemical analysis revealed Fe_2O_3 (1.3%), MnO (40.8%), SiO_2 (15.1%), Al_2O_3 (5.8%), MgO (3.1%), K_2O (0.8%), TiO_2 (0.1%), and CaO (4.1%) with a loss of 28.9% on ignition. Traces of Cr (761 ppm) and Ni (476 ppm) were also detected. The ore sample was chemically treated with 5 mol L^{-1} NaOH (without any thermal treatment) for 24 hours to concentrate iron oxide and related compounds. The powder X-ray diffraction (XRD) analysis of the sample, after (Fig. 1a) and before (Fig. 1b) chemical attack with 5 mol L^{-1} NaOH, was performed with a Siemens D500 powder diffractometer equipped with CuK_α radiation. XRD pattern of the untreated sample showed the presence of rhodochrosite and quartz (Fig. 1) as the principal constituent and also indicated the presence of minor proportions of rhodonite ($MnSiO_3$). The sample was concluded to be mainly rhodochrosite ($MnCO_3$) with some iron and calcium compounds. Measurement of the room temperature saturation magnetization was done with a portable soil magnetometer [2] and its value was found to be $\sigma_{sat} = 0.3$ J T^{-1} kg^{-1} at $B_0 \approx 0.3$ T. A complete room temperature magnetization curve $\sigma(B_{ext})$ of the chemically treated sample was obtained with a compact VSM conceived by Cugat et. al. [3]. The Mössbauer spectra were recorded in the standard transmission geometry with a Co^{57}/Rh source.

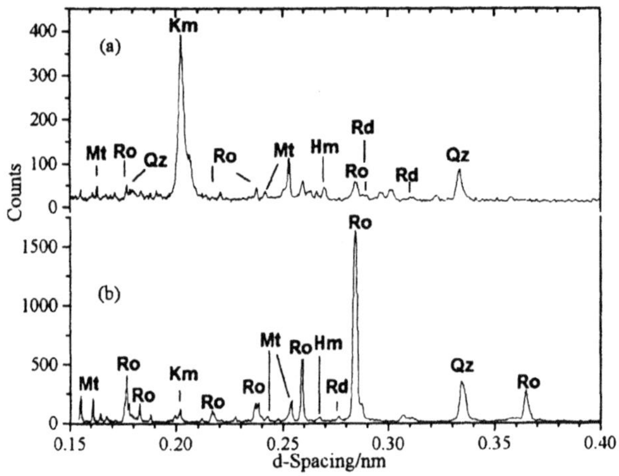

Figure 1: Powder X-ray patterns of the rhodochrosite sample (a) treated with 5 mol L^{-1} NaOH and (b) untreated.

3 Results and Discussion

Whole Sample

Room temperature Mössbauer spectrum (Fig. 2a) showed a broad and asymmetric doublet with some unresolved peaks inside. Quadrupole splitting (Δ) distribution could be fitted with two blocks, each one corresponding to an averaged isomer shift (δ) relative to α-Fe: one block (Fig. 2b, δ = 1.14 mm s^{-1}) with 80.6 % of the whole spectral area (WSA), relative to high spin Fe^{3+}, with P(Δ) maxima at Δ = 1.06 (4.9 %), 2.11 (21.9 %) and 2.94 (53.8 %) mm s^{-1}, and other (Fig. 2c, δ = 0.40 mm s^{-1}) with 19.6 % WSA, relative to high spin Fe^{2+}, with P(Δ) maxima at Δ = 0.24 (18.7 %), 0.78 (0.4 %) and 1.37 (0.3 %) mm s^{-1}. Dickson [4] studied four samples of rhodonite from Australia and England and found five paramagnetic components which had (I): δ = 1.08 - 1.09 mm s^{-1} and Δ = 1.27 - 1.33 mm s^{-1}; (II): δ = 1.17 - 1.21 mm s^{-1} and Δ = 1.93 - 2.02 mm s^{-1}; (III): δ = 1.208 - 1.214 mm s^{-1} and Δ = 2.50 - 2.53 mm s^{-1}; (IV): δ = 1.208 - 1.214 mm s^{-1} and Δ = 2.29 - 2.32 mm s^{-1} and (V): δ = 1.208 - 1.214 mm s^{-1} and Δ = 2.29 - 2.32 mm s^{-1}. In the present study, the main Fe^{2+} peaks in the P(Δ) distribution are centered in Δ = 2.11 and 2.94 mm s^{-1}, which approach the reported values of sites (II) and (V), respectively. In addition, Fe_2O_3, CaO and MnO contents from chemical analysis are comparable with two Dickson's samples from England, but SiO_2 is less than half, in the present sample.

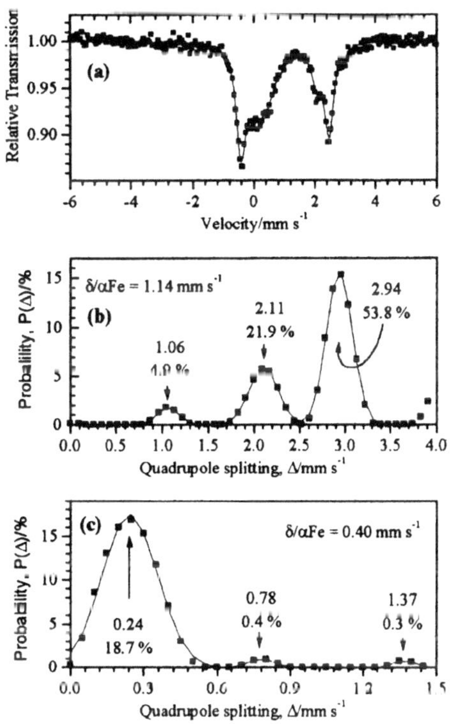

Figure 2: (a) Room temperature Mössbauer spectrum of the whole rhodochrosite-rich ore sample, least squares fitted with the two block quadrupole splitting distribution; (b) and (c): the profiles of each distribution block are decomposed in Gaussians

The whole sample Mössbauer spectrum at RT could exist because of replacement of Mn by Fe in $MnCO_3$ and/or $MnSiO_3$. It was not possible to assign any reflection from XRD pattern to silicate structure.

Chemically Treated Sample

The room temperature Mössbauer spectrum (Fig. 3a) shows a singlet (δ = 0.05 mm s^{-1}, 1.5 % WSA), a doublet (δ = 0.24 mm s^{-1}, Δ = 0.87 mm s^{-1}, 0.3 % WSA) and a broad hyperfine field distribution with an averaged δ = 0.03 mm s^{-1}. The profile of the hyperfine distribution can be decomposed in six Gaussians, centered at B_{hf} = 10.9 (Gaussian **1**, in Figure 4b, 2.1 % WSA), 13.7 (**2**, 3.9 %), 17.8 (**3**, 8.1 %), 22.8 (**4**, 45.4 %), 26.2 (**5**, 25.1 %) and 28.2 (**6**, 13.7 %) tesla.

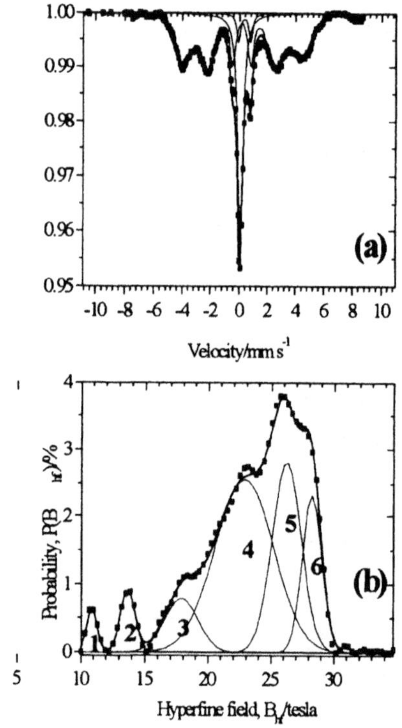

Figure 3: (a) Room temperature Mössbauer spectrum of the rhodochrosite sample treated with 5 mol L^{-1} NaOH, least squares fitted with a quadrupole splitting distribution. (b): numbers 1 to 6 indicate Gaussians in which the distribution profile was numerically decomposed.

Mössbauer spectra at 15 K (Fig. 4a) has essentially the same spectral features, with a singlet and a doublet totaling 2.6 % WSA, and also a broad hyperfine field distribution with 97.4 % of the whole spectrum, and an averaged δ = 0.11 mm/s. The profile of the hyperfine distribution can be decomposed in six Gaussians, centered at B_{hf} = 14.2 (Gaussian **1**, in Fig. 4b, 4.3 % WSA), 16.5 (**2**, 2.7 %), 20.1 (**3**, 16.5 %), 23.2 (**4**, 26.0 %), 25.7 (**5**, 30.3 %) and 27 4 (**6**, 17.6 %) tesla.

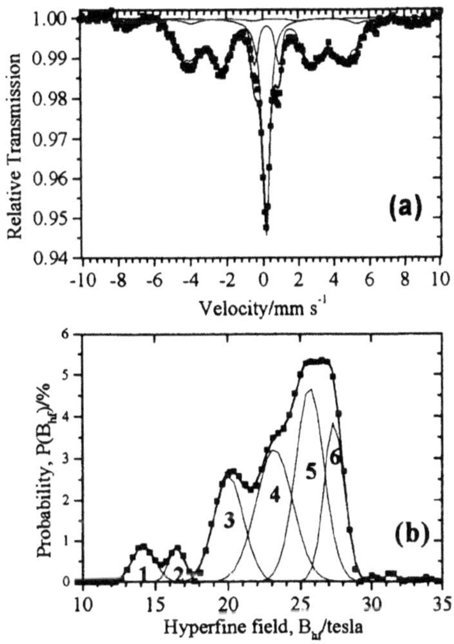

Figure 4: (a) 15 K Mössbauer spectrum of the rhodochrosite sample treated with 5 mol L^{-1} NaOH, least squares fitted with a quadrupole splitting distribution. (b): numbers 1 to 6 indicate Gaussians in which the distribution profile was numerically decomposed.

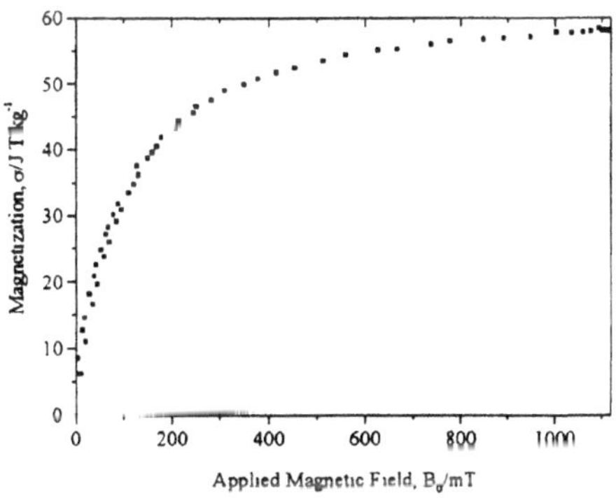

Figure 5: Room temperature magnetization curve of the rhodochrosite sample treated with 5 mol L^{-1} NaOH

The XRD (Fig. 1a) showed the principal components of kamacite (Fe,Ni), which has a ferromagnetic structure [5], and this may explain the relatively strong saturation magnetization (~53 J T^{-1} kg^{-1}) found (Fig 5) in the chemical extract.

Mössbauer results found in the present work approach but hyperfine fields are slightly lower than those reported by Scorzelli et al. [6], for Fe-Ni alloy from meteorites. As far as the chemical composition is concerned, a remarkable difference in this case is the relatively high Cr content given by microprobe analysis in some Fe,Ni-rich grains from the chemically treated sample, that revealed a highly heterogeneous powder, which corresponds to 0.6 % of the whole ore mass. Probed Fe-Ni-rich grains were found to have a mass ratio Fe:Ni of 8.8, together with minor proportions of Al and Si.

4 Conclusions

A sample of rhodocrosite-rich ore was found to contain additionally rhodonite and quartz. Room temperature Mössbauer analysis indicates a quadrupole distribution with one block due to high spin Fe^{2+} (δ = 1.14 mm s^{-1}) and other of high spin Fe^{3+} (δ = 0.40 mm s^{-1}). The chemical treatment of the sample with 5 mol L^{-1} NaOH for 24 hours renders a heterogeneous and magnetic powder (saturation magnetization, σ_{sat} ~53 J T^{-1} kg^{-1}), containing Fe-Ni-rich grains composed by about 8 mass% Ni, 70 % Fe and 19 % Cr together with minor proportions of Al and Si. Room temperature Mössbauer spectra of this chemically-treated sample was fitted mainly with a hyperfine field distribution ($P(B_{hf})$, $\delta_{averaged}$ = 0.03 mm s^{-1}), with six maxima centered at B_{hf} = 10.9 (2.1 % WSA), 13.7 (3.9 %), 17.8 (8.1 %), 22.8 (45.4 %), 26.2 (25.1 %) and 28.2 (13.7 %) tesla., which were assigned to kamacite.

Acknowledgments

The authors wish to thank Drs. Celso Guimarães and Ronald Vassimon Ferreira (Conselheiro Lafaiete Mine) for providing the sample and to Mr. Cristiano Soares and Ms Kélen Cristina Lopes Martins (EMBRAPA) for the technical support in the chemical analysis. FAP-DF, CNPq, FAPEMIG and CST for some financial support.

References

[1] C. Palache, H. Berman and C. Frondel. Dana's System of Mineralogy, Vol. II, 7th edition (1944) John Wiley & Sons, New York, USA.

[2]. J. M. D. Coey, O. Cugat, J McCauley and J. D. Fabris, Revista de Física Aplicada e Instrumentação 7 (1992) 25

[3] O. Cugat, R. Byrne, J. McCauley , and J, M, D. Coey, Rev. Sci. Instrum. 65 (1994) 3570.

[4] B. L. Dickson, The iron Distribution in Rhodonite, Am. Mineral. 60 (1975) 98-104.

[5] J. Danon, R. B. Scorzelli, I Souza Azevedo, and Christophe-Michael-Levy, Nature 281 (1979) 469

[6] R. B. Scorzelli, I. S. Azevedo, J. Danon and M A. Myers, J. Phys. F. Metal Phys. 17 (1987) 1993.

MÖSSBAUER STUDY OF Mg-DOPED MAGHEMITE

W. da N. Mussel[(1)]; *J. M. D. Coey*[(2)]; *J. D. Fabris*[(3)] *and M. F. de Jesus Filho*[(1),(†)]

[(1)]Departamento de Química, Universidade Federal de Minas Gerais, 31270-901 Belo Horizonte, MG, Brazil.
[(2)]Physics Department, Trinity College, University of Dublin, Dublin 2, Ireland.
[(3)]Visiting Researcher, Departamento de Química, UFMG, 31270-901 Belo Horizonte, MG, Brazil. Permanent address: EMBRAPA/CNPMS, Caixa Postal 151, CEP 35701-970 Sete Lagoas, MG; Brazil.

A set of five maghemite samples were prepared with increasing amounts of structural Mg. Electron microprobe analysis indicated that Mg was retained in the spinel lattice in a stoichiometric proportion of x = 0; 0.27; 0.49; 0.78 and 0.81 in the structural formula $Fe_{3-x-y}Mg_x\square_yO_4$, where $\square$ = vacancy. The lattice parameter of the tetragonal cell and Curie temperature were found to vary linearly with x. Mössbauer spectra obtained at 20 K were fitted with two magnetic subspectra ascribed to the octahedral $\{Fe^{3+}\}$ and tetrahedral $[Fe^{3+}]$ sites of spinel. The relative subspectral area ratio, $\{Fe^{3+}\}/[Fe^{3+}]$ is empirically related to x by a model based on the relative elemental occupancy of the spinel sites.

1. Introduction

Tropical soils developing on mafic lithology are commonly magnetic, in the sense that their saturation magnetization $\sigma_s > 1$ J T^{-1} kg^{-1}. Ti-rich maghemite (γFe_2O_3) has been reported as the most common spinel phase in those soils [1, 2, 3], but structural Al^{3+} and Mn^{2+} also replace iron in maghemite, as reported by Allan et al. [4] and Moukarika et al. [5], respectively. More recently, a natural Mg-rich maghemite has been identified and newly characterized [6, 7]. So far, the cation distribution among octahedral {B} and tetrahedral [A] sites and the effect of isomorphic replacement of iron by Mg^{2+} in the spinel lattice on the magnetic structure have not been well established. In an attempt to give more experimental support for data interpretation, Mg^{2+}-doped maghemites were synthesized and studied, in this work, by powder X-ray diffraction (XRD), Mössbauer spectroscopy and magnetic measurements.

2. Experimental

The synthesis procedure was based on a hydrothermal reaction. A solution containing ferrous sulfate and magnesium sulfate in variable proportion was added to a boiling twice-destilled and deionized water, fluxed with pure nitrogen in a closed system. An oxidant mixture of KNO_3/NaOH was added in drops to this solution to provide Fe^{3+}. The resulting magnetite samples were firstly washed with a solution of NaCl/NH_4OH to remove adsorbed ions, and then intensively washed with water and acetone, to remove impurities, and dried in air at room temperature. The pure and Mg-doped magnetites were heated at 240°C in air for 3 hours to convert them to the corresponding maghemites. Magnetization measurements were performed with a compact vibrating sample magnetometer conceived by Cugat et al. [8]. The powder X-ray diffraction (XRD) pattern was obtained with a Siemens D500 powder diffractometer and CuK_α radiation. A silicon standard was used for calibration.

(†)Professor Milton Francisco de Jesus Filho deceased on 02 January 1996.

Chemical compositions were obtained by probing polished section of grains, with a Leica Cambridge Stereoscan 360 scanning electron microscope, equipped with na energy-dispersive X-ray spectrometer. Mössbauer spectra were recorded in a conventional constant acceleration transmission spectrometer and a Co^{57}/Rh source. The Curie temperature was measured in a thermobalance coupled with a set of permanent magnets to give a nonuniform magnetic field of about 10 mT, with a heating rate of 5 °C min^{-1}.

3. Results and Discussion

The X-ray patterns (not shown) reveal reflections due to maghemite only. . Lines of superstructure at 23.91°, 26.22°, 32.35° and 34.09° 2θ indicate that the unit cell is tetragonal as result of vacancy ordering. Values of the *a*-dimension increase continuously (Figure 1(a)) with the magnesium content, taken as the index (x) in the spinel Formula 1, no second phase is seen in the x = 0.78 and 0.81 samples, where a central dublet appears in the Mössbauer spectra (Figure 2).

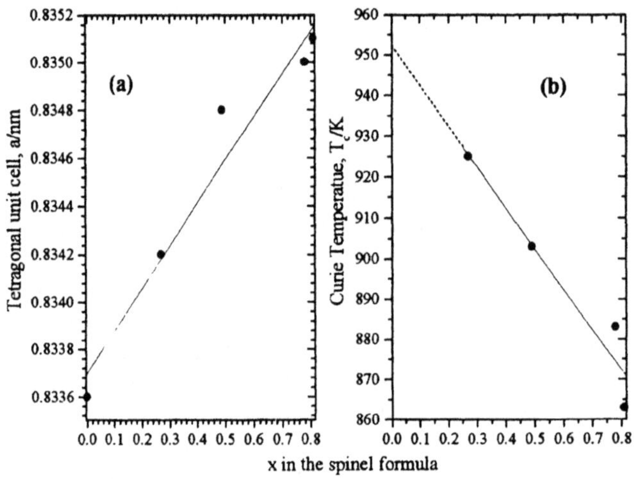

Figure 1. (a) Variation of a-dimension of the unit cell and (b) Curie temperature (T_c) with the magnesium content expressed as the index (x) in the spinel formula.

Formula 1

$$[Fe^{3+}]\{Fe^{3+}\Box_w\}Mg_xO_4$$

where { } and [ ] stand for octahedral and tetrahedral sites, respectively, and □ denotes cation vacancies.

Least squares fitting leads to the linear expression, as expected from Vegard's law ($R^2 = 0.95$),

Equation 1

$$a = 0.0018(2)x + 0.8337(1)$$

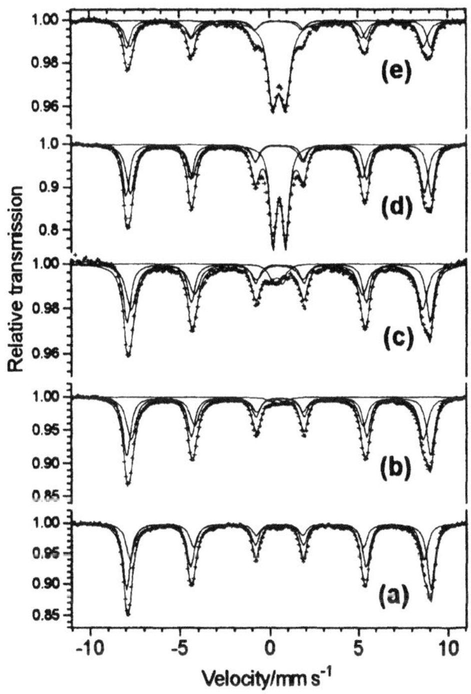

Figure 2. 20 K Mössbauer spectra of the (a) pure and (b) for x = 0.27; (c) x = 0.49; (d) x = 0.78 and (e) x = 0.81 Mg-doped maghemites.

In Equation 1, a is given in nanometers and numbers in parentheses are uncertainties in the last significant digit of the respective parameter. The a value deduced from Equation 1 for x = 0 is in good accord with the reported value for pure tetragonal maghemite (a = 0.834 nm, JCPDS card # 25-1402); the extrapolated a = 0.8355(2) nm for x = 1 approaches the indexed unit cell dimension of cubic magnesioferrite, Fe_2MgO_4 (a= 0.8375 nm, JCPDS card # 17-464). Differences in ionic radii (values taken from Kaye and Laby [9]) between both six-coordinated high spin $^{(VI)}Fe^{3+}$ (r = 65 pm) and $^{(VI)}Mg^{2+}$ (r = 72 pm) or four-coordinated high spin $^{(IV)}Fe^{3+}$ (r = 49 pm) and $^{(IV)}Mg^{2+}$ (r = 58 pm) may explain this variation.

The average particle sizes of the Mg-doped maghemites were calculated by the Warren-Averbach method with three independent determinations (d-variance and Fourier analysis both on distribution and slope) and range from 15 to 20 nm (Table 1).

The thermomagnetic curves of Mg-maghemites show characteristic reversibility at least up to 800 - 900 K. The pure maghemite undergoes an irreversible transformation at ~943 K due to a crystallographic change to the corundum structure. The thermal stability conferred by structural Mg^{2+}, as found in the present work, is in accordance with results reported by Fabris et al. [7] for a natural Mg-rich maghemite which presented a reversible magnetic behaviour up to 1170 K.

The Curie temperature (T_C) of the pure maghemite can be not directly obtained, due to the crystallographic phase change to hematite. However, T_C for Mg-doped maghemites is found to vary with the Mg^{2+} content (Fig 1(b)), according to the linear relation ($R^2 = 0.93$),

Equation 2

$$T_C = -100(19)x + 952(12)$$

Table 1
Lattice parameters, lattice strain and particle size dimensions (Warren-Averbach method) and unit cell dimension of the pure and Mg-doped maghemites.

Spinel formula	*a*/nm	*c*/nm	Strain	Particle size/nm	Volume of the unit cell/nm^3
$Fe_{2.67}\square_{0.33}O_4$	0.8336(1)	2.502(3)	$0^{(*)}$	$60^{(**)}$	1.738(2)
$Fe_{2.49}Mg_{0.27}\square_{0.24}O_4$	0.8343(2)	2.501(5)	0.0009	14.6(9)	1.741(3)
$Fe_{2.34}Mg_{0.49}\square_{0.17}O_4$	0.8349(4)	2.506(7)	0.0107	20.0(15)	1.747(4)
$Fe_{2.15}Mg_{0.78}\square_{0.07}O_4$	0.8352(3)	2.497(6)	0.0118	17.8(25)	1.742(4)
$Fe_{2.12}Mg_{0.81}\square_{0.06}O_4$	0.8354(3)	2.499(7)	0.0119	17.9(13)	1.744(4)

$^{(*)}$Used as standard.
$^{(**)}$Determined by the Scherrer method Klug and Alexander [13]. Used as standard for the other determinations by the Warren-Averbach method.

An estimated value can be obtained by extrapolating to x = 0. In the present case, T_C for pure γ-Fe_2O_3 is found to be 952 K, which is close to the reported T_C ~900 K and ~950 K, according to Coey [10] and Murad [11] respectively.

Mössbauer spectra obtained at 20 K are shown in Figure 2. Two magnetic subspectra assigned to [Fe^{3+}] and {Fe^{3+}} spinel sites along with a central doublet were least squares fitted. Hyperfine parameters are presented in Tables 2a and b. The relative area of the central doublet (Figure 2) increases with the Mg content but XRD patterns do not show any crystalline phase other than maghemite. It is possible that the doublet is due to amorphous iron oxide. The magnetic spectrum was decomposed in two subspectra ascribed to Fe^{3+} in the [A] and {B} spinel sites.

Table 2a
20 K Mössbauer parameters relative to iron in {B} sites of spinel structure of the pure and Mg-doped maghemites. δ = isomer shift relative to αFe; B_{hf} = hyperfine field; ε = quadrupole shift and RA = relative subspectral area.

	B Site			
Spinel formula	δ/mm s^{-1}	B_{hf}/tesla	ε/mm s^{-1}	RA/%
$Fe_{2.67}\square_{0.33}O_4$	0.435(2)	52.7	0.034(4)	62.5
$Fe_{2.49}Mg_{0.27}\square_{0.24}O_4$	0.459(9)	52 6	-0.006(2)	52.6
$Fe_{2.34}Mg_{0.49}\square_{0.17}O_4$	0.460(4)	52.5	0.001(6)	47.9
$Fe_{2.15}Mg_{0.78}\square_{0.07}O_4$	0 458(1)	52.8	0.018(4)	46.7
$Fe_{2.12}Mg_{0.81}\square_{0.06}O_4$	0 47(1)	52.6	0 03(1)	45.5

Table 2b
20 K Mössbauer parameters relative to iron in [A] sites of spinel structure of the pure and Mg-doped maghemites. δ = isomer shift relative to αFe; B_{hf} = hyperfine field; ε = quadrupole shift and RA = relative subspectral area.

	A Site				
Spinel formula	δ/mm s^{-1}	B_{hf}/tesla	ε/mm s^{-1}	RA/%	$RA_{\{B\}}/RA_{[A]}$
$Fe_{2.67}\square_{0.33}O_4$	0.390(3)	51.0	-0.065(6)	37.6	1.66
$Fe_{2.49}Mg_{0.27}\square_{0.24}O_4$	0.412(2)	50.7	-0.083(3)	47.4	1.11
$Fe_{2.34}Mg_{0.49}\square_{0.17}O_4$	0.400(8)	50.5	-0.08(1)	52.1	0.92
$Fe_{2.15}Mg_{0.78}\square_{0.07}O_4$	0.392(2)	50.9	-0.002(3)	53.3	0.88
$Fe_{2.12}Mg_{0.81}\square_{0.06}O_4$	0.424(4)	50.9	-0.001(1)	54.5	0.84

No other constraint was imposed to the fitting convergence routine except the linewidth of the {B} site, taken from free-fitting the spectrum of pure maghemite, which was kept fixed for all other samples, and the theoretical ratio 3:2:1 of line intensities. Hyperfine magnetic fields for this pure maghemite were found to be $B_{hf}[A]$ = 51.0 T and $B_{hf}\{B\}$ = 52.7 T, but they decrease with increasing Mg-content in both sites, except for x = 0.78 and 0.81. The higher doublet intensity in these samples lowered the relative sextet absorption and fitting accuracy might be somehow affected. Quadrupole shifts around zero are characteristic of Fe^{3+} in spinel lattice. The isomer shift differences $\delta_B - \delta_A$ range from ~0.05 to ~0.07 mm s^{-1}.

The tendency at higher x suggests that some elemental mechanism governs the isomorphic replacement of Fe^{3+} by Mg^{2+} in the oxide lattice. The continuous decrease of R_x with x (Figure 3) also suggests that the octahedral {B} site is preferentially occupied by the Mg^{2+}.

From the fraction x of Mg which is incorporated into the structure, a part {Mg} of it replaces a fraction {Fe} of iron in octahedral sites. However, this replacement tends to a saturation, in a exponential decay, imposed by the solubility limit of Mg in the octahedral site, as it can be expressed by Equation 3.

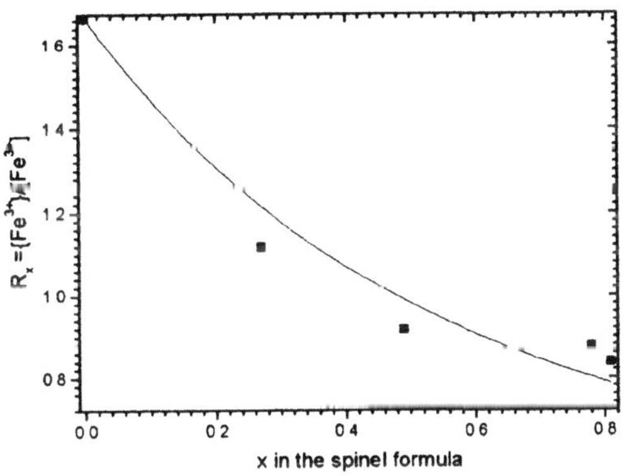

Figure 3. Variation of $R_x = \{Fe^{3+}\}/[Fe^{3+}]$ with the magnesium content expressed as the index (x) in the spinel formula.

Equation 3

$$-\frac{d\{Fe\}}{\{Fe\}} = \alpha \,.\, dx \ ,$$

where α is a partitioning constant, which accounts for the fraction ({Mg}) of the total structural Mg (x) occupying octahedral sites. By definition,

Equation 4

$$[Mg] = x - \{Mg\},$$

Equation 5

$$[Fe] = 1 - [Mg],$$

and

Equation 6

$$\{Mg\} = \{Fe\}_0 - \{Fe\},$$

where $\{Fe\}_0$ is the octahedral fraction of iron, for x = 0.

From Equation 3,

Equation 7

$$\{Fe\} = \{Fe\}_0 \, e^{-\alpha . x}.$$

Combining Equation 4 through Equation 7, and remembering that $\{Fe\}_0 = 5/3$ for pure maghemite,

Equation 8

$$R_x = (\{Fe\}_0 e^{-\alpha . x}) / (1 - (x - \{Fe\}_0 (1 - e^{-\alpha . x})) .$$

Equation 8 involves only one independent variable, x. The limits of x = {Mg} + [Mg] are zero (pure maghemite) and 1 (magnesioferrite, Fe_2MgO_4).

Sites A and B in 20 K Mössbauer spectra are only partially resolved, as indicated by the unequal intensities of lines 1 (leftmost) and 6 (rightmost). Iron occupancies in both sites taken from the relative subspectral areas are strongly affected by line widths estimation from the non-linear least squares of Mössbauer spectra fitting. Much better estimation would be achieved only by Mössbauer data from measurements with an externally applied strong magnetic field. Even though R_x-values taken from zero-field spectra bear somewhat high uncertainties, fitting the simple model expressed in Equation 8 with α = 0.88(4) leads to a limit R_x = 0.73(7), for x = 1.

The expected magnetization (σ) of any ferrimagnetic sample can be inferred from the unit formula (Formula 1) by

Equation 9

$$\sigma = F.(\{Fe^{3+}\} - [Fe^{3+}]).$$

F accounts for the magnetic moment of the Fe^{3+} ions in [A] and {B} sites and the molar mass of the spinel.

Figure 4 shows the relation between σ and S = $\{Fe^{3+}\} - [Fe^{3+}]/M$, where $[Fe^{3+}]$ and $\{Fe^{3+}\}$ are site occupancies of [A] and {B} sites as given by Mössbauer results, respectively, and M is the molar mass (kg mol^{-1}) of pure or Mg-doped maghemite. The experimental straight line reflects magnetization data obtained at 4.2 K and 5 T. Linear fitting gives ($R^2 = 0.99$)

Equation 10

$$\sigma = 13.2\,(3)S + 37.1(4).$$

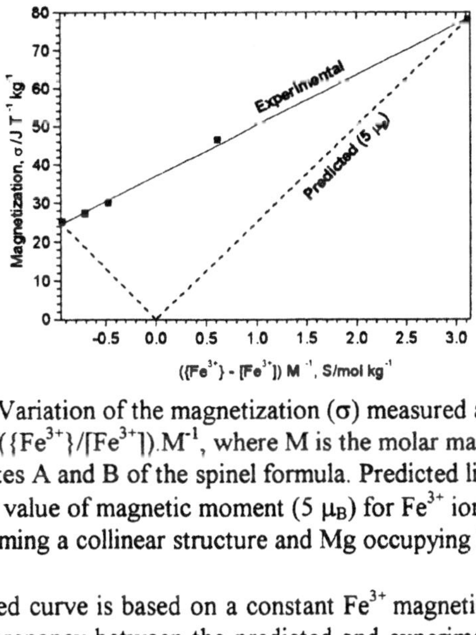

Figure 4. Variation of the magnetization (σ) measured at 4.2 K with $S = R_x.M^{-1} = (\{Fe^{3+}\}/[Fe^{3+}]).M^{-1}$, where M is the molar mass and Fe the iron index in sites A and B of the spinel formula. Predicted line is based on a constant value of magnetic moment (5 μ_B) for Fe^{3+} ion in both sites assuming a collinear structure and Mg occupying B sites.

The predicted curve is based on a constant Fe^{3+} magnetic moment of 5 μ_B, for both sites. The discrepancy between the predicted and experimental lines increases as the magnesium content increases (decreasing S, Figure 4). This discrepancy may be due to differences in magnetic moments of Fe^{3+}, with increasing Mg content, in A and B sites. Murthy et al. [12] reported magnetic moments by neutron diffraction as low as 2.48 μ_B for Fe^{3+} in B site and 3.89 μ_B in A sites, respectively, of magnesioferrite, based on polarized and unpolarized neutron diffraction measurements, and allocating 14 mol% Mg substitution on A sites. The low moments reflect spin canting in the ferrimagnetic structure.

4. Conclusions

A set of five maghemites with increasing Mg contents (x = 0; 0.27; 0.49, 0.78 and 0.81) were examined by chemical, XRD and Mössbauer analysis. Lattice parameter of the tetragonal cell and Curie temperature (T_c) were found to vary linearly with x according to the linear expression a/nm = 0.0018(2)x + 0.8337(1) and T_c/K = -

100(19)x + 952(12), respectively. Mössbauer spectra obtained at 20 K were fitted with two magnetic subspectra ascribed to the octahedral {Fe^{3+}} and tetrahedral [Fe^{3+}] sites of spinel. The relative subspectral area ratio, R_x = {Fe^{3+}}/[Fe^{3+}] is empirically related to x by an exponential function. Differences between experimental and predicted magnetizations may be due to variations of the ordered magnetic moments of Fe^{3+} ions in A and B sites with increasing Mg content. The reduction of A and B moments by Mg, which is too high to be explained only by covalence, must be due to spin canting.

Acknowledgments

Work financially supported by CNPq, FINEP and FAPEMIG (Brazil). CAPES (Brazil) granted a DSc fellowship to W. da N. Mussel.

References

[1] Resende, M., Allan, J., Coey, J. M. D., Earth Planet Sci. Letters, 78 (1986) 322.

[2] Resende, M., Santana, D. P., Franzmeier, D. P. and Coey, J. M. D., In: Proc. of VIII Intern. Soil Classification Workshop, Rio de Janeiro (1988) 78.

[3] Allan, J. E. M., Coey, J. M. D., Sanders, I.S., Schwertmann, U., Friedrich, G. and Wichowski, A., Min. Mag. 53 (1989) 299.

[4] Allan, J. E. M.; Coey, J. M. D.; Resende, M. and Fabris, J. D., Phys. and Chem. of Minerals 15 (1988) 470.

[5] Moukarika, A., O'Brien, F. and Coey, J.M.D., Geophysical Research Letters 18 (1991) 2043.

[6] Fabris, J. D.; Coey, J. M. D.; Jesus Filho, M. F.; Santana, D. P.; Goulart, A. T.; Fontes, M. F. and Curi, N., Hyperfine Interactions 91 (1994) 751.

[7] Fabris, J. D.; Coey, J. M. D.; Qi, Q. and Mussel, W. da N. American Mineralogist 80 (1995) 664.

[8] Cugat, O., Byrne, R., McCauley, J. and Coey, J. M. D., Rev. Sci. Instrum. 65 (1994) 3570.

[9] Kaye, G. W. C. and Laby, T. H., in: Tables of Physical and Chemical Constants and some Mathematical Functions, 14th edition (Longman, London and N.Y., 1975) p. 386.

[10] Coey, J. M. D., NATO ASI Series C: Mathematical and Physical Sciences 217 (1988) 397.

[11] Murad, E. Phys. Chem. Minerals 23 (1996) 248.

[12] Murthy, N. S. S., Natera, M. G., Begum, R. J., Youssef, S. I. in: Proceedings of the International Conference (1970) 60.

[13] Klug, H. P.and Alexander, L. E. in. X-Ray Diffraction Procedures for Polycrystalline and Amorphous Materials Second Edition, (John Wiley & Sons, New York, 1974).

PEDOGENIC INSTABILITY OF MAGNETITE FROM AMPHIBOLITE

Magda C. Ferreira Pinto[a], *Milton F. de Jesus Filho*[a(†)]; *Antonio T. Goulart*[b], *José D. Fabris*[c]; *Genilson P. Santana*[a]

[a]Departamento de Química, Universidade Federal de Minas Gerais (UFMG), 31270-901Belo Horizonte, MG, Brazil.
[b]Departamento de Química, Universidade Federal de Viçosa, 36571-000 Viçosa, MG, Brazil.
[c]Visiting Researcher, Departamento de Química da UFMG. Permanent address: EMBRAPA/CNPMS, Caixa Postal 151, CEP 35701-970, Sete Lagoas, MG; Brazil.

Iron oxide rich-portions obtained by magnetic separation, followed by selective chemical treatment with NaOH, from the fresh and altered parent rock samples of a pedosystem developing on amphibolite were studied by Mössbauer spectroscopy, powder X-ray diffractometry (XRD), magnetic measurements and chemical analysis. The fresh rock was found to contain magnetite, whereas only maghemite was found in the altered rock. The rock magnetite is therefore quite unstable to be preserved in the soil. Both fresh and altered rock samples contained also a ferric ilmenite. Unlike the amphibolite-magnetite this ferric ilmenite is preserved under the rock alteration processes representing the early stage of pedogenesis.

1. Introduction

Tropical soils are formed under relatively strong weathering conditions and typically concentrate large amounts of cation (mainly Fe^{3+}, but also Al^{3+}, Ti^{4+}, Mn^{2+}, among others) oxide minerals. The most common crystalline iron oxides are hematite (αFe_2O_3, hereinafter, Hm) and goethite ($\alpha FeOOH$, Gt), but magnetic soils may also contain maghemite (γFe_2O_3, Mh) or magnetite (Fe_3O_4, Mt). The characteristic spontaneous magnetization of these two iron oxides in the pure form are ~60 J T^{-1} kg^{-1} and 100 J T^{-1} kg^{-1}, respectively. However, much lower values are actually found for soil-Mh and soil-Mt, due to isomorphic replacement of Fe by other elements, particularly Al, Ti, Mg, and minor amounts of, e. g., Cr, Ca, Mn, Mo, Zn or Cu. Ferric ilmenite ($xFeTiO_3.(1-x)Fe_2O_3$, Im; $0.83 \leq x \leq 0.95$) is often associated with the spinel phase, in mafic systems [1], [2].

Basalt-magnetite is pedogenically unstable, being found only in the fresh parent rock but not in the soil, whereas ilmenite in such a lithosystem is much more stable during pedogenesis [3].

In the present paper we report some results on the study of iron-bearing minerals, particularly on the spinel phases and ilmenite, of a amphibolite from a weathering mantle found nearby Viçosa, Minas Gerais, Brazil.

2. Experimental

Cobbles of amphibolite exhibiting an altered zone (identified as a yellowish outer layer coating the fresh rock) were collected from a weathering mantle overlayed by a dusky red magnetic Alfisol located in a cut just beside the BR120, 3 km from Viçosa (20° 45' 20'' S 42° 52' 40'' W), Minas Gerais, Brazil. The fresh and altered rock samples were broken up with a steel hammer and powdered in a ball mill. The magnetic separates from the ground fresh and altered rock samples were obtained from a dilute slurry in water while stirring in a magnetic rod stirrer. The magnetic particles were then collected with a small hand magnet. The magnetic separates from the fresh and altered rock samples were treated with 100 mL of boiling NaOH 5 mol L^{-1} in three 3 h - periods to

(†)Professor Milton Francisco de Jesus Filho deceased on 02 January 1996.

remove silicates [4], [5]. The residual fraction was then treated with diluted HCl to remove any formed sodalite. The remainder (hereinafter referred to simply as extract) was analysed by powder X-ray diffractometry (XRD) with a Rigaku Geigerflex equiped with a graphite diffracted beam monochromator, using Cu(K_α) radiation and by room temperature and 80 K Fe^{57} Mössbauer spectroscopy, in a constant-acceleration transmission set up with a Co^{57}/Rh source. Magnetization measurements were made with a portable soil magnetometer conceived by Coey et al. [6]. The ilmenite-rich samples were obtained by successively attacking the NaOH-treated fraction with HCl 1:1 until no residual magnetization was easily detected. The chemical analysis was carried out by dissolving the sample in HF + HCl + HNO_3. The solution so obtained was then analyzed for iron by the standard dichromatometric method [7] and with a Spectroflame Analytical Instruments plasma emission spectrophotometer for the other elements.

3. Results and Discussion

The chemical composition of the whole fresh and altered amphibolite samples is presented in Table 1. From these data, this fresh rock corresponds to a high-Fe tholeiitic basalt, according to the Jensen [8] plot. Alteration processes in this case imply loss of silicates (lower contents of Si, Al, Ca and Mg, in the altered rock) and enrichment in iron and titanium compounds. XRD patterns (not shown) of untreated fresh rock reveal that augite is the main silicate and ilmenite and magnetite or maghemite are the only iron oxides in both fresh and altered amphibolite. Patterns of the magnetic separate both from fresh and altered rock samples are presented in Figs. 1a and 2a. Corresponding XRD of these samples treated with HCl, to selectively remove oxides other than ilmenite, are depicted in Figs. 1b and 2b. Cell edges for ilmenite for both fresh (a = 0.5074(9) nm; c = 1.411(8) nm) and altered rock (a = 0.5059(8) nm; c = 1.402(6) nm) samples are lower than those found by Doriguetto et al.[1] for samples of an altered amphibolite from the same pedological site.

Table 1
Chemical composition of the whole fresh and altered amphibolite samples.

	Fe_2O_3	Al_2O_3	TiO_2	CaO	MgO	MnO	P_2O_5	SiO_2	Na_2O	K_2O	H_2O	LOI(*)	Total
	mass%												
Fresh	14.60	12.23	4.87	7.42	4.15	0.16	0.66	47.15	2.67	2.77	1.24	1.60	99.5
Altered	29.40	9.37	6.38	0.62	1.23	0.09	1.15	33.95	1.27	1.72	4.96	8.66	98.8

(*)Loss on ignition.

No superstructure line was identified for the spinel phases in both fresh and altered samples. Values of the cubic unit cell dimension (a_0) of the spinel phases decrease on going from the fresh (a_0 = 0.8392(2) nm) to the altered rock magnetic (a_0 = 0.8313(1) nm) extracts. In the former, a_0 is closer to the characteristic value for magnetite (a_0 = 0.83967 nm, JCPDS card # 19-629), whereas in the later a_0 approaches that of maghemite (cubic, a_0 = 0.8350 nm, JCPDS card # 24-81). The saturation magnetization (σ) of the whole fresh (σ = 1.48(2) J T^{-1} kg^{-1}) and altered (σ = 1.78(3) J T^{-1} kg^{-1}) rock and their corresponding magnetic extracts (σ = 44.6(9) J T^{-1} kg^{-1} and σ = 36.9(4) J T^{-1} kg^{-1}, respectively) indicates that the magnetic separation is effective in concentrating the magnetic phase in the samples as much as 20 (altered rock) to 30 (fresh rock) fold. Mössbauer analysis provides clearer evidence of the nature of the spinel phase in the magnetic extracts from fresh and altered rock samples. Magnetite is found in the fresh rock sample (Figs. 3a and 3b). Room-temperature Mössbauer parameters (Table 2) indicate that the mixed-valence $Fe^{3+/2+}$ in the octahedral sites of magnetite represents 47 % of the whole spectral area. Considering that the recoilless fraction corresponding to this site (f_B) is 6 % lower than that of tetrahedral sites (f_A) [9], the relative sites occupancy would be 58.8 % in B and 41 2 % in A sites of the spinel structure. Hyperfine fields are significantly different from $B_{hf}(B)$ = 43.3 T and $B_{hf}(A)$ = 47.2 T found by de

Jesus Filho et al. [3], for the magnetite of a tholeiitic basalt. Mössbauer parameters of augite and ilmenite are presented in Table 3.

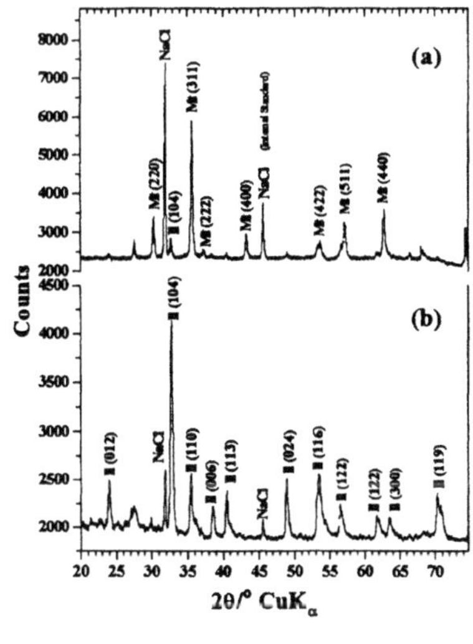

Fig. 1. Powder X-ray patterns of the (a) magnetic separate from the fresh amphibolite and (b) ilmenite-rich residue remained after attack of the magnetic separate with HCl. Mt = Magnetite, Il = Ilmenite. NaCl is the internal standard.

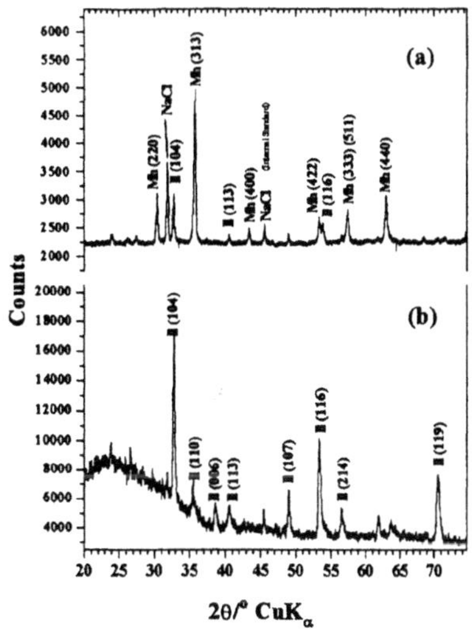

Fig. 2. Powder X-ray patterns of the (a) magnetic separate from the altered amphibolite and (b) ilmenite-rich residue remained after attack of the magnetic separate with HCl. Mh = Maghemite, Il = Ilmenite. NaCl is the internal standard.

A remarkable result is obtained for the magnetic extract of the altered rock. Maghemite can be identified from the RT and 80 K Mössbauer spectra (Figs.4a, 4b and 4c; Table 4) together with some silicate and ferric ilmenite. The central doublet due to ilmenite is resolved in the RT Mössbauer spectra in a narrower velocity scale (Fig. 4d and Table 5). The presence of only maghemite as spinel phase means that magnetite in this lithosystem is rather unstable during pedogenesis. At 80 K fitted Mössbauer spectra give 37 % of the subspectral area due to Fe^{3+} in tetrahedral and 42 % in octahedral spinel sites. Hyperfine parameters of this maghemite are presented in Table 6.

The Mössbauer spectra of HCl-treated sample obtained with a narrower velocity scale allow the separation of two Fe^{2+} and one Fe^{3+} doublets ascribed to ilmenite (Figs. 3c and 4d and Table 7). XRD pattern gives no evidence of significant amount of residual silicates. The subspectral Mössbauer areas (RA) can be corrected with the recoilless factor (f) by using the values reported by De Grave and Van Alboom [10] for f due to Fe^{2+} (f_{Fe}^{2+} = 0.650) and Fe^{3+} (f_{Fe}^{3+} = 0.837) for iron in stoichiometric ilmenite and hematite, respectively. Thus the molar fraction of ilmenite (x) in the solid solution

$$xFeTiO_3.(1 - x)Fe_2O_3$$

is related to RA by

$$2(1 - x)/x = (f_{Fe}^{2+} RA_{Fe}^{3+}/f_{Fe}^{3+} RA_{Fe}^{2+}).$$

Using RA_{Fe}^{3+} and total RA_{Fe}^{2+} from Mössbauer data (Table 7), x = 0.92 and 0.91, for the fresh and altered rock samples, respectively, which are consistent with results reported by Doriguetto et al. [1].

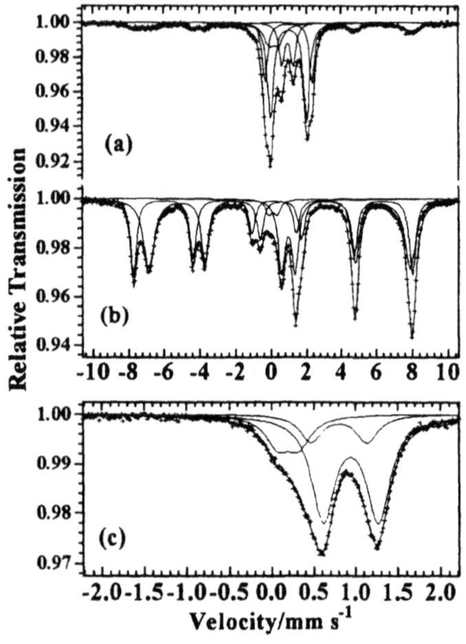

Fig. 3. Room temperature Mössbauer spectra of the (a) fresh amphibolite, (b) magnetic separate from the fresh rock and (c) ilmenite-rich residue remained after attack of the magnetic separate with HCl.

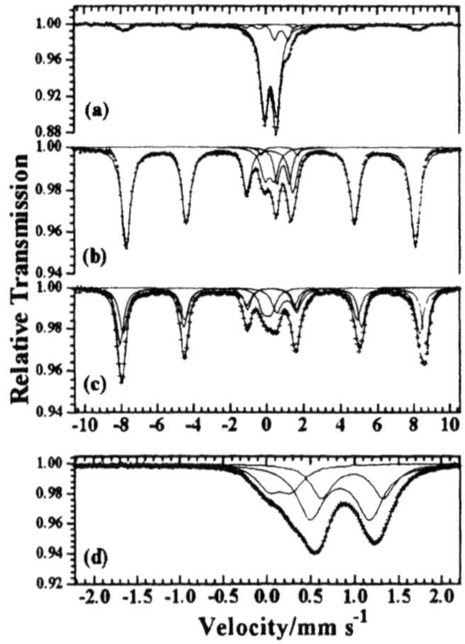

Fig. 4. Mössbauer spectra of the (a) whole fresh amphibolite at room tempeature (RT); of the magnetic separate from altered rock (b) at RT and (c) 80 K and (d) of the ilmenite-rich residue remained after attack of the magnetic separate with HCl.

Table 2

Least-squares fitted room temperature Mössbauer parameters for the residue remaining after 5 mol L^{-1} NaOH extraction from the fresh amphibolite. δ = isomer shift relative to αFe; Δ = quadrupole splitting; ε = quadrupole shift; B_{hf} = hyperfine magnetic field; RA = relative area of the subspectrum; { } and [ ] mean octahedral and tetrahedral sites, respectively, in the spinel structure.

Mineral	Site	δ/mm s^{-1}	ε, Δ/mm s^{-1}	B_{hf}/tesla	RA/%
Magnetite	{$Fe^{3+/2+}$}	0 66	-0.03	45 8	47
	[Fe^{3+}]	0 29	-0.02	48 8	35
Ilmenite	Fe^{2+}	1 09	0.75		15
Silicate(*)	Fe^{3+}	0 30	0.49		3

(*)Unresolved doublet. The resolved parameters for this subspectrum are presented in Table 3.

Table 3

Least-squares fitted room temperature Mössbauer parameters for the central doublets of the chemically untreated fresh amphibolite sample. δ = isomer shift relative to αFe; Δ = quadrupole splitting; RA = relative area of the subspectrum.

Mineral	Site	δ/mm s^{-1}	Δ/mm s^{-1}	RA/%
Augite	Fe^{2+}	1.14	2.02	47.9
	Fe^{2+}	1.13	2.67	22.8
	Fe^{3+}	0.41	0.83	14.3
Ilmenite	Fe^{2+}	1.03	0.72	15.0

Table 4

Least-squares fitted Mössbauer parameters room temperature for the central doublets of the chemically treated altered amphibolite sample. δ = isomer shift relative to αFe; Δ = quadrupole splitting; ε = quadrupole shift; B_{hf} = hyperfine magnetic field; RA = relative area of the subspectrum.

Mineral	Site	δ/mm s^{-1}	ε, Δ/mm s^{-1}	B_{hf}/tesla	RA/%
Maghemite	Fe^{3+}	0.32	-0.02	49.08	79.7
Ilmenite	Fe^{2+}	1,03	0.66		8.7
					9.8
	(*)	0.83	1.96		1.8

(*)Unresolved doublet. The resolved parameters for this subspectrum are presented in Table 5.

Table 5

Least-squares fitted Mössbauer parameters for the central doublets of the chemically untreated altered amphibolite sample. δ = isomer shift relative to αFe; Δ = quadrupole splitting; RA = relative area of the subspectrum.

Mineral	Site	δ/mm s^{-1}	Δ/mm s^{-1}	RA/%
Augite	Fe^{2+}	1.04	2.51	1.3
Ilmenite	Fe^{3+}	0.37	0.66	68.3
	Fe^{2+}	0.96	0.73	10.0

Table 6

Least-squares fitted 80 K Mössbauer parameters of the chemically untreated altered amphibolite sample. δ = isomer shift relative to αFe, Δ = quadrupole splitting; ε = quadrupole shift; B_{hf} = hyperfine magnetic field; RA = relative area of the subspectrum.

Mineral	Site	δ/mm s^{-1}	ε, Δ/mm s^{-1}	B_{hf}/tesla	RA/%
Maghemite	Fe^{3+}(A)	0.37	0(*)	50.20	37
	Fe^{3+}(B)	0.45	0(*)	51.82	42
	Fe^{3+}	0.22	0.41		10
	Doublet	1.25	1,08		11

(*)Fixed

Table 7

Least-squares fitted room temperature Mössbauer parameters of the ilmenite-rich residue remained after attack of the magnetic extract of the fresh and altered rock samples with HCl. δ = isomer shift relative to αFe; Δ = quadrupole splitting; RA = relative area of the subspectrum.

Sample source	Site	δ/mm s⁻¹	Δ/mm s⁻¹	RA/%
Fresh rock	Fe^{2+}	0.92	0.68	16
	Fe^{2+}	1.06	0.66	65
	Fe^{3+}	0.32	0.25	19
Altered rock	Fe^{2+}	0.96	0.67	53
	Fe^{2+}	1.09	0.70	27
	Fe^{3+}	0.28	0.25	20

4. Conclusions

Iron oxide rich extracts obtained by magnetic separation, followed by selective chemical treatment with NaOH, from a fresh and an altered portions of an amphibolite were examined in detail by chemical, powder XRD and Mössbauer anlaysis. The fresh rock was found to contain magnetite whereas only maghemite was identified in the altered rock sample. Ilmenite was the other iron-rich phase also present both in the fresh and altered rock samples. From Mössbauer analysis, this phase consists actually of a solid solution of the type $xFeTiO_3.(1 - x)Fe_2O_3$, with $x \sim 0.90$, in both fresh and altered amphibolite. These result imply that the rock-magnetite is rather unstable and ilmenite is somehow preserved under the rock alteration processes representing the early stage of pedogenesis.

Acknowledgments

Work financially supported by CNPq, FINEP and FAPEMIG (Brazil).

References

[1] A.C. Doriguetto, A.T. Goulart; M.F. de Jesus Filho; J.D. Fabris, G.P. Santana and J.M.D. Coey, Europ. J. Soil Sci. (1997). Submitted.
[2] A.T. Goulart, M.F. de Jesus Filho, J.D. Fabris and J.M.D. Coey, Hyp. Int. 91(1994)771-775.
[3] M.F de Jesus Filho,. J.D. Fabris, A.T. Goulart, J.M.D. Coey, B.A. Ferreira and M.C.F Pinto, Clays and Clay Minerals 43(1995)641-642.
[4] K. Norrish, R. M. Taylor, Journal Soil Science 12(1961)295-306.
[5] N. Kämpf, U. Schwertmann, Clays and Clay Minerals 30(1982)401-408.
[6] J.M.D. Coey, O. Cugat, J. McCauley and J.D. Fabris, Revista de Física Aplicada e Instrumentação 7(1992)25-30.
[7] P.G. Jeffery and D. Hutchison, *Chemical methods of rock analysis* (Pergamon, Oxford, 1981).
[8] L.S. Jensen, A new cation plot for classifying subalkalic volcanic rocks. Ontario Department of Mines, Miscellaneous Paper 66(1976).
[9] G. A. Sawatzky, F. van der Woude and A. H Morrish, Phys. Rev. 183(1969)747-757.
[10] De Grave and A. Van Alboom, Phys Chem. Minerals 18(1991)337-342.

MINERALOGICAL AND MAGNETIC STUDY OF AN AREA WHICH IS AFFECTED BY THE SPONTANEOUS COMBUSTION OF COAL.

L. D'ONOFRIO[1]; F.GONZALEZ-JIMENEZ[1]; N. ORIHUELA[2] and M.D. SOTO[2]
[1]Facultad de Ciencias, A.P.47586, [2]Facultad de Ingeniería, A.P.1053, Universidad Central de Venezuela, Caracas, Venezuela

Abstract

The area under study is located in the northwestern part of Lara Sate, Venezuela. Within this area a subvertical body of coal of the Late Tertiary crops out, with a size of approximately 4.5 m thick and 3 m height, being internally in combustion. In certain points of the topographical surface affected by this phenomenon the temperature reached up to 225 °C. This combustion affects thermally the shales adjacent to the coal, to the point of transforming them into natural brick. The thermal transformation undergone by these shales involves strong mineralogical changes which give new physical properties to them, in particular the magnetic ones. In the present work we have undertaken a detailed study of the modifications on the local geomagnetic field on the electromagnetic field emitted by remote stations, and also a mineralogical and magnetic characterization of different samples collected on the field, by Mössbauer spectroscopy (MS), magnetic susceptibility (χ) and powder X-ray diffraction (XRD) measurements. We have determined that the magnetic anomalies observed over the surface of the terrain are related to the presence of magnetite (Fe_3O_4) in the zones which have been affected both thermally and chemically by to the coal combustion.

1. Introduction

The process of subterranean coal combustion is known since the coal extraction activity began. However, this phenomenon has often been mistakenly considered as an igneous volcanic process. There are different causes for this combustion. The most important of these is the oxidation of coal by atmospheric oxygen [1,2,3,4], after a sudden exposure due to a landslide or human action.

The highest temperatures which the processes of subterranean coal combustion reach have been estimated by several authors [5,6,7], and in general they exceed 1000 °C. Usually the mineralogical products associated with these processes are classified into two large groups. In the first group are included all the minerals which are moved tens or even hundreds of meters, by the steams (pneumatolitics) that raise from the deep and hot zones up to the surface, and then crystallize at the edges of the chimneys. The second one includes the minerals that are produced by the *in situ* thermal alteration of the rocks adjacent to the coal in combustion (generally shales or sandstone), hold in a new rock known as porcelanite or clinker as result of the thermal and chemical alteration of the preexistent rocks. This process involves strong mineralogical changes which give new physical properties to the rocks, as color change, higher hardness and distinct magnetic properties. These magnetic properties depend highly on the presence of different iron oxides which can produce very important local magnetic anomalies [8,9,10,11].

A Venezuelan locality, known as "Las Monas Volcano", corresponds to the type of process that just has been described above. In the present work we attempted to explain the presence of a strong anomaly in the local geomagnetic field, by using Mössbauer spectroscopy, magnetic susceptibility and powder x-ray diffraction data analysis of several samples taken in the field.

2. "Las Monas Volcano"

"Las Monas Volcano" is located in the western part of Venezuela, northwest of Lara State, in the southern bank of the crick called Las Monas (10°21'23"North/70°30'19"West). This "Volcano" is developed from the combustion of the coal of the late Tertiary which only crops out subvertically to the level of the crick, as a remaining block of 4.5 m wide and 3 m height, where there are evidences of past combustion processes. The area that is active today is located up on the hill, above the outcrop of coal, in the southern direction. It covers about 800 m^2 of an uncovered and variegated area in red, orange, pastel, yellow and white, very different to the colors of the original shales. Also in this area stand out dour chimney structures through which the gases produced are expelled, and where are also synthesized the pneumolitic minerals. In the larger chimney it has been registered a temperature of 225°C on the topographical surface, but it is expected to be considerable higher deep inside and near to the coal in combustion. The largest thermal alteration in this area is evident in the shales adjacent to the coal (stratigraphically above the coal) which are transformed into natural brick (porcelanites or clinker).

3. Qualitative Analysis of the Total Magnetic Isointensity Map

With the purpose of evaluating the influence of the coal combustion of the "Las Monas Volcano" over the geomagnetic field, we performed a magnetic survey that comprised an area of 550 m^2 within the total active area. The measurement stations were distributed along six profiles, 1 m between them, and with 7 to 10 m between profiles (Fig. 1). A proton precession magnetometer with a sensitivity of 1 gamma ($1\gamma = 1nT$) was used. Time intervals smaller than half an hour were used for the measurements in order to control the drift during these intervals.

The magnetic isointensity map (Fig 1) presents contours with a variation of 650 γ from the northeast to the southeast, with a clear tendency to increase in the southeast part, where high magnetic fields stand out, being associated with the three chimneys located in this area. The high magnetic fields observed in the central and southern regions, lead us to think that these fields are related to the altered shales located in the active area. This interpretation has been confirmed by the susceptibility measurements made in several samples taken from the coal outcrop, in the bank of the crick and in the region nearby the profiles. There is also the possibility that the high magnetic fields of the central and southern area be related to an ensemble of small longitudinal cracks observed in the active area of this locality, just in between the third and fourth transversal profiles, and parallel to the main profile. In the extreme southeast part, the density of the isointensity contours tend to increase. The field gradient distribution goes smoothly down to a value of 50 γ/m. In this area also the local closures diminishes. This tendency can be interpreted as the approaching of the body of coal to the area of measurement, which would increase the magnetic behaviour associated to the presence of minerals produced by the coal combustion. This interpretation would also indicate that the dimensions of the body of coal increases towards the south, according to the topography of the active area under study.

In order to complement the information obtained by the total magnetic survey, a survey of the electromagnetic field signals emitted by a remote station (VLF) was also made. Likewise, this signal is altered by the local magnetic contribution, given results similar to those ones given by the previous survey. There is an increase of the VLF phase component in the areas following the increase of the magnetic response, with a corresponding decrease in the out of phase response. This can be explained by the increment of the secondary field around the areas altered by the combustion, and it is complemented with the appearance of anomalies with low wave number associated to the chimneys edges.

4. Sample Collection and Powder X-Ray Diffraction Data Analysis

Four samples where collected in the shale above the coal that generates the combustion process (Fig. 2). One of them (sample *d*) was taken in the active area at only 2 meters from the larger chimney. The other three samples were taken from the shales above the coal near to the bank crick, where the statigraphic order is conserved. When the physical characteristics of these three samples and its relation with the remnant coal are observed, it can be confirmed that the sample with the higher degree of thermal alteration is the one which is nearest to the coal (sample *c*). On the contrary the sample with the lowest or absent degree of thermal alteration is the one obtained away from the coal by nearly 1 m (sample *a*). The intermediate degree of thermal alteration is in between them (sample 2*b*). The XRD analysis performed in the samples *a*, *b* y *c*, is covalent with the thermal degree of alteration proposed before ($c>b>a$), on the basis of the minerals identified. These measurements were made in a Philips 1130 equipment with a Cu tube operated at 40 KV and 20 mA. For each sample two types of diffractograms were obtained. In the first one the sample was randomly oriented and scanned from 2.5-70°(2θ) with a step size of 0.02°(2θ) and counting time of 0.5 seconds per step. In the second diffractogram the sample was oriented in order to emphasize the signals from the clays, scanned from 5-35°2θ with a step size of 0.01° and counting time of 0.5 seconds per step. The results are shown in Table 1.

5. Mössbauer Spectroscopy Results

The Mössbauer spectra were obtained with a home made spectrometer running in the triangular symmetric mode for the velocity, with 512 channels to storage the data that then is summed up symmetrically in 256 channels The radioactive source is Co^{57} diluted in Palladium. In Fig. 3 are presented the Mössbauer spectra obtained at room temperature for the four samples. The points are the experimental data and the continuos line are the least-squares fit curve, being the free parameters: the isomer shift with respect to metallic iron (IS), FWHM (Γ), the electrical quadrupole splitting (QS), the hyperfine magnetic field (CH) and the relative spectral area (P) of each subspectrum which contributes to the total spectrum of the sample. These fitted parameters are shown on the Table 2, where also the iron site or iron valence state responsible for the magnetic behaviour is given.

6. Magnetic Susceptibility Results

The total magnetic susceptibilities χ of the samples as a function of temperature between 77 K and room temperature are shown in Fig. 4. These measurements were performed in a Faraday balance with applied magnetic field of 2854 Oe for *a*, *b* and *d* samples and 1425 Oe for the sample *c*. It can be observed that the order of magnitude of the room temperature χ for samples *a*, *b* and *d* are comparable, 5.5×10^{-6}, 8.5×10^{-6} and 11×10^{-6} emu/g respectively, although the values for samples *b* and *d* are larger than sample *a* due to the presence of hematite. At first sight it can be said that these three samples have a paramagnetic Curie-Weiss behaviour. One the contrary, the sample *c*, which is one that has been in contact with the coal, have a value of χ much larger than in the previous samples (4.1×10^{-4} emu/g at room temperature), and also it presents a magnetic transition below 130 K which is typical of magnetite (Fe_3O_4), the Verwey transition.

7. Discussion

The results of the characterization are summarized as follows:

1) From Mössbauer and χ measurements the sample *a* presents a paramagnetic Curie behaviour which corresponds to a mixture of clays with Fe^{2+} and Fe^{3+} ions well separated.
2) The samples *b* and *d* although come from different regions, have similar behaviour, which may reveal similar thermal oxidation treatment. From Mössbauer hematite (Fe_2O_3) is clearly identified with 55.5% of relative spectral area in sample *b* and 33.2% of relative spectral area in sample *d*. The paramagnetic Fe^{2+} y Fe^{3+} ions are characteristic of clays heated in the presence of oxygen, which preserve its crystalline structure [12].
3) The sample *c* contains hematite (17% of relative spectral area) and magnetite (31% of relative spectral area) This fact is a good indicator that the minerals present in, e.g., samples *b* and *d* have been reduced by gases as CO and hydrogen produced by the coal combustion with eventual presence of water. Furthermore, the paramagnetic Fe^{3+} (24%) and Fe^{2+} (29%) present in sample *c* can be contained in the remnant clays "squeletons" that have been first heated to high temperature and then partially reduced [13].

It can be emphasized that the magnetic anomalies observed on the surface are due to presence of magnetite in two specific subterranean environments where the reduction processes are dominant. The first of these environments is surely located in the contact interface between the coal that is (or was) in combustion and the altered shale. The second one is probably located in the walls of the chimneys and cracks were the reduction gases flow up to the surface. This fact explains the close relationship observed in the magnetic isointensity map (Fig. 1) between the location of the high magnetic field and the structure of cracks and chimneys present in the terrain.

Acknowledgments. We want to acknowledge the collaboration of Miss Oneida Leon and Mr. Edgar Jaimes in the susceptibility measurements and the figures.

References.

[1] DONALD, J & HUNT, J *Coal mining.* 3° edition. School of Mineral Industries. The Pennsylvania State College. Pennsylvania. (1952).

[2] KIN, A. *Laboratory studies on spontaneous heating of coal. A summary of information in the literature.* United States Department of the Interior. Bureau of Mines. Information Circular 8756. (1975) 13 p.

[3] RAMLU, M *Mine disasters and mine rescue.* Edit Balkema/Rotterdam. (1991) 400p.

[4] LEÓN, P *Fuegos en minas de carbón. Colección de tecnología y seguridad minera.* Ministerio de Industria, Comercio y Turismo. Secretaría General de Energía y Recursos Minerales. Instituto Geominero de España. (1992) 171 p.

[5] CHURCH, B; MATHESON, A & HORA, Z *Combustion metamorphism in Hot Creek area, British Columbia.* Canadian Journal of Earth Sciences. V 16. N 6. (1979) 1883-1887

[6] COSCA, M; GEISSMAN, J; SIMMOS, W & COATES, D *Pyrometamorphic rocks associated with naturally burned coal beds, Powder River Basin, Wyoming.* American Mineralogist. V 74. (1989) 85-100

[7] CLARK, B & PEACOR, D *Pyrometamorphism and partial melting of shales during combustion metamorphism: mineralogical, textural and chemical effects.* Contribution to Mineralogy and Petrology. V 112. (1992) 558-568

[8] MOTISCKA, P. Memorias V Congreso Geológico Venezolano. T. II. Caracas. (1977) 663-691

[9] HASBROUCK, M and HADSELL, F. *Geophysical exploration techniques applied to western United States coal deposits.* Edit Miller Freeeman First International Coal Exploration Symposium. San Francisco. (1976) 256-287

[10] CISOWSKY, S and FULLER M. Geological Society of America Bulletin. 99 (1987) 21- 29

[11] MORALES, V.; ESCOBAR, Y.; SOTO, M. and ORIHUELA, N. Acta Científica Venezolana. 46 Supl. N° 1.XLV Convención anual AsoVAC.USB. Caracas.(1955) 222

[12] H. Constant, F. Gonzales-Jimenez y M. Rosa Brussin. Hyperfine Interactions, 57 (1990) 1789-1794

[13] F. Gonzales-Jimenez, H. Constant, R. Iraldi, E. Jaimes and M. Rosa-Brussin. Hyperfine Interactions, 27 (1986) 927-930

Tables an figures

Table 1: XRD analysis The amount of minerals present is given in decreasing order.

Sample	Degree of Thermal Alteration	Minerals present
a	Low or absent	Low quartz 46%; smectite 22%; muscovite 22%; kaolinite 10%
b	Intermediate	Low quartz 54%; muscovite 36%; kaolinite10%; smectite; hematite
c	Large	Low and high quartz ; tridimite; mullite; maghemite?
d	?	Low quartz 62 %; alunite 38%

Table 2: Mössbauer parameters from spectra of Figure 3.

Samples	DI (mm/s)	Γ (mm/s)	QS (mm/s)	CH (kG)	P (%)	Compounds or iron valence
a	0.29	0.52	0.61	-	59.9	Fe^{3+} in clays
(Figure 3a)	1.25	0.38	1.92	-	40.1	Fe^{2+} in clays
b	0.38	0.36	- 0.22	516	55.5	Hematite
(Figure 3b)	0.25	0.40	0.48	-	34.1	Paramagnetic Fe^{3+}
	0.36	0.40	1.30	-	10.4	Distorted Fe^{3+}
c	0.38	0.44	- 0.21	514	16.8	Hematite
(Figure 3c)	0.72	0.56	- 0.01	460	17.9	Magnetite
	0.33	0.54	- 0.04	490	12.8	Magnetite
	0.35	0.52	0.73	-	23.7	Fe^{3+} in clays "skeleton"
	0.93	1.24	1.71	-	28.8	Fe^{2+} in clays "skeleton"
d	0.37	0.36	- 0.23	521	33.2	Hematite
(Figure 3d)	0.25	0.40	0.48	-	17.5	Paramagnetic Fe^{3+}
	0.36	0.40	1.31	-	41.3	Distorted Fe^{3+}

Figure 1. Magnetic isointensity map of the "Las Monas Volcano"
The geographical unit are in meter and the
magnetic ones are in gammas (1γ = 1 nT)

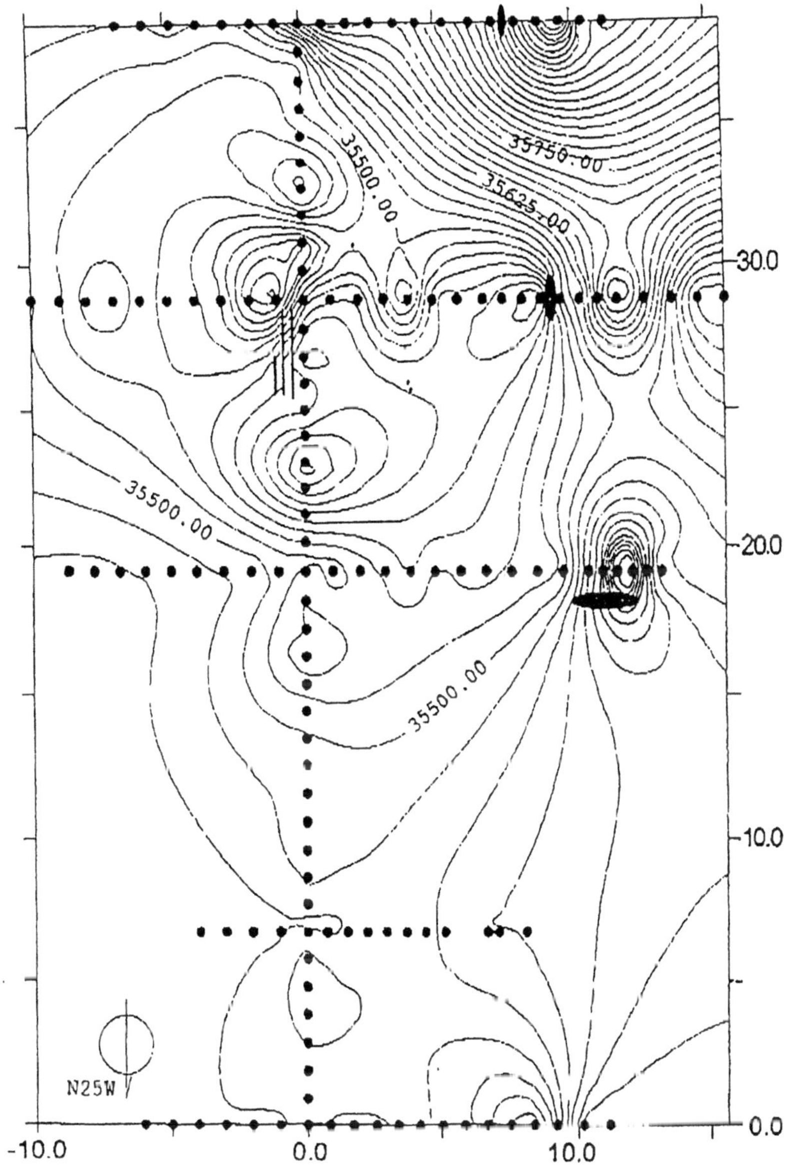

Figure 2. Schematic representation of the places where the samples were taken

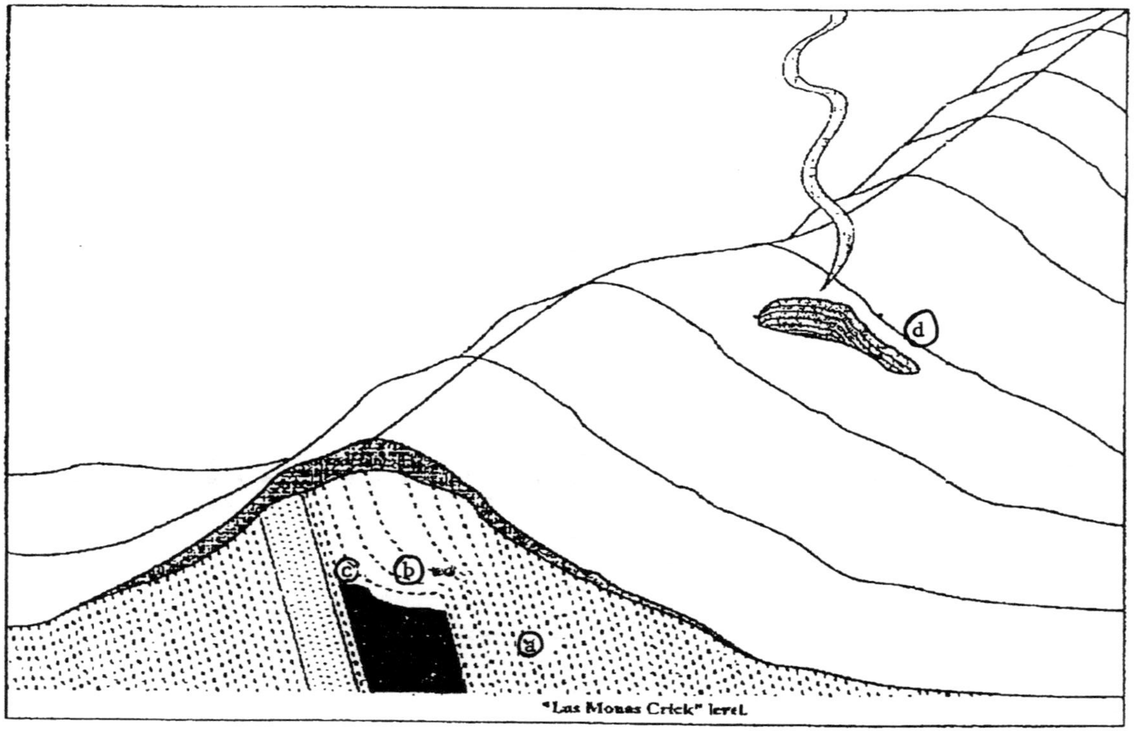

Figure 3. Mössbauer spectra of the samples at room temperature

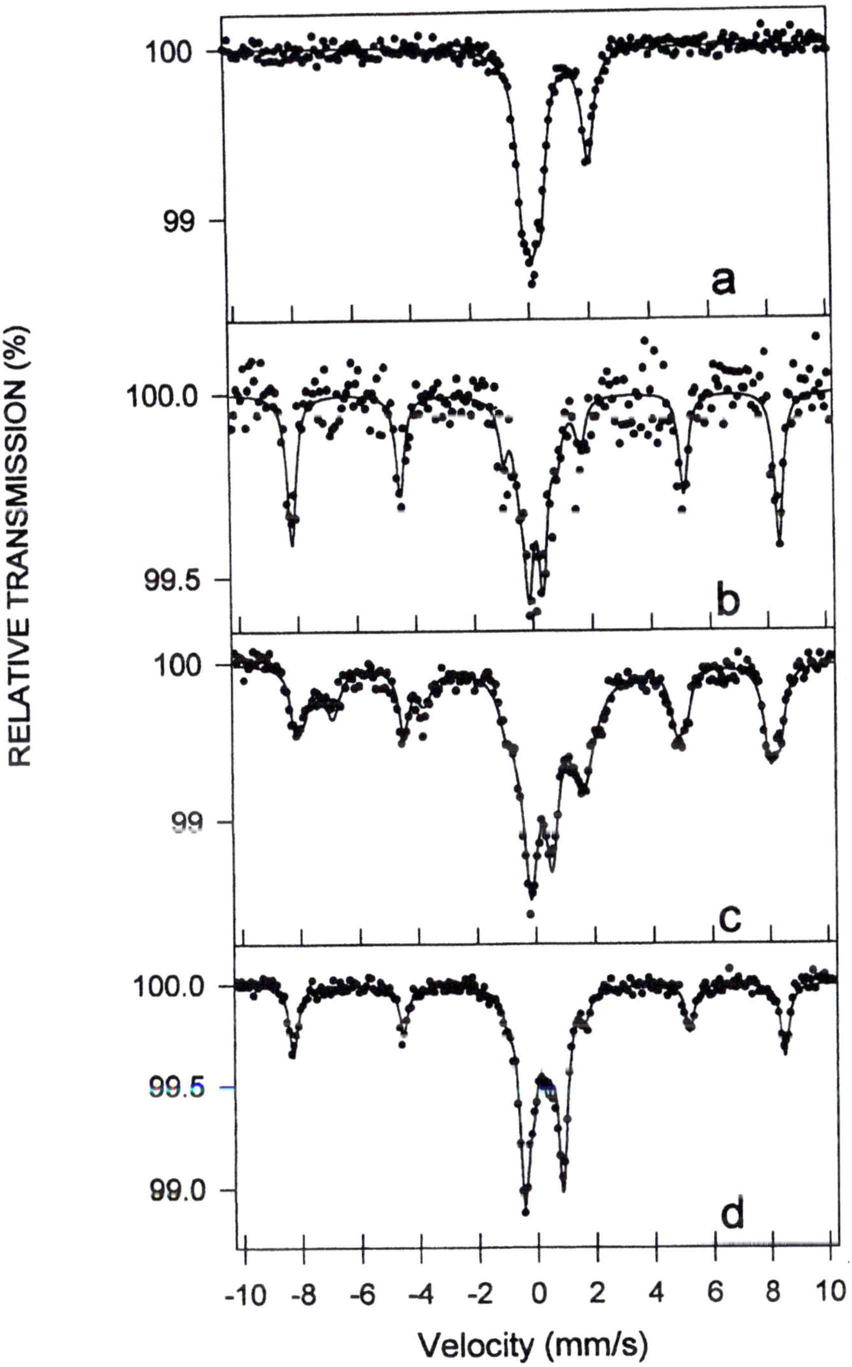

Figure 4. Magnetic susceptibility of samples as a function of temperature

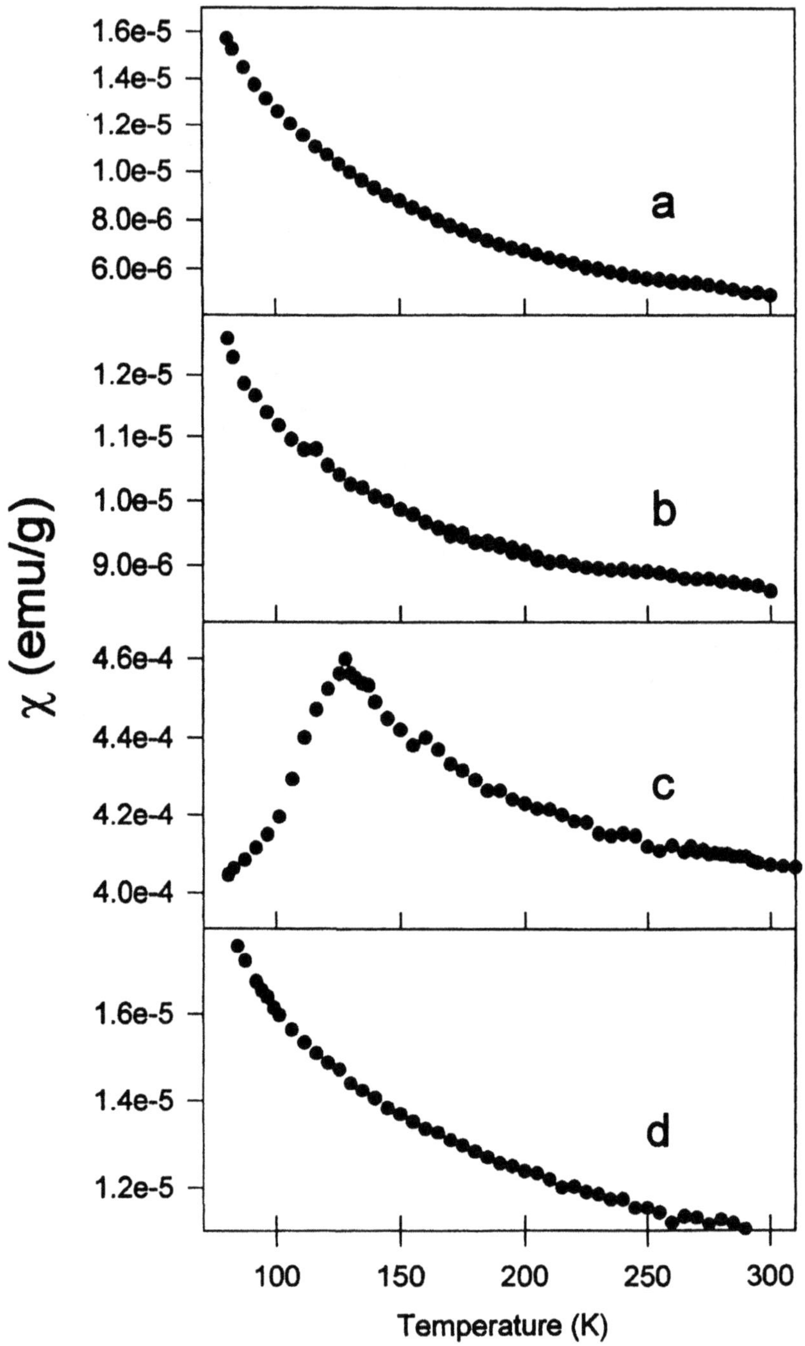

STUDIES IN THE TERNARY SYSTEM Fe-Nb-O [1]

E.C. Torres-Tapia, L.F. Moreira, M.R. do Amaral Jr, P.H. Domingues and J.M. Neto

Instituto de Física, Universidade Federal do Rio de Janeiro, Cidade Universitária
C.P. 68.528, 21945-970, Rio de Janeiro, RJ. Brasil

Homogeneous mixtures of Fe_2O_3 and Nb_2O_5 for Fe:Nb stoichiometries 1:1 and 2:1 were calcined at 980°C and 1200°C and then hydrogenated at 450°C until all traces of hematites have disappeared.. The samples with stoichiometry 2:1 were subsequently annealed at 1000°C. The reaction products are analysed in terms of Mössbauer spectroscopy and x-ray diffraction.

1. Introduction

One of the main interests in the study of the niobium-iron-oxygen diagram is the determination of the oxygenated phases that may occur, for example, in alloys and some special steels containing niobium. This phase diagram is, in general, very complex due to the coexistence of several phases. It was shown in the literature that the system Nb_2O_5-FeO reacts for temperatures above 1100°C giving phases like $FeNb_2O_6$ and $Fe_4Nb_2O_9$, while the system Nb_2O_5-Fe_2O_3 gives compounds like $FeNbO_4$ and $FeNb_{11}O_{29}$ [1,2]. At 1180°C [3] besides the last two iron-niobium oxides also $Fe_4Nb_2O_9$ can be produced. At this temperature the rutile structure of the NbO_2 (Nb_2O_4) is capable of taking large amounts of $FeNb_2O_6$ into a solid solution by the equilibrium reaction $Nb_2O_4 \leftrightarrow Fe_{0.67}Nb_{1.33}O_4$ forming a solid solution with the $FeNb_2O_6$. For the system Fe-Nb-O, at 1200°C was observed the formation of several phases, including the phases $FeNb_2O_6$, $Fe_4Nb_2O_9$, and was also shown that the spinel Fe_3O_4 is capable of incorporating small quantity of niobium ions in its structure giving a solid-solution of magnetite doped with niobium ions whose end composition was determined as $Fe_{2.43}Nb_{0.54}O_{4.00}$ ($Fe_7Nb_2O_{12}$) [4]. So, the reaction between oxides of niobium and iron at high temperatures depends on the path of the reaction. Our intention in this paper is to investigate the reaction products when niobium oxide and magnetite react through a different path, that is, produced under reducing atmosphere, where non-stoichiometric mixtures of Fe and Nb are pre-calcined at high temperatures and then, before annealing, reduced under H_2 flow at a lower temperature to produce magnetite, and follow the changes the phases undergo by the use of Mössbauer spectroscopy that is an excellent technique to detect the occurrence of iron in different environments and different states of oxidation, and x-ray diffraction.

2. Experimental

Homogeneous mixtures of Nb_2O_5 (CBMM, 99.9%) and Fe_2O_3 (analytical grade Riedel de Häen AG) with stoichiometries Fe:Nb=1:1 and 2:1 were calcined in air atmosphere at both 980°C for 48hs and 1200°C for 36hs. In both processes the materials were cooled down, and then hydrogenated at 450°C at a rate of one hour per gram of material in order to reduce the non-reacted hematite to magnetite, and then cooled again to room temperature in argon atmosphere. The materials were then ground and analysed by x ray and Mössbauer and the hydrogenation repeated until all traces of hematite have disappeared. Half of the samples with stoichiometry Fe:Nb=2:1 were then annealed at 1000°C for three hours, also in argon atmosphere. The following convention was adopted along this work: the samples calcinated at 980°C and 1200°C are identified as "A" and "B" respectively. The number following this letter indicates the molar ratio Fe:Nb: the molar ratio 1:1

---

[1] This work was partially supported by CNPq, FINEP and FAPERJ.

and 2:1 are designed by "1" and "2", respectively. An "A" at end of the designation means that the sample was annealed.

The Mössbauer spectra were all taken at room temperature with a conventional constant acceleration spectrometer (CMTE Elektronic) and a $^{57}Co/Rh$ source. The isomer-shifts were refered to metallic iron, and the spectra were fitted using the program Normos, and x-ray diffraction data were taken with a Siemens type F diffractometer using Co K_α and Cu K_α radiation.

3. Results and Discussion

Figures 1 and 2 show the Mössbauer spectra for these samples and Table I the hyperfine parameters obtained from its least-squares fit. The Mössbauer spectrum of the sample B1 was fitted with only one doublet of Fe^{3+}, and was identified by x-rays diffraction (XRD) as the monoclinic phase of $FeNbO_4$ (P-II in Table I). XRD diffractogram of this sample also identified traces of the monoclinic compound $FeNb_{11}O_{29}$ (P-VII) [2], Fig. 3, not seen in the Mössbauer spectrum due to its very small iron content. As it can be observed for the sample A1, pre-calcined at the temperature of 980°C, this temperature was not sufficient to complete the reaction as occurring in sample B1 and magnetite was still observed contributing with about 33% of the total Mössbauer spectrum that shows, besides magnetite and $FeNbO_4$ (P-II), another broadened Fe^{2+} phase (P-I) contributing with about 27% of the total spectrum. If we consider, as a first approximation, that all iron present in the samples are seen by Mössbauer and that no niobium enters the magnetite structure or interstitial sites, a simple analysis of iron and niobium content in the chemical formulae of the materials indicated that for the sample A1 the phase P-I should have about 60% of niobium, that is, the analysis of this spectrum indicates a proportion of Fe:Nb=0.45 for this phase, suggesting that it could be associated with the $FeNb_2O_6$. However, this phase could not be clearly determined by XRD due

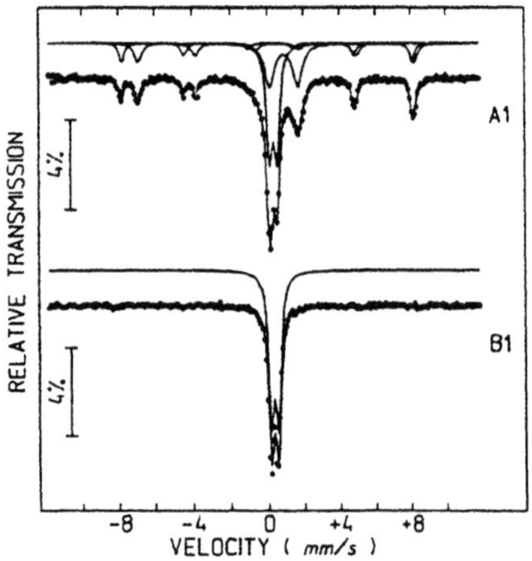

Fig. 1. Mössbauer spectra of samples with molar ratio Fe:Nb equal to 2:1.

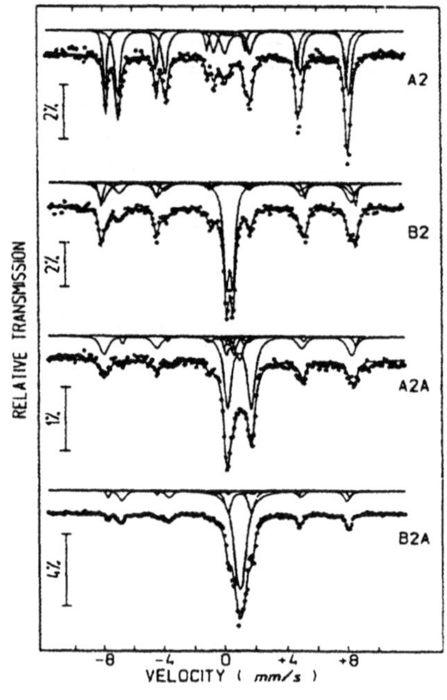

Fig. 2. Mössbauer spectra of samples with molar ratio Fe:Nb equal to 1:1.

to the fact that its diffraction pattern and those of magnetite and $FeNbO_4$ are similar, and the superimposition of these patterns hinder an unambiguous determination. Furthermore, as this material was analysed as-cast, the number of defects in the structure may also render difficult its identification by XRD. On the other hand, if we consider that part of niobium ions enter in the magnetite structure or in interstitial sites, the same analysis would lead to a Fe:Nb relation between

Table I - Hyperfine parameters of the fitted Mössbauer spectra. (A) and (B) are the magnetite tetrahedral and octahedral sites respectively, P-I is $FeNb_2O_6$, P-II is $FeNbO_4$, P-III is würstite, P-IV is $Fe_4Nb_2O_9$ and P-V is $FeNb_2O_6$. H is the hyperfine magnetic field, QS is quadrupolar splitting and IS is the isomer shift relative to metalic iron (^{57}Co/Rh).

		Magnetite		P-I	P-II	P-III	P-IV		P-V	P-VI
		A site	B site	Fe^{2+}	Fe^{3+}	Fe^{2+}	Fe^{3+}	Fe^{2+}	Fe^{2+}	Fe^{3+}
A1	Area(%)	13.7	19.7	26.0	39.8	-	-	-	-	-
	H(kOe)	493.0	468.5	-	-	-	-	-	-	-
	QS(mm/s)	-0.065	0.076	1.573	0.410	-	-	-	-	-
	IS(mm/s)	0.167	0.614	0.915	0.308	-	-	-	-	-
A2	Area(%)	31.4	54.4	12.4	1.9	-	-	-	-	-
	H(kOe)	489.5	468.8	-	-	-	-	-	-	-
	QS(mm/s)	-0.085	0.069	1.573	0.411	-	-	-	-	-
	IS(mm/s)	0.134	0.594	0.915	0.308	-	-	-	-	-
B1	Area(%)	-	-	-	100.00	-	-	-	-	-
	H(kOe)	-	-	-	-	-	-	-	-	-
	QS(mm/s)	-	-	-	0.430	-	-	-	-	-
	IS(mm/s)	-	-	-	0.343	-	-	-	-	-
B2	Area(%)	28.9	16.4	2.8	38.4	-	-	-	-	13.5
	H(kOe)	504.0	472.0	-	-	-	-	-	-	518.2
	QS(mm/s)	-0.215	0.179	1.598	0.430	-	-	-	-	-0.175
	IS(mm/s)	0.288	0.631	0.904	0.343	-	-	-	-	0.304
A2A	Area(%)	28.9	5.9	-	-	8.4	10.2	4.0	42.6	-
	H(kOe)	503.2	472.2	-	-	-	-	-	-	-
	QS(mm/s)	-0.132	0.206	-	-	0.386	0.514	1.113	1.532	-
	IS(mm/s)	0.200	0.793	-	-	1.088	0.350	0.843	1.073	-
B2A	Area(%)	5.7	16.9	-	-	61.4	1.0	1.4	13.7	-
	H(kOe)	491.9	457.1	-	-	-	-	-	-	-
	QS(mm/s)	0.030	-0.031	-	-	0.329	0.501	1.113	1.532	-
	IS(mm/s)	0.154	0.577	-	-	1.088	0.309	0.742	0.962	-

1:2 and 1:1, equivalent to $(Fe^{2+},Fe^{3+})Nb_2O_6$, that is, $FeNbO_4$ partially reduced. The presence of this phase is also to be expected when $FeNbO_4$ is produced at lower temperatures in reducing atmosphere [5], as was our case. X-ray shows also traces of another unidentified phase, Fig. 4.

For the samples with stoichiometry 2:1 it is natural always to expect the presence of magnetite due to the excess of iron. The phases obtained for the sample A2, as seen by Mössbauer, are similar to those for sample A1, except for the proportions. For the firing temperature of 1200°C (sample B2), the hydrogenation was not able to reduce all hematite to magnetite, and the hematite phase (P-VI) is still present with 13% even after sucessive cycles of hydrogenation, suggesting that this temperature is sufficiently high to allow some Nb ions to migrate into the hematite structure, inhibiting a further reduction. The presence of hematite in the Mössbauer spectrum can be explained considering that the amount of Nb diffusing into the hematite for the as-cast samples is sufficiently small to not alter its crystalline structure nor its hyperfine interactions appreciably, so that both Mössbauer, Fig. 2, and XRD only see the hematite sextet. This behaviour seems to be common to all mixtures calcinated at 1200°C independently of the starting proportions of Fe and Nb [6]. It can also be observed for this sample that there is a deficiency of iron in sites B (octahedral) of the magnetite. Normally, in the magnetite the contribution of sites B should be about the double of that of A sites (tetrahedral). In our case, due to the high contribution of Nb in the phases P-I and P-II (≈40%) and a small contribution in the hematite, the amount of Nb ions entering the structure of magnetite will be very small. So, the deficiency of iron in this sample is due essentially to vacancies in the B-sublattice, a result quite normal in view to the fact that the material is as-cast. From our data is estimated that no more than 5% of Nb ions are allowed to enter the magnetite.

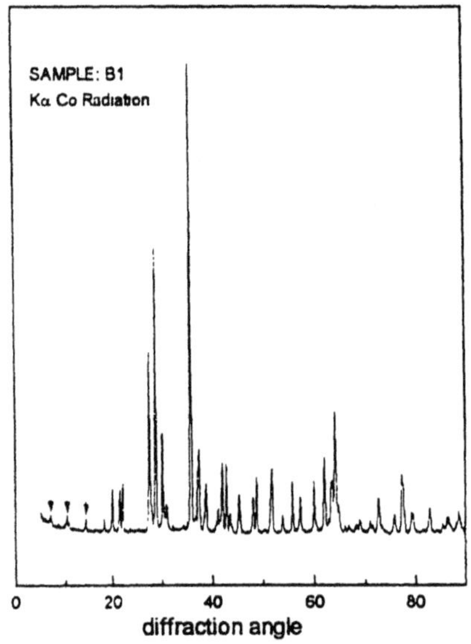

Fig. 3. X-ray diffraction pattern of B1 sample. The arrows indicate the small lines of $FeNb_{11}O_{29}$.

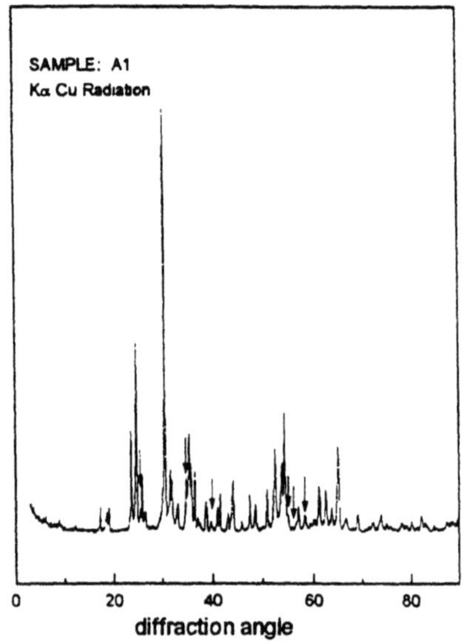

Fig. 4. X-ray diffraction pattern of A1 sample. The arrows indicate an unidentified phase.

When the samples A2 and B2 were submitted to a subsequent annealing at 1000°C, the Mössbauer spectra, besides the magnetite showed that the phases P-I, P-II and P-VI disappeared giving place to the new phases P-III (wustite), $(Fe^{2+},Fe^{3+})Nb_2O_6$ P-IV, and P-V, $Fe_4Nb_2O_9$ P-V, also confirmed in the XRD diffractograms.

There are some interesting aspects to be noted in the annealed materials. According to Tokikazi [5], in reducing atmosphere conditions, the Fe in phase P-IV is always present as Fe^{2+} and Fe^{3+} whose relative contribution depends on the reducing conditions. In fact, the Mössbauer spectrum also sees a doublet of iron in the oxidation state 3+ with 10% of the total area. Another point is that the phase P-V that appeared when both A2 and B2 samples were annealed may be considered as coming from the hematite since, as stated by Turnock [3], this phase coexists with hematite that can incorporate Nb into its structure forming the phase $Fe_4Nb_2O_9$. When the samples were annealed the diffusion of Nb ions into the hematite structure increases with the resulting stabilization of the phase $Fe_4Nb_2O_9$ [3]. According to Bertaut et al. [7] the iron in this phase is always present in its oxidation state 2+, and its contribution to the total Mössbauer spectra is only

4% for the sample A2A and 1.4% for B2A, suggesting that calcination at lower temperatures favours its stabilization. There are no Mössbauer results reported in the literature on this substance.

Finally, there is the wustite phase, P-III, observed in the Mössbauer spectrum as a very broadened doublet. The x-ray analysis for this phase detected two phases, generally presented in the literature as $Fe_{1-x}O$ (one with stoichiometry x=0 and the other with x=0.087). The phases identified by x-ray diffraction are shown in Table II.

Table II - Phases identified by x-ray diffraction. P-VII corresponds to the compound $FeNb_{11}O_{29}$ not seen in the Mössbauer spectra

Sample	Magnetite	P-I	P-II	P-III	P-IV	P-V	P-VI	P-VII
A1	X	X	X					X
B1			X					X
A2	X	X	X					X
B2	X	X	X				X	
A2A	X			X	X	X		
B2A	X			X	X	X		

X-ray results showed that the lattice parameter for the magnetite for A-samples does not change appreciably. For the sample A2A, as the Mössbauer spectrum presents a very broadened linewidth and an increased value for the hyperfine field, the contribution for the A-site of magnetite in Table I may be due to the presence of another splitted unidentified Fe^{3+} sextet due, probably, to the low reaction temperature.

The sample B2A presents a completely different behaviour. For this sample about of 12% of niobium are present in the Mössbauer spectrum, remaining ≈38% to be accomodated. Considering the limit of 7% for the diffusion of niobium into the magnetite forming the stoichiometric compound $Fe_4Nb_2O_{12}$ **[3]**, the difference (≈30%) may be attributed to the defficiency of iron in the Fe-Nb phases because of the small annealing time or even to the existence of interstitial Nb ions.

4. Conclusion

The solid state reaction in the system Fe-Nb-O with Fe:Nb ratios of 1:1 and 2:1 depends critically on the temperature of calcination. So far, no pure phase was obtained. Instead, compounds of niobium and iron with oxidation states 2+ and 3+ were found. In particular, and only for 1200°C and with Fe:Nb ratio 1:1, a nearly pure phase of $FeNbO_4$ (with traces of $FeNb_{11}O_{29}$, not seen in the Mössbauer spectrum) was obtained.

References

[1] J.C.A. Burdese and M.L. Bolera; La Metallurgia Italiana. 4(1965)150.
[2] V.K. Tronov, L.M. Kovba and Z.Y. Polshchikova; Russ. J. Inorg. Chem. **13**(1968)786
[3] A.C. Turnock; J. Amer. Ceram. Soc., **49**(1966)177.
[4] K. Kitayama; J. Sol. St. Chem., **69**(1987)101.
[5] E. Tokizaki, Y. Sugitani and K. Nagashima, Mat. Res. Bull. **21**(1986)231
[6] E.C. Torres-Tapia, L.F. Moreira, M.R. Amaral Jr, P.H. Domingues and J.M.Neto, to published
[7] E.F. Bertaut, L. Corliss, F. Forrat, R. Aleonard and R. Pauthenet, J.Phys.Chem. Solids, **21**(1961)234

HYPERFINE INTERACTIONS IN LAYERED SILICATES

Joice Terra and D.E.Ellis*
Department of Physics and Astronomy and Materials Research Center, Northwestern University,Evanston, Il 60208, USA.
* On leave from: Centro Brasileiro de Pesquisas Físicas, Rio de Janeiro, R.J., Brazil
Correspondence to: Joice Terra

Density functional self-consistent spin-polarized calculations with the first-principles Discrete Variational embedded cluster method were performed to obtain the electronic structure of metal sites in 2:1 layer silicates. The hyperfine parameters isomer shift, quadrupole splitting and magnetic hyperfine fields at octahedral Fe^{2+} sites in annite were computed and compared with experimental measurements derived by Mössbauer spectroscopy. Cluster size effects were investigated by systematically expanding the number of coordination shells included in the variational region.

I. INTRODUCTION

There have been many theoretical investigations in solids silicates in order to improve the comprehension about sorption and catalytic processes [1-4].

In the present work, we are concerned with the iron-bearing silicate annite that is typically found as a random solid solution. It has a mica layer silicate[5] which is a symmetric arrangement of two tetrahedral layers attached to opposite faces of a single octahedral sheet(T-O-T structure), as illustrated in Fig.1a.

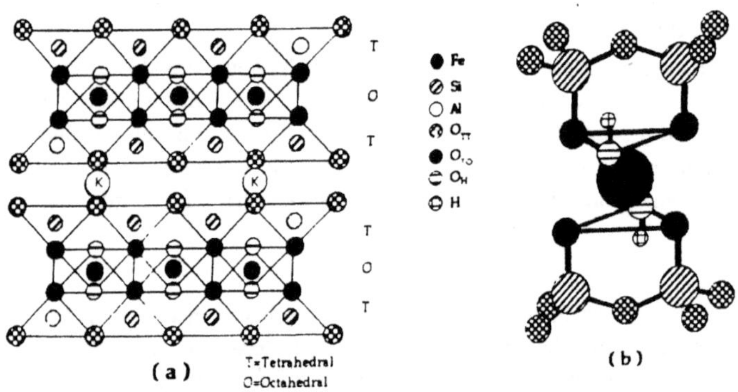

Figure 1- (a) Schematic illustration of annite composed by stacking the tetrahedral and octahedral sheets. (b) The largest cluster selected to represent annite with Fe^{2+} at the origin.

The molecular formula for ideal annite may be written as:

$$KFe_3(AlSi_3)O_{10}(OH)_2 . \quad (1)$$

In Eq.1, the tetrahedral sites are occupied by Si^{4+} and Al^{3+} in the ratio 3:1. Fe - atoms are nominally in the bivalent state octahedrally surrounded by six fourfold coordinated oxygens of two types, termed O_{TO} and O_H. All of these anions are shared by three octahedra; however, each O_{TO} is also the apex of one tetrahedron whereas O_H, which is bonded to one H atom forming the OH^- group, is unshared by tetrahedra. The hydroxyls are arranged around Fe in different positions, giving rise to *cis* and *trans* Fe^{2+} sites respectively in the ratio 2:1 along the octahedral layer. There are two other types of O , which are twofold coordinated to tetrahedral cations (Al or Si). In the present calculations they are considered to be equivalent and defined as type O_{TT}. Finally, the mobile K cations occupy the interlayer sites and ensure the electrical neutrality of the structure.

Here we report self-consistent embedded cluster calculations employing the discrete-variational (DV) method and density functional theory for two clusters representing annite. The forty atomic positions in the unit cell were derived from the position coordinates obtained by Guggenheim et al.[6] for the sample MOC-biotite, together with the lattice parameters from a synthetic annite[7] to model ideal annite as expressed in Eq.1. The clusters were embedded in the charge density of many layers of external atoms and the potential created was made consistent with the cluster potential, as outlined below.

II. THEORETICAL METHOD

The full details of the discrete variational (DV) method have been well-documented elsewhere[8]. The main task is to solve the set of one-electron Kohn-Sham[9] equations of density functional theory[10], in the local density approximation (LDA), for the cluster(in hartree a.u.):

$$h\phi_{i\sigma}(\bar{r}) = [-\frac{1}{2}\nabla^2 + V_c(\bar{r}) + V^{\sigma}_{xc}(\bar{r})]\phi_{i\sigma}(\bar{r}) = \varepsilon_{i\sigma}\phi_{i\sigma}(\bar{r}). \tag{2}$$

In Eq.2, the Coulomb potential V_c includes both the electron-nucleus attraction and electron-electron repulsion, and V^{σ}_{xc} describes the exchange and correlation effects of spin σ as derived by von Barth and Hedin[11]. According to LDA, the molecular potentials are functional of the electron density ρ_σ of spin σ:

$$\rho_\sigma(\bar{r}) = \sum_i n_{i\sigma}|\phi_{i\sigma}(\bar{r})|^2 \tag{3}$$

where the summation in (3) runs over the molecular spin orbitals $\phi_{i\sigma}$ with occupation $n_{i\sigma}$ which are expanded on a basis of atomic numerical orbitals (LCAO) centered at the symmetrically equivalent atoms. According to Eq.3, the charge and spin densities are:

$$\rho_c(\bar{r}) = \rho_\uparrow(\bar{r}) + \rho_\downarrow(\bar{r}) \tag{3a}$$

$$\rho_s(\bar{r}) = \rho_\uparrow(\bar{r}) - \rho_\downarrow(\bar{r}). \tag{3b}$$

The discrete variational method leads to the secular equations, which are solved self-consistently on a three-dimensional numerical grid. This grid is chosen to be regular inside a sphere of radius 1.8 au around the central Fe atom for an accurate evaluation of hyperfine parameters on Fe [12]; outside this sphere and around other atoms , a pseudo random diophantine scheme is used to define the points[8]. The number of points employed for the self-consistent calculations is of the order of 6000 for the Fe sphere and 14000 and 26000 points respectively for the smaller and larger clusters to describe the rest of the clusters' space. To facilitate the calculation of the electron-electron repulsion, densities are projected onto a superposition of overlapping spherical densities[13] centered on each nucleus:

$$\rho(\bar{r}) \cong \rho^{model} \equiv \sum_{nlI} d^{I}_{nl}\rho^{I}_{nl}(\bar{r}) \tag{4}$$

with

$$\rho^{I}_{nl}(\bar{r}) = \sum_{q}\left|R^{q}_{nl}(r_q)\right|^2 Y^{0}_{0}(\hat{r}_q). \tag{5}$$

Here r_q is the local coordinate relative to site q and the q-summation is only over the set, represented by superscript I, of the atoms equivalent by symmetry to the atom q. The O atoms are not equivalent in the solid, as discussed in the previous section. They are defined by three different sets labeled as O_{TO}, O_{TT} and O_H . The coefficients d^{I}_{nl} are obtained variationally by a least-squares[13] fit to the "true" density of Eq. 3, subject to the condition that ρ^{model} integrates correctly to the total number of electrons.

Optimized atomic basis functions are obtained from numerical self-consistent atomic DF calculations, with configurations determined in turn self-consistently from the embedded clusters. For the smallest cluster , all orbitals are kept in the variational space. For the largest cluster, the variational basis functions used are *3d* ,*4s*, *4p* for Fe, *3s*, *3p* for Si, and *2s*, *2p* for the O atoms, with low-lying states represented as a 'frozen core' . All core and shallow-core orbitals not included in the variational basis were orthogonalized to the valence basis functions in the first iteration and kept "frozen" subsequently.

An embedding scheme was adopted to take into account effects of atoms in the solid external to the cluster. Numerous layers of atoms around variational cluster are represented by numerical atomic charge and spin densities, from which the total charge and spin densities are assembled and the DF potentials constructed. The potential is truncated at the core region of exterior atoms (i.e. a repulsive pseudopotential is added) to simulate effects of the Pauli exclusion principle[14]. The long range Coulomb interaction is taken into account by Ewald summation[15].

III. RESULTS AND DISCUSSION

Annite is described by two clusters centered at the Fe^{2+}. The largest cluster is depicted in Fig.1b. The smallest cluster is defined by considering only the central $FeO_4(OH)_2$ fragment of the largest one. In the present calculations all hydroxyls occupy the trans octahedral positions. As may be noticed in Fig.1b , the Fe-centered large cluster is made up without any Al^{3+} around the central cation. Such a procedure in modeling annite is adopted to obey the proportion of 3:1 exhibited respectively between Si and Al in the tetrahedral layer(Fig.1a).

Hyperfine parameters as measured by Mössbauer spectroscopy can give important information about electronic structure and chemical bonding. The purpose here is to calculate these parameters at Fe^{2+} nucleus by using the two iron-centered clusters(Fig.1b) representing annite end-member. The results are analyzed considering the effect of cluster size in the hyperfine properties and are compared to the experiment.

The isomer shift δ, as defined in Mössbauer Spectroscopy is given by[16]

$$\delta = \alpha[\rho_A(\mathbf{0}) - \rho_B(\mathbf{0})], \tag{6}$$

where the calibration constant α involves only nuclear parameters and its value is fixed for each Mössbauer isotope, while the subscripts A and B refer to the absorber and source respectively. The electron-charge density $\rho_c(0)$ at nuclear site of Fe in the absorber A is defined in Eq.3a, where only orbitals belonging to the totally symmetric representation(a_g) of the point group of the cluster(C_{2h}) will have finite probability at the origin in a non-relativistic approach. To determine δ, we first performed calculations for free Fe atoms and ions, for which δ was measured in frozen gas matrices[17], employing the atomic density functional self-consistent method in order to obtain nuclear constant $\alpha = -0.228\text{mm}/\text{s}.a_0^{-3}$ and thus calculate the isomer shifts, using this value of α and the self-consistent values of $\rho_c(0)$, derived from Eq.3a at the origin.

For the γ-ray transition of 14.4keV of ^{57}Fe, the interactions between the nuclear quadrupole moment Q and the electric-field-gradient of exterior charges V_{zz} splits the Mössbauer resonance line into a doublet, whose separation in mm/s is given by:

$$\Delta EQ = 10.109QV_{zz}(1+\frac{\eta^2}{3})^{1/2} \tag{7}$$

with the so-called asymmetry parameter η defined as

$$\eta = \frac{V_{xx}-V_{yy}}{V_{zz}}, \tag{7a}$$

V_{xx}, V_{yy} and V_{zz} are the diagonal components of the electric-field-gradient(EFG) tensor in the system of the principal axis and are labeled according to the rule:

$$|V_{zz}| \geq |V_{yy}| \geq |V_{xx}| \tag{7b}$$

From Eq.7b, the values of η belong to the range $0 \leq \eta \leq 1$. In Eq.7, Q and V_{zz} are expressed in barn(b) and atomic units of a_0^{-3} respectively.

In the present calculation, the components V_{jl} of the EFG tensor, before applying the diagonalization procedure, are given(in atomic units):

$$V_{jl} = -\int \frac{\rho_c(\vec{r})(3x_j x_l - \delta_{jl} r^2)dv}{r^5} + \sum_q \frac{Z_q^e(3x_{qj}x_{ql} - \delta_{jl} r_q^2)}{r_q^5}. \tag{8}$$

The first term in Eq.8 is the electronic contribution, calculated in the DV method as a sum over three-dimensional grid with $\rho_c(\vec{r})$ given by Eq.3a; the second term is the point-charge contribution of the surrounding O or(and) Si nuclei of the two Fe-centered clusters, which are described by the effective charge Z_q^e defined as the nuclear charge minus the frozen-core electrons; the q-sum is over the nuclei of the cluster excepted Fe-nucleus, where the electric-field gradient is calculated.

The Fermi or contact hyperfine field is given by the expression

$$H_c = (\frac{8\pi}{3})g_e\mu_B\frac{1}{2}[\rho_\uparrow(0) - \rho_\downarrow(0)], \tag{9}$$

where g_e is the electronic g-factor, μ_B the Bohr magneton and the term in brackets is the difference between the electronic density at the central Fe - nucleus, located at the origin of the cluster coordinate system, for spin up and spin down $\rho_s(0)$, with $\rho_s(0)$ given by Eq.3b. The dipolar term H_D can be written as

$$H_D = g_e\mu_B M_{zz}, \tag{10}$$

where M_{zz} is the largest component of H_D tensor in the system of the principal axis; the elements M_{ij} of H_D tensor are derived from the first term in Eq.8 however, in this case, $\rho_c(\vec{r})$ is replaced by $\rho_s(\vec{r}) = \rho_\uparrow(\vec{r}) - \rho_\downarrow(\vec{r})$(Eq.3b) and the convention expressed in Eq.7b is also applicable here; in the present work, the orbital contribution H_L of the hyperfine field is not considered. Therefore, the magnetic hyperfine field at Fe-nucleus H_F is given by

$$H_F = H_c + H_D. \tag{11}$$

In Table I are given the values found for the hyperfine parameters for the iron-central clusters representing annite end-member. The total charge density $\rho_c(0)$ is larger for the smallest cluster resulting in a minor isomer shift(δ), as expected since the nuclear constant α is negative for ^{57}Fe. The experimental δ for Fe^{2+} lies at +1.365(5)mm/s at T=4K with respect to α-iron[18]; it is closer to δ=1.23mm/s derived from the largest cluster.

The principal components and corresponding directions of the electric-field gradient(EFG) are shown in Table I. It can be verified that the EFG-matrix elements are profoundly affected by the cluster size. It is seen in this table that the value of V_{zz} is negative with the corresponding eigenvector aligned into the z axis for $FeO_4(OH)_2$, while $Fe(Si_4O_{14})(OH)_2$ has a positive V_{zz} in the x direction. The value calculated for asymmetry parameter η=0.22(Eq.7a) shows a small deviation from axial symmetry around Fe for $FeO_4(OH)_2$; however, the value derived from the largest cluster, η=0.98, lies to the upper limit of η . By using Eq.7, where the value of nuclear quadrupole moment is that derived by Dufek et al.[19](Q=+0.16b), we found ΔEQ=-4.62 and +3.73mm/s for $FeO_4(OH)_2$ and $Fe(Si_4O_{14})(OH)_2$ respectively. Experimental value for comparison is not available at 4K; however, the values at liquid nitrogen and room temperatures reported: for octahedral Fe^{2+} it is observed a sizable increase in ΔEQ from 2.40 to 2.88mm/s as the temperature is lowered[7]. The ΔEQ value derived from the largest cluster is better agreement with such trend than that one calculated for $FeO_4(OH)_2$. Due to the comparable magnitudes of ΔEQ and hyperfine magnetic field(H_F), the spectrum of the synthetic annite is not resolved at 4K.

Table I
Hyperfine interaction parameters calculated for central Fe of $Fe(Si_4O_{14})(OH)_2$ representing annite-mica structure.

		$FeO_4(OH)_2$	$Fe(Si_4O_{14})(OH)_2$
$\rho_c(0)^a$		142.69	142.33
δ(mm / s)		+1.15	+1.23
EFGtensor(a_0^{-3})	V_{xx}	+1.11 (1.0,0.0,0.0)	-0.02 (0.0,1.0,0.0)
	V_{yy}	+1.73 (0.0,1.0,0.0)	-2.33 (0.0,0.0,1.0)
	V_{zz}	-2.84 (0.0,0.0,1.0)	+2.35 (1.0,0.0,0.0)
η		0.22	0.98
ΔEQ(mm / s)		-4.62	+3.73

$\rho_{s(0)}{}^{a}$	-0.78	-0.69
H_c(kOe)	-411	-362
H_D(kOe)	-181	+151
H_F(kOe)	-592	-211

[a]In units of a_0^{-3}. *[b]Related to* α *iron.*

The sum of H_c and H_D gives the magnetic hyperfine field H_F. The value calculated for the smallest cluster is H_F=-592kOe and H_F=-211kOe for the other one. For calibration, we consider the definitive result obtained by Preston et al. [20] for α-Fe at room temperature: H_F=330kOe corresponds to line 1 to line 6 splitting of 10.657mm/s and determine the Fe^{2+} sextets related to $FeO_4(OH)_2$ and $Fe(Si_4O_{14})(OH)_2$ with line 1 to line 6 splittings of 19.10 and 6.82mm/s respectively. The difficulty in resolving properly high-spin Fe^{2+} spectrum is due the comparable magnitudes of H_F and ΔEQ[21]. However, it is possible to establish a upper limit value of H_F. According to the results from Ref.21, the synthetic annite sample, whose lattice parameters are considered here, exhibits Fe^{3+} in both octahedral and tetrahedral sites. The Fe^{3+} sextets are resolved into two distinct spectra with line 1 to line 6 splittings of 17.5 and 14.5mm/s related respectively to Fe^{3+} fourfold and sixfold coordinated to O. The positions of some of the lines pertaining to Fe^{2+} spectra are determined and indicate that H_F due to Fe^{2+} is less than those measured for Fe^{3+} in both sites; therefore, it is acceptable to define 14.5mm/s as a upper limit for line 1 to line 6 splitting(or H_F=440.kOe). In this case, the value derived from the largest Fe-central cluster(H_F=-211kOe) is consistent with the experimental trend.

IV. CONCLUSIONS

We have investigated the electronic structure of annite end-member by means of self-consistent spin-polarized density functional Discrete Variational embedded cluster method. The ideal annite structure(Eq.1) is modeled by expanding MOC-biotite sample[6] into the synthetic annite lattice[7]; both atomic positions of MOC-biotite and lattice parameters of synthetic annite are experimentally well determined. Two clusters centered at Fe are defined to describe the solid. Calculations of Mössbauer hyperfine parameters present a good quantitative agreement with the experimental trend. Predictions for the signs of ΔEQ and H_F are made and the importance in including the next near neighbor is focused.

ACKNOWLEDGMENTS

We thank Prof. S. Guggenheim and Prof. D.G. Rancourt for helpful discussion. This work was supported by the MRL Program of the National Science Foundation, at the Material Research Center of Northwestern University, under Award No. DRM-9120521. The author J.T. thanks the support received from Brazilian agency CNPq under contract No. 200124/95-1.

REFERENCES

[1] B. J. Teppen, D. M. Miller, S. Q. Newton and L. Schäfer, J. Phys. Chem. 98 (1994) 12545.
[2] H. V. Brand, L. A. Curtiss and L. E. Iton, J. Phys. Chem. 96 (1992) 7725.
[3] J. C. White and A. C. Hess, J. Phys. Chem. 97 (1993) 6398.
[4] Yong-nian Xu and W. Y. Ching, Phys. Rev. B44 (1991) 11048.

[5] R. W. G. Wyckoff, *Crystal Structures*(Wiley, New York, 1968) vol.4.
[6] S. Guggenheim et al., unpublished.
[7] D. G. Rancourt, I. A. D. Christie, M. Royer, H. Kodama, J. -L. Robert, A. E. Lalonde and E. Murad, Am. Mineral. 79 (1994) 51.
[8] D. E. Ellis and G. S. Painter, Phys. Rev. B2 (1970) 2887; D. E. Ellis, Int. J. Quantum Chem. S2 (1968) 35; E. J. Baerends, D. E. Ellis and P. Ros, Chem. Phys. 2 (1973) 41.
[9] W. Kohn and L. J. Sham, Phys. Rev. A140 (1965) 1133.
[10] R. G. Parr and W. Yang,*Density Functional Theory of Atoms and Molecules* (Oxford University Press, New York, 1989).
[11] U. von Barth and L. Hedin, J. Phys. C5 (1972) 1629.
[12] D. E. Ellis, D. Guenzburger and H. B. Jansen, Phys. Rev. B28 (1983) 3697.
[13] B. Delley and D. E. Ellis, J. Chem. Phys. 76 (1982) 1949.
[14] D. E. Ellis, G. A. Benesh and E. Byrom, Phys. Rev. B16 (1977) 3308.
[15] J. C. Slater, *Insulators, Semiconductors and Metals* (New York: Mc Graw- Hill, 1967) vol.3.
[16] N. N. Greenwood and T. C. Gibb, Mössbauer Spectroscopy, (Chapman and Hall, London, 1971).
[17] See, J. Terra and D. Guenzburger, J. Phys. Chem. 99 (1995) 4935, and references therein.
[18] D. G. Rancourt, private communication.
[19] P. Dufek, P. Blaha and K. Schwarz, Phys. Rev. Lett. 75 (1995) 3545.
[20] R. S. Preston, S. S. Hanna and J. Heberle, Phys. Rev. 128 (1962) 2207.
[21] D. G. Rancourt, I. A. D. Christie, G. Lamarche, I. Swainson and S. Flandrois, J. Mag. Magnetic Mat. 138 (1994) 31.

II. AMORPHOUS MATERIALS AND NANOCRYSTALLINE MATERIALS

MÖSSBAUER SPECTROSCOPY AND YOUNG'S MODULUS MEASUREMENTS IN THE STUDY OF THE CRYSTALLIZATION PROCESSES OF SEVERAL FeSiB BASED GLASSES.

J. Moya [1,2], *M. Vázquez* [1], *V. Cremaschi* [2], *B. Arcondo* [2] *and H. Sirkin* [2].
1. Instituto de Magnetismo Aplicado (Salvador Velayos) UCM-RENFE-CSIC, Madrid.
2. Facultad de Ingeniería, UBA, Buenos Aires.

Fe-Si-B alloys with the addition of small quantities of Cu and Nb allow the development of a particular kind of nanocrystalline material characterized by an ultrafine structure with crystalline size of about 10 to 50 nm embedded in an amorphous phase. In some cases the key of the nanocrystallization is attributed to Cu while the stability of the remnant amorphous matrix is attributed to Nb. The alloy known by the trade name FINEMET ($Fe_{73.5}Si_{13.5}B_9Nb_3Cu_1$) is one typical example. These structures are obtained by means of one hour isothermal annealings of the amorphous precursor material at temperatures about 500 °C. The interest on these materials obeys to the potential applications of their outstanding magnetic properties.

The crystallization processes of Fe-Si-B-Nb-Cu and Fe-Si-B-Sn alloys with different compositions are studied employing Mössbauer spectroscopy and x-rays diffraction while their magnetoelastic behavior, where a ΔE effect is observed, is analyzed by means of Young's modulus measurements.

INTRODUCTION

Fe-Si-B is a well known amorphous alloy (1) being one of its outstanding properties its soft magnetic behavior. In 1988 Yoshizawa et al. (2) discovered that the addition of small quantities of Nb and Cu to this alloy produces, during crystallization from the amorphous state, an ultra fine structure made of Fe(Si) randomly oriented grains of around 10-12 nm size, embedded in an amorphous matrix. These new kind of nanocrystalline materials improve remarkably the soft magnetic properties of its precursor Fe-Si-B, showing values of saturation magnetization ($\mu_0 M_S$) next to 1.3 T, coercive forces smaller than 1 A/m and a magnetostriction nearly zero ($\lambda \approx 1\times10^{-6}$). In this system, the nanocrystallization is attributed to Cu clusters (3) while Nb seems to stabilize the remnant amorphous matrix.

The addition of other elements with low diffusion coefficients and inmiscible in α-Fe can also induce nanocrystallization being not necessary the presence of Cu (4). In particular, Sn fills the first conditions so we analyze its role in the crystallization process.

In a previous work (5) we have investigated the relaxation process up to nanocrystallization in the FeSiBNbCu system. In the present research, new results are presented and they are compared with those obtained in similar studies carried on the FeSiBSn system.

EXPERIMENTAL TECHNIQUES

Amorphous ribbons of composition $Fe_{79-x}Si_{11}B_{10}Sn_x$ (x=1 and 2,6; samples A1 and A2,6 respectively) and $Fe_{73.5}Si_{22.5-x}B_xNb_3Cu_1$ (x= 6, 8, 9, 10 and 12; samples B6 to B12) with typical sections of 1mm×20μm, were obtained by a melt spinning device. Isothermic annealings of one hour were performed under vacuum (10^{-5} mbar) at different temperatures in the range between 350 and 600 °C to induce relaxation and crystallization on the samples. The evolution of the structure from amorphous to crystalline state was observed by X-ray diffraction (XRD), Mössbauer spectroscopy (MS) and Young's modulus measurements (ΔE effect).

Fig. 1 shows XRD patterns of samples A1, A2.6, B9 and B12 at different annealing temperatures (T_a) obtained in a Rigaku diffractometer using Cu Kα radiation.

Mössbauer spectroscopy studies at room temperature were performed with ^{57}Co source in transmission geometry and a NORMOS Program was employed to fit the spectra. This program uses an histogram distribution of hyperfine magnetic field (Bhf) and the subspectra relative depths were calculated using a linear regression plus damping. A linear dependence between isomer shift (IS) and magnetic field was supposed and a bimodal distributions were obtained from the fitting. Fig. 2 shows the Mössbauer spectra of samples A1, A2.6, B9 and B12 after different annealing temperatures.

Young's modulus measurements were made in 2 cm long samples before and after the heat treatment (E_0 and E respectively) using the vibrating reed technique (6). The relative variation $\Delta E/E_0$ were obtained for different annealing temperatures. In the case of ferromagnetic alloys with high magnetostriction constants (λ), the Young's modulus value may diminish as a result of the additional magnetostrictive strain caused by the magnetic domains movement that induces the applied stress (ΔE effect). This fact is very important in amorphous ferromagnetic alloys and it could take place also with an annealing that increases the magnetic domains mobility (7) The relation between these parameters are given by formulae [1] (8).

$$\frac{\Delta E}{E_0} = -\frac{2\lambda}{5\sigma_i} E_S \qquad [1]$$

where E_S is the Young's modulus at the magnetic saturation state, E_0 the unmagnetized one, λ the magnetostriction constant and σ_i the internal stress.

RESULTS AND DISCUSSION

FeSiBSn System

Fig. 1(a) shows typical XRD patterns of annealed samples. It can be seen that crystallization starts at 450 °C with the α-Fe phase. At 500 °C the peak corresponding to α-Fe phase shifts to greater values of 2θ. This could be attributed to the enhancement by Si of the solid solution α-(FeSiSn) where Si and Sn are located in sites of the α-Fe cell. Fe_2B phase also begins to precipitate at this temperature. The presence of the Fe_3B metaestable phase was not observed. Sample containing 2.6% Sn shows, at 450 °C, a greater crystallization than the sample with 1% Sn. Banerji et al. (9) have still found amorphous structure in $Fe_{78}Si_9B_{11}$ alloy annealed during 16 hour, at 450 °C. The diminution of crystallization temperature may be associated to Sn presence.

Young's modulus measurements are shown in Fig. 3 for sample A2.6. From the as quenched state up to 400 °C, a diminishing of $\Delta E/E_0$ is observed. As no new phase or atomic reorganization are noticed neither in XRD patterns nor in Mössbauer spectra (concerning Fe element), this diminishing may have a magnetostrictive genesis (ΔE effect). In the annealing process, the internal stresses frozen in during quenching are lessen (see equation [1]). This fact is in good agreement with results obtained in a previous work (5,10). At 450 °C the increase of $\Delta E/E_0$ value is related with the beginning of crystallization, which continue up to 550 °C.

Mössbauer spectra of as-quenched (a-q) samples A2.6 and annealed A2.6 below 450 °C show a distributed magnetic hyperfine field (Bhf). The spectra were fitted employing a histogram distribution which Bhf peak value is 25.2 T and the standard deviation σ 4 T. The IS peak value is 0.03 mm/s.

The spectra of samples A2.6 annealed at 500 °C show an amorphous phase (14%) plus five interactions corresponding to non stoichiometric Fe_3Si which Bhf are 31.7, 30.5, 28.3, 24.1 and 19.1 T. The spectra of sample A1 annealed at 500 °C show an amorphous phase (28%) plus four interactions corresponding to Fe_3Si non stoichiometric which Bhf are 31.7, 28.3, 24.1 and 19.4 T.

It may be mentioned that sample of 2.6% of Sn presents more crystallization than sample with 1% of Sn annealed at the same temperature. Both samples present the same crystallization phase Fe_3Si non stoichiometric. Other phase as Fe_2B cannot be completely discarded taking into account that a small amount of it would be hardly observed among the reported interactions.

The Mössbauer spectra of the sample obtained from T_a= 450 °C show an amorphous phase and four magnetic interactions corresponding to different sites in the crystalline net. The Mössbauer parameters are in agreement with those corresponding to non-stoichiometric Fe_3Si in Fe-Si-B-Nb-Cu system (11) but the fields are smaller than those corresponding to large size Fe_3Si grain. Taking into account that the hyperfine field distribution do not depend on the amount of Si in solution, this reduction could be attributed to the small grain size (11,12).

Sample A2.6 annealed at 600 °C show completely crystallization. In this spectra the values of the Bhf fields are increased and this values resemble those of Fe_3Si large grain size.

<u>FeSiBNbCu System</u>

XRD patterns of Fe-Si-B-Nb-Cu alloys in different steps of relaxation are shown in Fig. 1(b) (sample B9 and B12) and Mössbauer spectra of sample B12 are shown in Fig.2(a). The difference with Fe-Si-B-Sn system is that crystallization of α-Fe_3Si phase starts at higher temperature (T_a= 480 °C for B12 and about 490-500 °C for B9). In this case not significant shift of the peaks are observed.

According to both techniques (XRD and MS), crystallization of α-Fe_3Si starts at T_a= 480 °C in the sample B12. At lower T_a only amorphous phases can be observed. Between room temperature and 460 °C the Mössbauer distributions show no difference, so the Fe environment do not change in this interval of temperatures (5). Mössbauer studies of samples B9 and B12 performed up to T_a= 460 and 480 °C are shown in Fig. 2(b). Spectra of sample B12 at T_a= 480 °C show in the external sides the presence of a Bhf= 31.8 T corresponding to Fe_3Si non stoichiometric. The fitting also shows Bhf of 28.2 and 24.1 T from the same phase.

The Young's modulus measurements show a similar behavior than that observed in series A, specially in the low boron contents (B6-B8) samples (5): a diminution of E during amorphous relaxation and an increase when crystallization starts. As shown in fig. 3 an abrupt increase of E is observed in samples B9 and B12 at T_a= 440-480 °C which can not be related with crystallization. This behavior may be caused by the clustering of Cu, previous nanocrystallization, that takes place near T_a= 450 °C as it was reported by Hono et al. (3). We suggest that this clustering introduces internal stresses making the magnetic dipoles movement more difficult. In sample B9 it can be observed a new decrease in the $\Delta E/E_0$ values which can be attributed to a release of those stresses. In sample B12 this last decrease is not present due to the beginning of total crystallization.

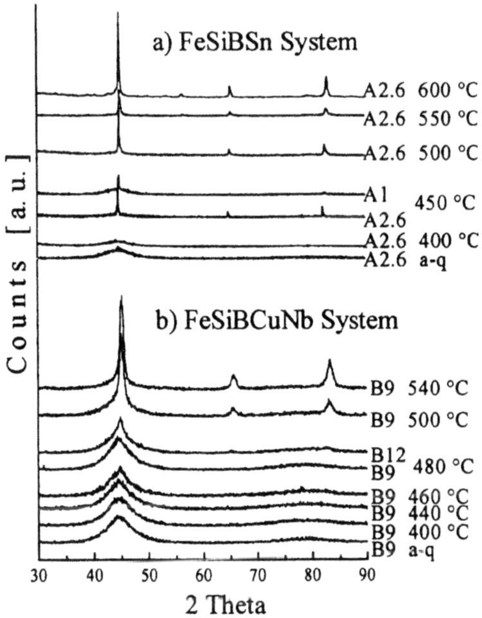

Fig. 1: XRD of samples A1, A2.5, B9 and B12 with different anneling temperatures, for FeSiBSn (a) and FeSiBCuNb (b) systems.

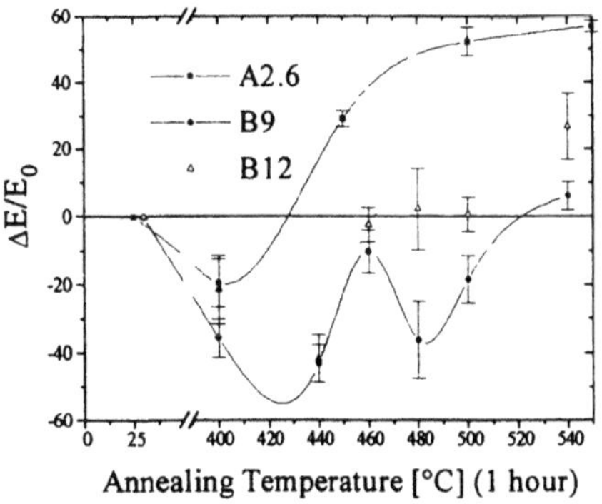

Fig. 3: $\Delta E/E_0$ measurements of samples A2.6, B9 and B12.

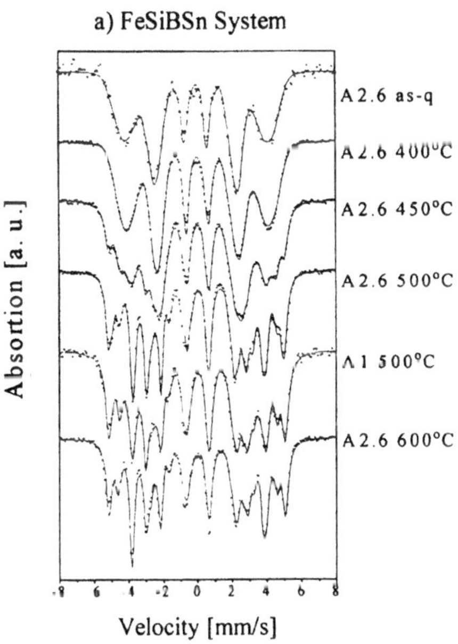

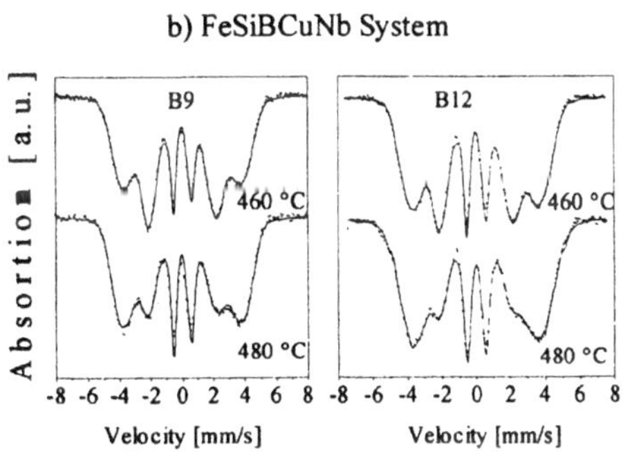

Fig. 2: a) Mössbauer spectra of samples A1 and A2.6. b) Comparison of samples B9 and B12 of amorphous phases and the first crystallization stages only obtained in B12 at T_a= 480 °C.

REFERENCE

[1] F.E. Luborky, in: Amorphous Metallic Alloy, ed. F.E. Luborsky. (Butterworths Publ. 1983) p. 360-380.
[2] Y. Yoshizawa, S. Oguma and K. Yamauchi. J Apply. Phys. 64 (10) (1988) 6044-6046.
[3] K. Hono, J.L. Li, Y. Ueki, A. Inoue and T. Sakurai. Appl. Surface. Science 67, (1993) 398-406.
[4] R.Yavari, in: IV International Workshop on Non- Crystalline Solids, Madrid 1994, eds. M.Vázquez and H.Hernando (World Scientific 1995), p. 35-43.
[5] J. Moya, M. Vázquez, V. Cremaschi, B. Arcondo and H. Sirkin. To be Published.
[6] K. Bothe, H. Neuhäuser; J. Non-Cryst. Solids 56 (1983) 279.
[7]J.Perez, F.Fouquet and G.Lormand in: Les Amorphes Metalliques;(les editions de physique 1983) p. 541.
[8] S. Chikazumi in: Physics of Magnetism, (Krieger Publishing Company. Malabar, Florida 1978).
[9] N. Banerji, U.C.Johri, V. N. Kulkarni and R. M. Singru, Journal of Material Science 30 (1995) 417-425.
[10] J. Moya, F. Audebert, B. Arcondo and H. Sirkin. To be Published.
[11] G. Rixecker, P. Schaaf and U. Gonser, J. Phis.Matter 4 (1992)10295-10310.
[12] M. Arita, S. Nasu and F. E. Fujita, Trans. Jap. Inst. Met 26 (1985) 71-20.

CRYSTALLIZATION OF AMORPHOUS Ga-Te-Fe ALLOYS WITH LOW Fe CONTENT.

M.Fontana[†], *B. Arcondo*[†], *M.T. Clavaguera-Mora*[‡] *and N. Clavaguera*[‡].
[†] Departamento de Física, Facultad de Ingeniería, Universidad de Buenos Aires.
Paseo Colon 850 - (1063) Buenos Aires - Argentina.
[‡] Grupo de Física dels Materials, Universidad Autónoma de Barcelona.

The Ga-Te system, when quenched from the melt, has a small glass forming composition range centered in the $Ga_{20}Te_{80}$ composition. When a small amount of Fe is added to the original alloys, partially amorphous samples are obtained. A ternary amorphous phase plus $FeTe_2$ traces are observed by means of x-rays diffraction. The crystallization process is analyzed employing differential scanning calorimeter, x-ray diffraction and Mössbauer spectroscopy. The activation energy, crystallization temperature, crystallization enthalpy and crystallization products are reported. A small addition of iron to Ga20Te80 binary composition generates more stable amorphous phases.

1. Introduction

Metal-chalcogenides systems and among them IIIb-Te alloys, have been widely studied because of their typical semiconducting behavior. The GaTe liquid alloys present a deep electronic conductivity minimum simultaneously with both a maximum in the thermoelectric power [1] and a significant excess heat capacity [2,3] for compositions close to the Ga_2Te_3 stoichiometry.

Amorphous $Ga_{20}Te_{80}$ alloy has been obtained by rapid quenching from the melt [4,5], and Luo et al. [6] have reported a glass forming composition range from 10 to 30 at. % Ga. Amorphous $Ga_xTe_{(100-x)}$ samples obtained by vapor deposition in a composition range x from 20 to 60 have also been reported [7]. The study of the crystallization kinetics of the $Ga_{20}Te_{80}$ amorphous alloy has been reported [8].

Another interesting metal-chalcogen amorphous alloy is Fe-Te. Amorphous Fe-Te thin films have been obtained by rf sputtering in the ranges 15 to 60 and 70 to 90 at.% Te[9].

Both equilibrium systems have been widely studied [10,11,12,13,14,15,16,17]

The antiferromagnetic compound $FeTe_2$ is a equilibrium phase of the binary system FeTe. It is an orthorhombic anomalous marcasite like $FeSe_2$ with iron in octahedral coordination. The Mössbauer spectrum of this phase at room temperature shows a asymmetrical doublet with an isomeric shift IS = 0.38±0.03 mm/s and a quadrupolar splitting QS = 0.48±0.03 mm/s. This value indicates a high spin Fe^{+2}, with a sp^3d^2 and the additional effect of π-bonding also. The absence of magnetic ordering may be explained by the relaxation phenomena. [14]

A small glass forming range has been observed in the Ga-Te-Fe system when the piston and anvil technique is employed [18]. The key of the amorphization process is the presence, along a short range of temperature, of the Ga_2Te_5 phase

The aim of this work is to analyze the crystallization of amorphous Ga-Te-Fe alloys with low Fe content, to determine the crystallization products, to observe the changes in the environment of the iron atom and to compare with the binary amorphous alloy $Ga_{20}Te_{80}$.

2. Experimental

Master alloys of compositions $Ga_{20}Te_{80}$ (alloy A), $Ga_{13.3}Te_{81.7}Fe_5$ (alloy B) and $Ga_{19.6}Te_{78.4}Fe_2$ (alloy C) have been prepared by melting mixtures of elemental (99.99 % purity) Gallium, Tellurium and Iron at 600 °C, in graphite crucibles under Ar atmosphere. The samples have been rapidly quenched by the piston and anvil technique[19] , under Ar atmosphere. Amorphous samples with almost circular form of 30 μm average thickness have been obtained. Their thermal evolution on heating has been studied in a differential scanning calorimeter (DSC) Perkin Elmer DSC-7 under dynamic Ar atmosphere. Continuous heating experiments have been performed at scan rates ranging between 5 and 80 K/min and isothermal experiments have been carried at several temperatures (sample A: 457, 460 and 463 K ,sample B: 413, 418, 423 and 428 K, and sample C: 505, 510 and 515 K). The measuring temperature has been attained by previous heating of the amorphous samples from room temperature to the annealing temperature at a rate of 80 K/min.

The rapidly quenched samples, their crystallization steps and as-cast samples have been analyzed by X-ray diffraction (XRD) using monochromatized Cu (Kα) radiation and by means of Mössbauer spectroscopy (MS), at room temperature, using a ^{57}Co (Rh) source transmission geometry. Powder for the measurements of XRD and MS was obtained interrupting the DSC scan at a temperature above each crystallization peak. The isomeric shift (IS) is reported relative to Fe(Rh). In order to avoid textural effects when analyzed by means of MS, powdered cast materials have been mixed with graphite and then sealed in plastic holders. Rapidly quenched samples have been measured as obtained, thermally heated samples have been measured as powder.

3. Experimental results

Figure 1 shows the DSC curves (dH/dT versus T, where H is the enthalpy and T the temperature in K) obtained along the crystallization of samples A, B and C, the heating rates were of 5 to 80 K/min.

The X-ray spectra of A, B and C rapidly quenched samples, their spectra after each crystallization process and the as-cast samples spectra are showed in Figure 2, 3 and 4, respectively.

The rapidly quenched sample A has been obtained completely amorphous, Te and Ga_2Te_3 have been identified in the as-cast sample A.

B and C rapidly quenched samples are partially amorphous with FeTe2 crystalline phase. Te, Ga_2Te_3 and $FeTe_2$ have been identified in B and C as-cast samples.

Mössbauer spectra of as-cast, rapidly quenched samples B and C, and their crystallization products after each DSC peak show asymmetrical doublets (spectra of sample B are showed in Figure 5). The Mössbauer data were fitted using two independent Lorentzian lines allowing different line widths. Fitted parameters are reported in Table 1. As $FeTe_2$ is an antiferromagnetic material with a polymeric type network, these asymmetries, in powdered samples are product either of relaxation, or Goldanskii Karyagine effect or both [14]. In as quenched samples texture effects [15] as well as stresses are also considered.

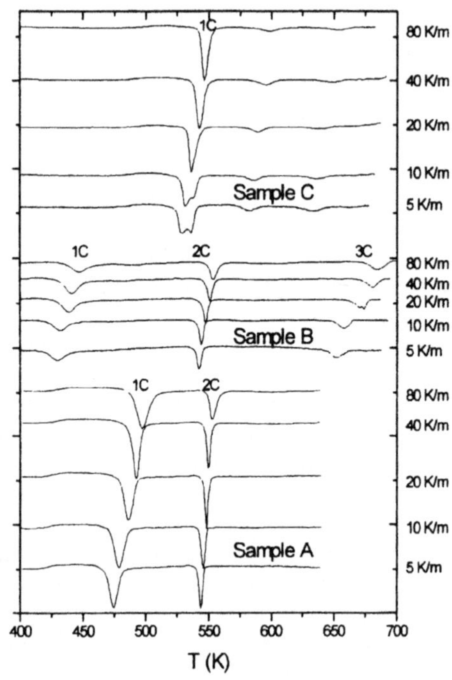

Figure 1 DSC curves (dH/dT versus T, where H is the enthalpy and T the temperature in K) obtained along the crystallization of samples A, B and C for heating rates of 5, 10, 20, 40 and 80 K/min. The main peaks of each sample are indicated (first (1c), second (2c), and third (3c) crystallization).

The main parameters of the crystallization kinetic: crystallization temperature, transformation enthalpy and the phases appeared in each crystallization process are showed in Table 2. The crystallization temperature has been determined for the 20 K/m scan.

The value of the apparent activation energy is quoted in Table 2, whereas the plots , for the first peak, are showed in figure 6. The apparent activation energy is obtained from the peak temperature in the continuous heating processes (Kissinger method [20]), and from the isothermal processes at the peak time (Arrhenius method).

The isothermal measurements have been performed to analyze completely the first crystallization peak in the three samples. The Avrami exponent, a parameter of the crystallization kinetic, has been obtained by Johnson-Mehl-Avrami-Erofeev expression [21], this is reported in Table 2.

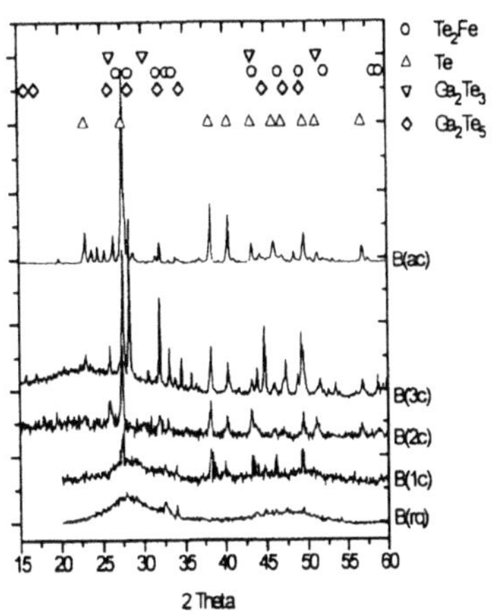

Figure 3. X-ray (Cu Kα) spectra for B sample, rq : rapid quenched sample, ac : cast sample; 1c, 2c and 3c: first, second and third crystallization.

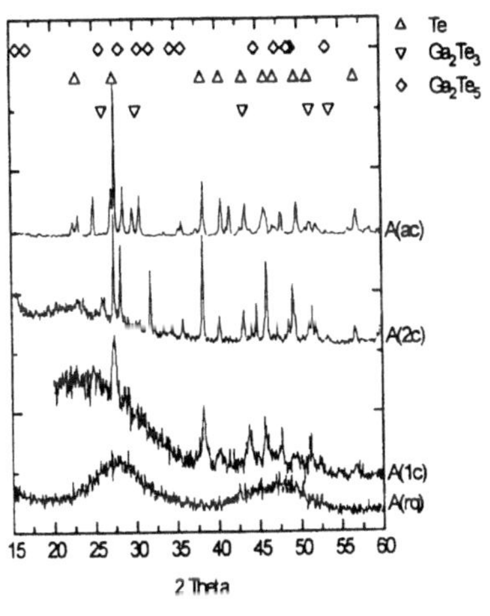

Figure 2. X-ray (Cu Kα) spectra for A sample, rq : rapid quenched sample, ac : cast sample; 1c and 2c: first and second crystallization.

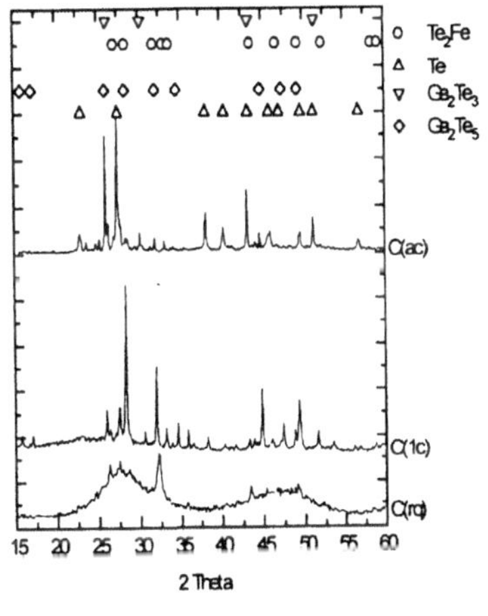

Figure 4. X-ray (Cu Kα) spectra for C sample, rq : rapid quenched sample, ac : cast sample and 1c: first crystallization.

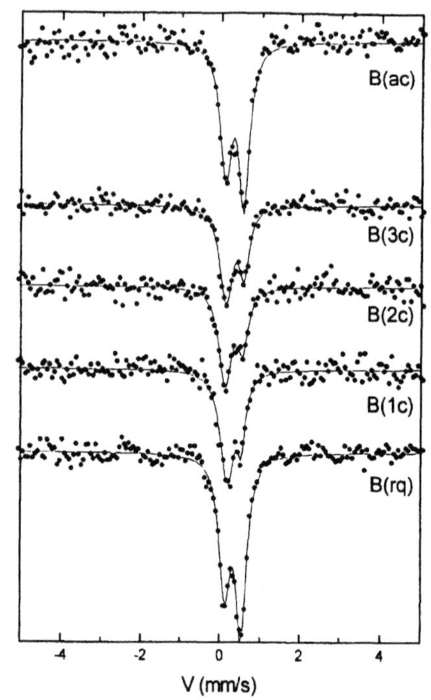

Figure 5. Mössbauer spectra obtained at room temperature for sample B with a ^{57}Co (Rh) source, rq : rapid quenched sample, ac : cast sample; 1c, 2c and 3c: first, second and third crystallization.

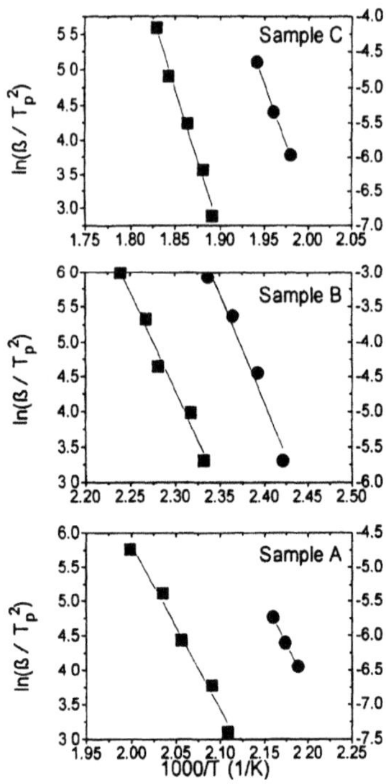

Figure 6. Plot of both $\ln(\beta/T^2)$ with β/T^2 in units of K^{-1} min^{-1}, T being the peak temperature obtained under continuos heating at scan rates β, and ln(t) , t being peak time (in s) obtained under isothermal annealing versus inverse of temperature, for the first crystallization peak of the samples A, B and C.

4. Discussion

From the results obtained from MS, it can be concluded that the iron atom is bonded to the tellurium atom as an association $FeTe_2$ in all samples: amorphous and crystalline. The isomeric shift values remain constant, while the quadrupolar splitting in the amorphous is smaller than in the crystalline samples. Then , in the alloy B, $FeTe_2$ crystallizes in the second crystallization peak.

From the calorimetric scan of samples A and C, it can be observed that the addition of iron defers the crystallization, superposes the first and second crystallization and stabilizes the amorphous phase. Crystallization temperature and activation energy are higher. The Avrami exponent value indicates that the crystallization process is controlled by interface [21] in both samples A and C.

The iron is always in $FeTe_2$ association, so the remaining material is a binary matrix Ga_xTe_{1-x}. The crystallization temperature in sample B is lower that in sample A. This is attributed to a higher amount of Te in the binary matrix Ga_xTe_{1-x}. This different composition favours the early crystallization of Te, the Avrami exponent value indicates a crystallization process controlled by diffusion. This results are due to two causes: 1) diffusion is favoured by temperature in samples A and C, so the crystallization processes are controlled by interfase, 2) a higher Fe concentration, associated to Te, plus a lower temperature in the amorphous phase in sample B difficulties the diffusion.

The crystallization products are determined by the type of crystallization process. The matrix of the amorphous phase consist in Ga_2Te_3 like clusters plus Te. In interfase controlled process Ga_2Te_5 and Te are crystallized at the end of a favoured diffusion process. In the diffusion controlled process Te and Ga_2Te_3 crystallizes almost without diffusion.

5. Conclusions

A small addition of iron to $Ga_{20}Te_{80}$ binary composition generates more stable amorphous phases. The temperature of crystallization and the activation energy are greater. The kinetic of crystallization is equal: the process is controlled by interfase.

In alloy B, the binary matrix Ga_xTe_{1-x} is more rich in Te and the crystallization temperature lowers. As Fe concentration increases, $FeTe_2$ clusters already existing in the amorphous phase obstruct the diffusion, associated to the lower temperature, and change the crystallization process and crystallization products.

Table 1. Mössbauer parameters from rapid quenched (rq), cast(ac) and their crystallization steps (1c, 2c and 3c: first, second and third crystallization) of B and C samples. The isomeric shift (IS), the quadrupolar splitting (QS), the line widths (Γ1, and Γ2) and the area of the second line relative to the first one (D21) are reported.

sample	IS(mm/s) ±0.02	QS(mm/s) ±0.03	Γ1 (mm/s) ±0.02	Γ2 (mm/s) ±0.02	D21
B(rq)	0.33	0.39	0.29	0.33	1.10
B(1c)	0.35	0.37	0.47	0.23	0.53
B(2c)	0.36	0.46	0.42	0.27	0.59
B(3c)	0.37	0.48	0.42	0.27	0.67
B(ac)	0.36	0.48	0.34	0.34	1.00
C(rq)	0.37	0.35	0.47	0.32	0.72
C(1c)	0.36	0.48	0.30	0.30	1.00
C(ac)	0.35	0.46	0.60	0.34	1.06

6. References

1. Y. Waseda, *The Structure of Non-Crystalline Materials* (Mc Graw Hill Inc., USA 1980) 168.
2. S. Takeda, S. Tamaki, A. Takano and Okazaki, J.Phys.C **16** (1983) 467.
3. E. Irle, B. Gather, R. Blachnik, U. Kattner, H. Lukas and G. Petzow, Z.Metallkde **78** 8 (1987) 535.
4. S.K. Hsiung and R. Wang , J Appl.Phys. **49** (1) (1978) 280.
5. G. Parthasarathy, S. Asokan, S.S.K. Titus and R.R. Krishna, Phys. Letters A **131** (7,8) (1988) 441.
6. H.L. Luo and P. Duwez, Appl. Phys. Lett. **2** (1963) 21.
7. R. Wang, Bulletin of Alloy Phases Diagrams vol.2, N°3 (1981) 269.
8. N.Clavaguera, M.T. Mora and M.Fontana , Accuracy in the Experimental Calorimetric Study of the Crystallization Kinetics and Predictive Transformation Diagrams: Application to a Ga-Te Amorphous Alloy, send to J.Material Research.
9. K. Chiva, K. Tokumitsu and H. Ino, J. Appl.Phys. **60** (3) (1986) 1019.
10. M. Hansen and K. Anderko, Constitution of Binary Alloys (Mc Graw-Hill, New York 1958) 458.
11. M. Julien Pouzol, S. Jaulmes et F. Alapini, Acta Cryst. B**33** (1977) 2270.
12. J.G. Antonopoulos, TH. Karakostas, G. Bleris, N.A. Economou, J. Mat. Sci. **16** (1981) 733.
13. R. Blachnik und E. Irle, J. of Less-Common Metals **113** (1985) L1.
14. K.V. Reddy and S.C. Chetty, Phys Stat. Sol. (a) **37** (1976) 687.
15. J. Ward and V. McCann, J.Phys.C **12** (1979) 873.
16. H. Verma and G. Rao, Phys.Review B **22**, (7) (1980) 3152.
17. V. Fano and I. Ortalli, Phys. Stat. Sol. (a) **10** (1972) K121.

18. M. Fontana, M. Bormioli and B. Arcondo; IV International Workshop on Non- Crystalline Solids, ed. M.Vázquez and H.Hernando (Madrid, 1994) World Scientific, 91.
19. G. Quintana, H. Sirkin, M. Rosen, D. Kurlat, E. Frank; Revista Brasileira de Fisica 9 (1979) 1.
20 Kissinger H. E., Anal. Chem., 29 (1957) 1702.
21 Crhistian J. W., "Physical Metallurgy", ed R.W.Cahn (North-Holland), New York, (1970), chap 10.

Table 2: Main parameters of the crystallization kinetic: crystallization temperature (measured for 20 K/m), transformation enthalpy, phases appeared in each crystallization process, apparent activation energy (obtained by Kissinger and Arrhenius methods) and the Avrami exponent.

Sample	**Ga20Te80 (alloy A)**	**Ga13.3Te81.7Fe5 (alloy B)**	**Ga19.6Te78.4Fe2 (alloy C)**
First Peak			
Phases	Te	Te	Te and Ga2Te5
Tc (K)	474.4 ± 1.0	438.4 ± 1.0	536.4 ± 1.0
ΔH (J/g)	23 ± 2	10 ± 1	21 ± 2
Activation Energy (KJ/mol)			
Kissinger	201 ± 13	232 ± 19	341 ± 20
Arrhenius	205 ± 7	256 ± 33	288 ± 10
Avrami Exponent	5.3 - 6.2	1.5 - 2.2	4.0 - 5.5
Second Peak			
Phases	Ga2Te5	Te and Ga2Te3	-
Tc (K)	549.4 ± 1.0	543.5 ± 1.0	-
ΔH (J/g)	10 ± 0.5	7 ± 0.5	-
Activation Energy (KJ/mol)			
Kissinger	616 ± 37	565 ± 25	-
Third Peak			
Phases	-	Ga2Te5	-
Tc (K)	-	673.0 ± 1.0	-
ΔH (J/g)	-	5 ± 0.3	-
Activation Energy (KJ/mol)			
Kissinger	-	257 ± 33	-

CRYSTALLIZATION PROCESS OF $Fe_{38.7}Ni_{38.7}B_{17.4}Sn_{5.2}$ AMORPHOUS RIBBONS.

A. Ureña, V. Cremaschi, B.Arcondo and H. Sirkin.
Departamento de Física, Facultad de Ingeniería, Universidad de Buenos Aires.

SUMMARY

Amorphous ribbons of $Fe_{38.7}Ni_{38.7}B_{17.4}Sn_{5.2}$ were prepared by means of the melt spinning technique. The samples were annealed at a constant rate from room temperature up to different temperatures to induce crystallization. The structural evolution from the amorphous to the crystalline state was analyzed at different steps of this process. Results obtained by X-ray diffraction and Mössbauer Effect are reported.

INTRODUCTION

Amorphous ribbons obtained by the Melt Spinning technique remain mechanically stressed due to the production process. This stress produces anisotropy in the material which can affect the elastic properties. Annealing processes performed on the ribbons deliver such stresses and, depending on the temperature, induce structural relaxation or crystallization. In a previous work [1], the relaxation process of the FeBSn system has been studied by means of the relative Young's modulus variation. A strong change has been observed previous crystallization. Similar changes has not been detected in the FeNiBSn system by this method. In order to understand these different behaviors, a systematic study along the annealing process up to crystallization is necessary. In this work X ray diffraction and Mössbauer spectroscopy are employed in order to begin the analysis of the crystallization process on samples of FeNiBSn obtained by the Melt Spinning technique.

EXPERIMENTAL PROCEDURE

An alloy of composition $Fe_{38.7}Ni_{38.7}B_{17.4}Sn_{5.2}$ has been prepared using 99.99% purity materials in a high frequency furnace under Argon atmosphere. Metallographic and X-ray Cu Kα radiation analyses have been carried out in the as-cast sample in order to confirm its homogeneity. As results obtained during these studies suggest the presence of a τ phase composed of Fe-Ni-B-Sn not reported previously, another alloy of composition $Fe_{3.15}$ $Ni_{17.85}$ Sn_2 B_6 was prepared in a similar way, in order to analyze this possibility

Ribbons of the $Fe_{38.7}Ni_{38.7}B_{17.4}Sn_{5.2}$ alloy quenched from the liquid state have been obtained by the Melt Spinning technique (17μm thickness and 1 mm width). The amorphous structure was confirmed by X-ray diffraction analyses. These ribbons were annealed under high vacuum (10^{-5} mbar) from room temperature (RT) up to different temperatures in the range between 100°C and 550°C. A heating rate of 4.8° C/min was employed. At different steps of the annealing process, the system was analyzed employing X-ray diffraction (Cu Kα radiation) and RT Mössbauer effect (ME). The ME experiment were performed by using a $^{57}Co(Rh)$ source in transmission geometry. The sample $(Fe_{15}Ni_{85})_{21}B_6Sn_2$ was analyzed at RT and at 15K.

RESULTS AND DISCUSSION

<u>*X-Ray Diffraction Analyses*</u>

The X-ray diffraction pattern of the as-quenched ribbons consists in two wide peaks, one at 2θ= 44.22° and the other at 2θ= 79.54° corresponding to an amorphous structure.

Annealed samples remain amorphous up to 390°C. The first crystalline structure corresponding to bcc (Fe,Ni) (Kamacite) is observed in samples annealed between T= 410°C and T= 450°C. The diffraction pattern corresponding to this phase can be observed in Fig. 1 (T= 430°C and 450°C). In samples annealed at higher temperatures the XR spectra show a set of lines that correspond to a τ phase $Cr_{21}W_2C_6$ (D84) type (Fig. 1; T= 470°C). This structure is stable in the Ni-B-Sn system ($Ni_{21}Sn_2B_6$) [3] and in the Ni-Fe-B system ($Ni_{18,5}Fe_{4,5}B_6$) [2] and had not been observed previously in the $(Fe_xNi_{(1-x)})_{21}Sn_2B_6$ stoichiometry. In order to analyze if a similar phase can be obtained in the Fe-Ni-B-Sn system, a new alloy was prepared starting from $Ni_{18,5}Fe_{4,5}B_6$ and changing 1 % of Fe and 1 % of Ni by 2 % of Sn.. The new alloy presents a XR diffraction pattern (Fig. 2) which is in agreement with both reported works [3,4].

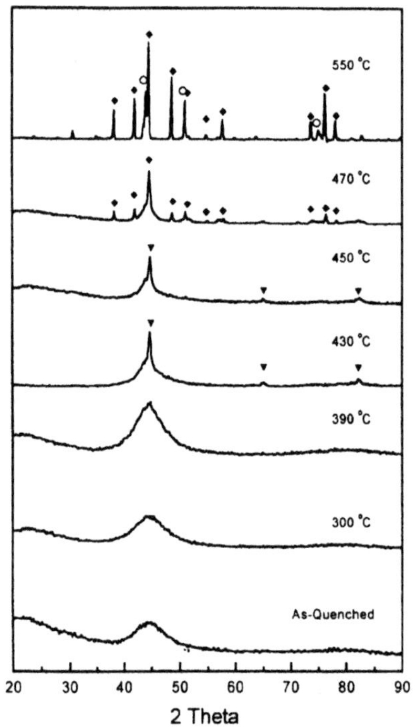

Figure 1. X ray diffraction patterns obtained with Cu Kα radiation of as quenched and annealed samples. ▼: Kamacite patterns. ◆· $Fe_{3\,15}$ $Ni_{17\,85}$ Sn_2 B_6 patterns, o: fcc (Fe,Ni) patterns.

Metallographic and scanning electron microscopy (EDAX) inspection show slabs with the same morphology of the τ phase (Ni-B-Sn) and the presence of iron. These facts make the hypothesis of a τ phase in a quaternary system more credible. Further studies are necessary to confirm this.

In samples annealed at T= 470°C and higher, this phase becomes dominant. Samples annealed up to 550°C are almost completely crystalline. In this step, the Kamacite phase has practically disappeared and new phases, among them fcc (Fe,Ni), were observed.

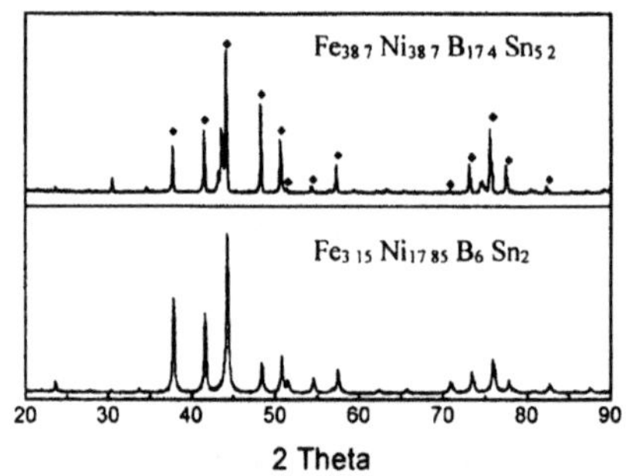

Figure 2. X ray diffraction patterns of as cast $Fe_{38\,7}$ $Ni_{38\,7}$ $Sn_{5\,2}$ $B_{17\,4}$ and. $Fe_{3\,15}$ $Ni_{17\,85}$ Sn_2 B_6. ◆· $Fe_{3\,15}$ $Ni_{17\,85}$ Sn_2 B_6 patterns.

Mössbauer spectroscopy results

Mössbauer spectra of the as quenched and thermally treated samples were fitted employing the NORMOS program [3], where a histogram distribution of hyperfine field (Bhf) is considered. A linear dependence of the isomer shift (δ_{IS}) with the magnetic field is admitted. In all the cases the δ_{IS} values are referred to Fe (Rh).

Table 1: Hyperfine parameters from as quenched and annealed samples.

Sample	As-Q	300°C	390°C	430°C	450°C	470°C	550°C	Phase
Bhf (T)	21.8	23.2	24.3	22.5	21.5	19.5		**Amorphous**
δ_{IS}(mm/s)	0.03	0.03	-0.01	0.04	-0.05	-0.05		
Area (%)	100	100	100	74	50	8		
Δ (mm/s)				0.32	0.48	1.16	1.29	**Site I**
δ_{IS} (mm/s)				0.06	-0.06	0.08	0.04	
Area (%)				3	3	10	12	
Bhf (T)				34	34	34		**bcc-(Fe,Ni) (Kamacite)**
δ_{IS} (mm/s)				-0.03	-0.08	-0.04		
Area (%)				18	19	15		
Bhf (T)						29	29	**fcc-(Fe,Ni) (Taenite)**
δ_{IS} (mm/s)						-0.04	-0.07	
Area (%)						16	35	
Bhf (T)				28.4	27.3			**Site II**
δ_{IS} (mm/s)				0.09	0.09			
Area (%)				5	9			
Bhf (T)					21.7			**Site III**
δ_{IS} (mm/s)					0.24			
Area %					19			
Bhf (T)						23	23	**Site IV**
δ_{IS} (mm/s)						0.03	0.0	
Area (%)						38	30	
Bhf (T)						21.1	19.9	**Site V**
δ_{IS} (mm/s)						0.04	-0.03	
Area %						13	20	
Bhf (T)							26.1	**Site VI**
δ_{IS} (mm/s)							-0.16	
Area (%)							3	

To fit the spectra of partially crystallized samples, Bhf distributions plus discrete interactions were employed while for the sample annealed at 550°C a multiple sites fitting program was employed. Hyperfine parameters are reported in Table 1, the fitted spectra and the Bhf distributions are plotted in fig. 3 and 4 respectively.

Table 2: $Fe_{3.15}$ $Ni_{17.85}$ Sn_2 B_6 hyperfine parameters (at different temperatures).

	300 K				**15 K**			
	I	II	III	IV	I	II	III	IV
Δ_{QS}	0.23	0.29	0.92	1.15	0.24	0.33	0.83	1.16
δ_{IS} (mm/s)	0.17	0.06	0.03	0.09	0.24	0.08	0.11	0.14
Area (%)	20	21	23	36	21	20	13	46

Bhf distributions of samples annealed at 300°C and at higher temperatures exhibit a contribution at very low hyperfine field (Fig.4). In addition, the Bhf peak value increases while the higher field limit in the histogram moves, from 30.6T in the as quenched sample, to 33T in that annealed at 390°C . Both facts denote the presence of clustering before crystallization.

In samples annealed at 430°C the crystallization of a distorted Kamacite phase is observed. The δ_{IS} does not agree with its literature's value [4]. This fact is probably due to the presence of Sn in solid solution. Other crystalline environments are observed at this temperature. One of them (site I) corresponds to the clusters already present at 300°C while site II may be attributed to a magnetic phase.

In samples annealed at 450°C an evolution in the Kamacite phase is observed. Its δ_{IS} increases towards the values reported in [4]. More crystalline Fe environments become observable (II and III).

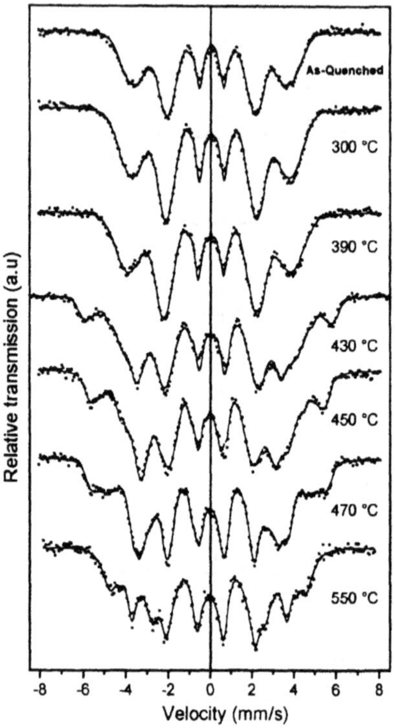

Figure 3. Mössbauer spectra obtained at room temperature with $^{57}Co(Rh)$ source of as quenched and annealed $Fe_{38.7}Ni_{38.7}B_{17.4}Sn_{5.2}$ samples.

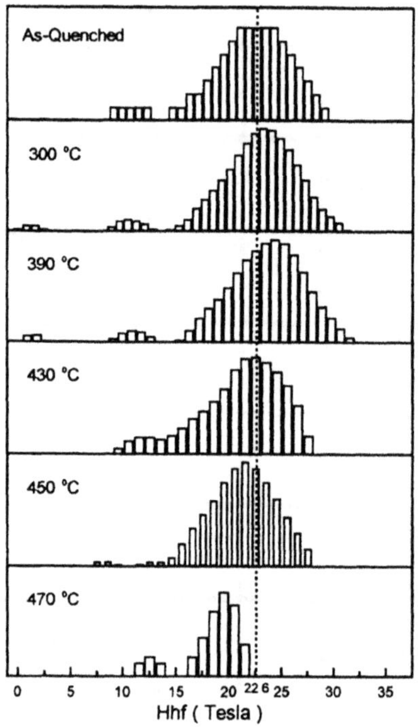

Figure 4. Hyperfine magnetic field distributions of as quenched and annealed $Fe_{38.7}Ni_{38.7}B_{17.4}Sn_{5.2}$ samples.

In samples annealed at 470°C a further evolution in the Kamacite bcc structure towards fcc-(Fe,Ni) (Taenite) [5] is observed while sites IV and V appear at the expenses of sites II and III.

In samples annealed at 550°C Taenite plus several crystalline environments are observed (I, IV, V and VI). Despite it is very difficult to attribute them to any phase, X-ray diffraction results suggest that most of them belong to the τ phase.

Mössbauer spectra of $Fe_{3.15}Ni_{17.85}Sn_2B_6$ sample obtained at room temperature and at 15K are shown in Fig. 5. Both spectra are very similar. The best fit consist in four Fe sites reported in Table 2.

Comparing the previously analyzed spectra with that corresponding to $Fe_{3\,15}$ $Ni_{17\,85}$ Sn_2 B_6 the crystalline site I in the annealed samples may be attributed to this τ phase. It is evident the drastic effect of Fe concentration on the magnetic behavior of this phase.

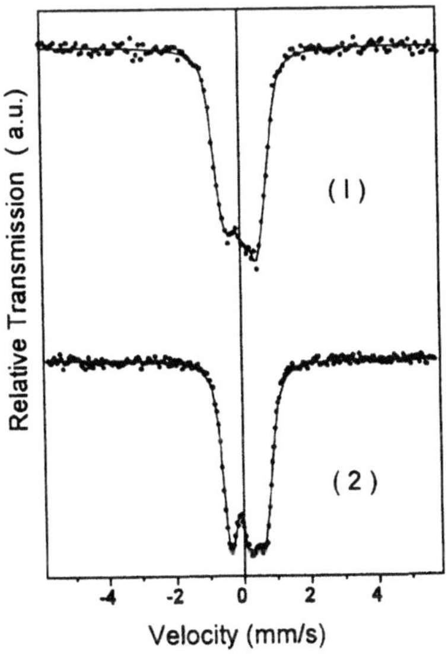

Figure 5. Mössbauer spectra obtained at room temperature (1)and at T=15K (2)with ^{57}Co(Rh) source of $Fe_{3\,15}$ $Ni_{17\,85}$ Sn_2 B_6 sample.

CONCLUSIONS

Rapidly quenched $Fe_{38.7}Ni_{38.7}B_{17.4}Sn_{5.2}$ samples remain amorphous along annealings performed up to 390°C. Well below this temperature clustering is observed (since T=300°C in this work). At least two different type of clusters are present. Once that crystallization begins with Kamacite at 410°C the evolution of the crystalline phases denotes strong diffusive processes.

AKNOWLEDGEMENTS

We thank Dr. Celia Saragovi for providing the low temperature facilities.

REFERENCES

[1] H.H. Stadelmaier and L. T. Jordan, Z. Metallkde 53 (1962) 719.
[2] M. Ayel, R. Rivière and G. Mounier, C. R. Acad. Sc. Paris, t.264 C (1967) 1756.
[3] R. A. Brand,(1987) NORMOS Programs, Internal Report Angewandte Physik, Universität Duisberg.
[4] R. B. Scorzelli, I. Souza Azevedo, I. Ortalli, G. Pedrazzi and A. Bonazzi, Hyp. Int. 83 (1994) 479.
[5] R. B. Scorzelli in: Proceedings of the first Latin American Conference Applications of the Mössbauer effect, Rio de Janeiro 31 de Oct-4 Nov 1988, eds. E. Baggio-Saitovitch, E Galvão da Silva and H. R. Rechenberg. (World Scientific, Singapore), p. 65.

ABOUT THE FITTING PROCEDURE OF MÖSSBAUER SPECTRA ON NANOCRYSTALLINE ALLOYS

Jean-Marc Greneche[1], Marcel Miglierini[1]*, and Anna Slawska-Waniewska[2]

1 Laboratoire de Physique de l'Etat Condensé, URA CNRS 807
Faculté des Sciences, Université du Maine, F72017 Le Mans Cedex France
2 Institute of Physics, Polish Academy of Sciences, Al. Lotnikow 32/46, 02-668 Warszawa, Poland

Abstract: We summarise a novel fitting procedure of Mössbauer spectra obtained on Fe-based nanocrystalline alloys: it consists of one sextet and two independent distributions of hyperfine field attributed to bulk bcc-Fe crystalline grains, interface zone and amorphous residual matrix, respectively. We essentially report in the present communication on the different stages which permitted to establish this present fitting model, in conjunction with the structure of nanocrystalline alloys.

Introduction

During the first crystallization stage of amorphous FeCuMB(Si) alloys (M=Nb, Cr, Zr, Ta, Ti, ...), a nanoscale structure appears with the emergence of bcc-Fe(Si) crystalline grains 10-20nm diameter, embedded in the remaining amorphous matrix: this structural behaviour originates very soft magnetic properties [1, 2]. Large effort has been devoted to study the structural and magnetic properties of some nanocrystalline alloys. In addition to diffraction techniques and electron microscopy, ^{57}Fe Mössbauer spectrometry as local probe is an excellent tool to elucidate the nature of the different hyperfine interactions and to quantize the presence of non equivalent atomic surroundings in these heterogeneous Fe-based materials [3]. In Si-containing nanocrystalline alloys, the complexity of the hyperfine structure of Mössbauer spectra which is due to the presence of many non equivalent iron sites in the bcc-FeSi crystalline phase, prevents from an accurate and detailed analysis of the intergranular phase which plays an important role in the magnetic behaviour. On the contrary, Fe(Cu)MB nanocrystalline alloys are better candidates, because bcc-Fe emerging grains are characterised by one single well-resolved sextet. Unlike studies reported in literature, we present in this communication a novel fitting procedure which can be applied on Fe(Cu)MB nanocrystalline alloys, established in conjunction with the topology of hyperfine interactions within Fe-based nanocrystalline alloys (details are given in [4]).

Experimental section

A special interest was given to Fe(Cu)ZrB and Fe(Cu)TiB nanocrystalline alloys in order to model the intergranular phase. The samples result from melt-spun ribbons annealed for different temperatures and times. Mössbauer experiments were performed in transmission geometry in the temperature range 4K-500K. The ribbon plane is oriented perpendicular to the γ-beam. The hyperfine parameters were refined by using both NORMOS DIST [5] and MOSFIT [6] programs: the discrepancy of respective results is found lower than the experimental error.

Description of the fitting procedure

Low temperature Mossbauer spectra obtained on Fe(Cu)MB nanocrystalline alloys exhibit clearly one sextet with well-resolved sharp lines and one sextet with broad lines while the sextet is superimposed to a central subspectrum which progressively collapses to an asymmetrical doublet with increasing temperatures (T>400K). The sextet is unambiguously assigned to bcc-Fe grains and the other component to the intergranular phase. A first fitting procedure of low temperature spectra consists in considering one single discrete distribution of hyperfine fields $P(H_{hyp})$ with following assumptions: the values of isomer shift (IS) and quadrupolar shift (2ε) for all sextets are constrained equal and free, the value of linewidth (Γ) constant. The profile of $P(H_{hyp})$ exhibits a narrow peak at 33-34T unambiguously assigned to bcc-Fe crystalline grains and three pronounced peaks and nearly zero $P(H_{hyp})$ values around 20T. One of these peaks (around 17-18T) has no physical meaning within $P(H_{hyp})$: indeed, it corresponds to the positions of the intermediate lines (fifth and second) of high field components ($H_{hyp} \approx 30T$) which intensity is strongly dependent on the orientation of the magnetisation of crystalline grains. Consequently, for the bcc-Fe component, the value of angle β which defines the direction of the hyperfine field respects to that of the γ-beam will be free in next fits. The first block (<20T) is comparable to $P(H_{hyp})$ estimated from the as-quenched amorphous alloy spectrum: consequently, it can be unambiguously assigned to the amorphous matrix of the nanocrystalline alloy. The upper block exhibits H_{hyp} values smaller than those characteristics of the crystalline Fe phase but higher than those of the amorphous matrix: we may first attribute this component to iron atoms which are located neither in crystalline grains, nor in amorphous matrix, a kind of « interface zone ». So, in a second stage, one can split $P(H_{hyp})$ into two independent blocks and one single component which corresponds to Fe grains, i.e. three different values of IS, 2ε maintained free during the fitting procedure. The adjustment is better than that obtained with the previous fitting model and one peak of $P(H_{hyp})$ (that located around 17-18T) disappears and an in-plane magnetisation is evidenced from the crystalline grains. A third stage consists in considering a linear correlation between IS and H_{hyp} in the lower field distribution block in order to account for the slight asymmetry of the magnetic spectrum; this assumption is consistent with correlated distributions encountered in FeZr(B) based amorphous systems. This new hypothesis leads to well fitted low temperatures spectra.

One can mention the following preliminary conclusions: (i) the refined values of hyperfine parameters of the sextet attributed to bcc-Fe crystalline grains are in excellent agreement with expected ones; (ii) concerning the high field distribution, IS and H_{hyp} values are slightly smaller than those of bcc-Fe phase, 2ε value is close to zero, which suggests that these iron nuclei are located in the close vicinity of crystalline grains; (iii) the third component comprises two parts which relative intensities are temperature dependent: one with low fields which may be assigned to paramagnetic or weakly magnetically coupled Fe which increases with temperature and the upper part to magnetic Fe moments, both corresponding to the residual amorphous phase. Nevertheless, there remains some ambiguities: (i) the criteria to define the lower limit of the high hyperfine field distribution block and the upper limit of the low one, including solutions with overlap of both blocks and (ii) the orientation or the preferred orientation (magnetic texture) of the hyperfine field within the intergranular phase which is given by angle β.

The appropriate limits of $P(H_{hyp})$ can be optimised by following the physical evolution of hyperfine parameters obtained after fitting series of Mössbauer spectra as a function of the temperature. To liberate from texture effects, Mössbauer spectra can be registered either using the magic angle configuration leading to a texture-free spectrum (3:2:1:1:2:3) [7] or in presence of an external field applied to the ribbon in order to saturate the magnetisation: when the hyperfine field is parallel (perpendicular) to the ribbon plane, one can expect 3:4:1:1:4:3 (3:0:1:1:0:3) configuration. The alignment of the hyperfine field is parallel or perpendicular to the tensile strengths induced during the quenching process when the magnetostrictive behaviour of these systems is positive or negative, respectively. The lower contributions of the $P(H_{hyp})$ block assigned to the amorphous phase can be eventually replaced by a distribution of quadrupolar splittings linearly correlated to that of an isomer shift, consistently with a paramagnetic doublet: one observes an increase of this latter component with temperature. Consequently, taking into account such considerations, the final fitting procedure can be applied to any Mössbauer spectra obtained on Fe(Cu)MB nanocrystalline alloys, whatever the temperature: one example is illustrated on fig.1 with the total spectrum, the three subcomponents and the distribution of hyperfine fields in the case of $Fe_{86.5}Cu_1Zr_{6.5}B_6$ alloy annealed at 600°C for 1 hour. The hyperfine data and the shape of $P(H_{hyp})$ can be discussed with respect to the expected topology of hyperfine interactions within Fe-based nanocrystalline alloys [4, 8-11].

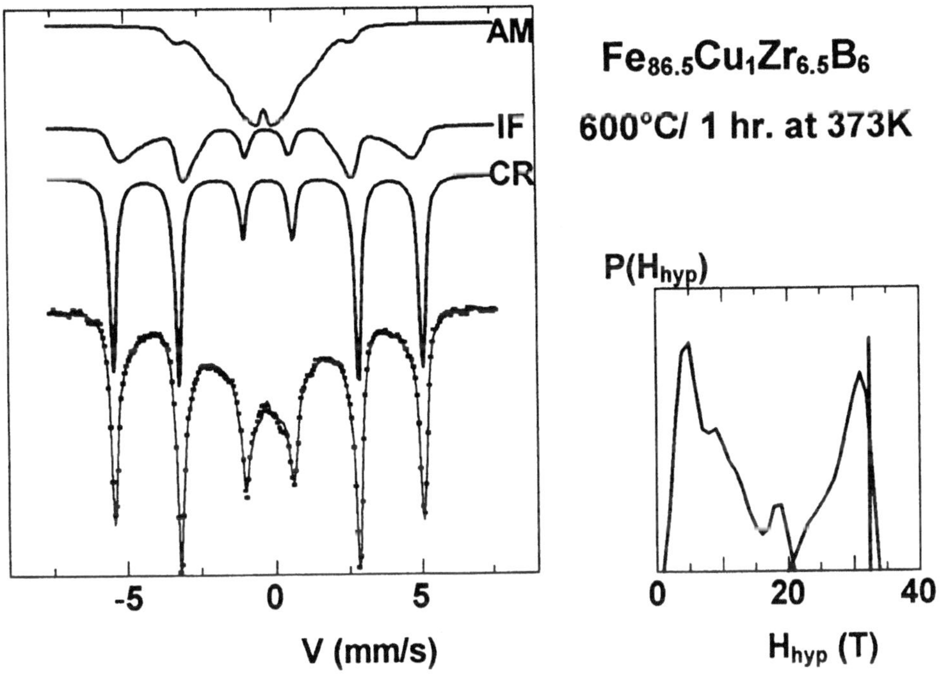

Fig. 1: Mössbauer spectrum recorded at 373K on $Fe_{86.5}Cu_1Zr_{6.5}B_6$ nanocrystalline alloy annealed at 600°C for 1h with its decomposition into 3 subcomponents attributed to the crystalline grains (CR), the amorphous matrix (AM) and the interface (IF), and the distribution of hyperfine fields $P(H_{hyp})$.

Discussion

The component represented by the outer single sextet corresponds clearly to iron atoms located in the bulk of crystalline grains, because of the nice agreement of the hyperfine parameters. The lower $P(H_{hyp})$ block is attributed to the amorphous remainder: indeed, IS and 2ε values and the linear correlation between IS and H_{hyp} are close to those encountered in as-quenched amorphous alloys spectra while $P(H_{hyp})$ peak value shifts towards lower H_{hyp} values, because of the smaller iron content. Then, the thermal evolution of $P(H_{hyp})$ which exhibits an increase of low fields (i.e. quadrupolar interactions) with temperature, is consistent with non homogeneous concentrations of Zr within the matrix: indeed, the Curie temperature in FeZr(B) amorphous alloys is strongly dependent on Zr content. This probably results from the diffusion mechanisms during the crystallization process. According to the hyperfine data, the second block may be attributed to iron atoms located in the outer surface of crystalline grains and its thickness can be estimated at ≈ 1 nm from their absorption areas [9, 11]. In that sense, Mössbauer spectrometry is able to elucidate the nature of hyperfine interactions in iron-based nanocrystalline alloys, providing that great attention is given on the fitting procedure on series of spectra recorded under different experimental conditions.

*) On leave from: Department of Nuclear Physics and Technology, Slovak Technical University, Ilkovièova 3, 812 19 Bratislava, Slovakia

References:

[1] Y. Yoshizawa, S. Oguma, and K. Yamauchi, J. Appl. Phys. 64 (1988) 6044

[2] K. Suzuki, A. Makino, A. Inoue, and T. Masumoto, J. Appl. Phys. 70 (1991); 74 (1993) 3316

[3] J.M. Greneche, this conference

[4] M. Miglierini and J.M. Greneche, J. Phys.: Condens. Matter 9 (1997) at press

[5] R. A. Brand, unpublished NORMOS programs

[6] J. Teillet, and F. Varret, unpublished MOSFIT program

[7] J.M. Greneche and F. Varret, J. Physics C: Solid State Physics, 15 (1982) 5333-5344

[8] M. Miglierini, Y. Labaye, N. Randrianantoandro, and J.M. Greneche, Proceedings of the 9th Conference on Rapidly Quenched and Metastable Materials, Mater. Sci. Engin. A (1997) at press

[9] J.M. Greneche, and A. Slawska-Waniewska, Proceedings of the 9th Conference on Rapidly Quenched and Metastable Materials, Mater. Sci. Engin. A (1997) at press

[10] A. Slawska-Waniewska, K. Brzozka and J.M. Greneche, Acta Phys. Pol., (1997) at press

[11] A.Slawska-Waniewska, A.Roig, E.Molins, J.M.Greneche, and R. Zuberek, J. Appl. Physics, (1997) at press

III. Chemical Structure and Bonding

MÖSSBAUER STUDIES OF HOST:GUEST INTERACTIONS IN ENCAPSULATED SnF_2

A. Abras[a], *R. D. Sinisterra*[b *], *C. A. L. Filgueiras*[b], *C. A. Menezes*[b], *C. A. Alves de Carvalho*[c], *M. E. Cortés Segura*[d], *N. Tortamano*[e]

[a] *Departamento de Física, Universidade Federal de Minas Gerais, C.P. 702, 30161-970 Belo Horizonte, MG, Brazil*
[b] *Departamento de Química, Universidade Federal de Minas Gerais, 30161-970 Belo Horizonte, MG, Brazil*
[c] *Departamento de Química, Universidade Federal de Ouro Preto, Ouro Preto, MG, Brazil*
[d] *Departamento de Odontologia Restauradora, Universidade Federal de Minas Gerais, Belo Horizonte, MG, Brazil*
[e] *Área de Clínica Integrada, Faculdade de Odontologia, Universidade de São Paulo, São Paulo, SP, Brazil*

Abstract

The encapsulation of chemicals in a host cage-like ligand has the property of changing the chemical and physical properties of the guest compound. Cyclodextrins have been shown to be promising hosts in a number of situations, leading to new species with wide technological and biological applications.

We have studied the interaction between SnF_2 and α- cyclodextrin. The compounds were prepared either in solution or in solid state, and were characterized by Mössbauer spectroscopy, powder X-ray diffraction, i.r. spectroscopy, and thermal analysis (TG, DTG, DSC).

Mössbauer spectroscopy proved to be a remarkable analytical tool to determine the amounts of included and free SnF_2, which were clearly identified, as well as the extent of oxidation of the guest compound to SnO_2. The method also allowed the monitoring of the several techniques used to perform inclusion.

1. Introduction

Cyclodextrins are non reducing oligosaccharides composed of least six (1→4) linked α D glucosyl residues, built up from six, seven or eight glucopyranose rings, respectively known as α, β, and γ-cyclodextrins, which have the shape of a hollow, truncated cone. They form inclusion compounds with suitably sized molecules which fit wholly or partially within their central cavity. Its most important property is that the interaction host:guest occurs without any covalent bonding. Current applications research involving cyclodextrin compounds cover a very broad spectrum of

---

*Author to whom correspondance should be addressed

industries, e.g. agrochemical, chemical processing, personal care, foodstuffs and pharmaceuticals [1,2].

We have investigated the interaction between SnF_2 and α-cyclodextrin (α-cd). The precursor and the inclusion compounds were studied by Mössbauer spectroscopy, powder X-ray diffraction, i.r. spectroscopy and thermal analysis (TG, DTG and DSC).

The type of inclusion investigated here is of great importance as means of protecting the highly reactive SnF_2 species and producing a material capable of slowly releasing the SnF_2 moiety, highly important in dentistry.

2. Experimental

α-cyclodextrin hydrate (α-cd) was purchased from Aldrich and SnF_2 from Strem, and both were used without further purification.

The inclusion compounds were prepared by three methods. The first one consisted of mixing an aqueous solution of α-cd and SnF_2 in a 1:1 molar ratio under N_2 and stirring for 15 minutes (preparation **P1**) and for 24 hours (preparation **P2**). In both cases the water was evaporated under vacuum and the host:guest compound obtained. The second method involved a reaction in the solid state between SnF_2 and α-cd in a 1:1 molar ratio under N_2 (preparation **P3**). The third preparation consisted of dry-freezing an aqueous solution of SnF_2 and α-cd, also in a 1:1 molar ratio (preparation **P4**). In order to compare the results a physical mixture of the reactants was also prepared with the same molar ratio as the host:guest compound (**P5**)

Several physical methods were used to characterize these products. Mössbauer spectra were obtained using a conventional constant acceleration spectrometer, moving a $CaSnO_3$ source kept at room temperature. The samples were measured at liquid nitrogen and room temperatures. All spectra were fitted by means of Lorentzian line shapes. Isomer shift values are given with respect to the $CaSnO_3$ source. X-ray powder diffraction (XRD) was carried out with a Rigaku diffractometer, fitted with a LiF monochromator and employing Cu Kα radiation with 5-40 ° (2θ) step scanning. TG and DTG curves were obtained from a Mettler-TA4000 system, using a dynamic N_2 atmosphere and a heating rate of 10°/min. The DSC curves were obtained from a Shimadzu DSC-50 apparatus with the same heating rate. I.r. spectra were recorded from a Perkin-Elmer spectrometer model 283B, using CsI pellets.

3. Results and Discussion

Inclusion compounds are more stable to air compared to tin(II) fluoride. The XRD results suggest a highly amorphous character for the inclusion product, in stark contrast with either the physical mixture, SnF_2 or α-cyclodextrin. This points to the existence of a supramolecular host-guest compound. Lowering of the crystalline character upon inclusion is likely to result from loss of water molecules from the host [3].

Formation of the inclusion product is accompanied by a sharpening of all the bands in the i.r. spectrum, compared to those of free α-cyclodextrin and the physical mixture. In the OH stretching region this sharpening can be associated with a decrease in hydrogen-bonding, which is consistent with the XRD results. Likewise, sharpening of the C-O-C bands around 1050 cm^{-1} is also indicative of less hydrogen-bonding in the material upon inclusion.

The TG and DSC curves of the inclusion products show the loss of water molecules in the range 70-100 °C, in accordance with the other data. The TG curve of SnF_2 shows its high thermal stability up to 506°C, suggesting intense association in the solid state. The TG curve of the host-guest compound, on the other hand, is markedly different and shows decomposition only at 259°C, which prevents any interpretation as a simple mixture. When a physical mixture of SnF_2 and α-cyclodextrin is subjected to TG under N_2 , the residue is only Sn (13.8%), whereas the inclusion compound gives a complex residue (33.2%), indicating strong chemical interaction in the supramolecular species.

The Mössbauer spectra of SnF_2 and of the inclusion compounds obtained by different methods are given in Figure 1. The best fitting parameters are summarized in Table 1, which includes literature data for comparison.

It can be seen from Figure 1(A) and Table 1 that the SnF_2 used in this work accounts for 96% of the relative area of Sn(II), which is in excellent agreement with literature data for α-SnF_2 [4]. The remaining 4% relative area is due to impurities, characteristic of tetravalent tin octahedrally coordinated by fluorine. This is seen from the good agreement between our parameters and those reported for $(H_3O)_2SnF_6$ [5] and $[SnF_6]^{2-}$ octahedra [6]. The Mössbauer spectrum of SnF_2 was also obtained at room temperature and showed 97% of the relative area due to Sn(II) with parameters δ = 3.42(1) mm/s, Δ= 1.56(1) mm/s and Γ = 1.03(1) mm/s, in accordance with literature data [7].

As expected the Mössbauer parameters for the physical mixture (**P5**) are identical with those of free SnF_2 (Table 1). This result shows that there is no host:guest interaction in this case.

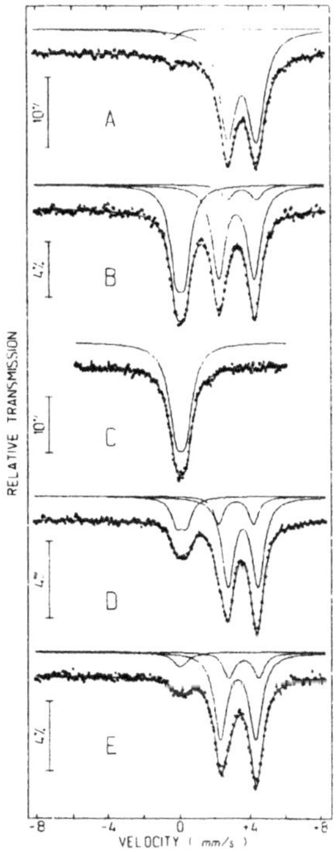

Figure 1. Mössbauer spectra at 85 K of SnF_2 (A) and inclusion compounds: aqueous solution for 15 minutes (B), 24 hours (C), solid state (D) and dry-freezing (E).

Table 1. Mössbauer parameters at 85 K for SnF_2 and their inclusion compounds with α-cyclodextrin.

Compound	Site 1 (free SnF_2)				Site 2 (included SnF_2)				Site 3 (Sn(IV))			
	δ^b	Δ^c	Γ^d	RA[e]	δ	Δ	Γ	RA	δ	Δ	Γ	RA
SnF_2	3.49(1)	1.62(1)	1.13(1)	96%					-0.45(4)	0	0.81(4)	4%
αSnF_2[a]	3.49(5)	1.61(5)	1.00(5)									
P1	3.47(2)	1.64(4)	0.82(6)	8%	3.15(1)	2.01(1)	0.90(2)	53%	-0.04(1)	0.50(1)	0.81(2)	39%
P2									-0.04(1)	0.54(1)	0.96(1)	100%
P3	3.46(1)	1.65(1)	0.87(2)	68%	3.14(3)	2.01(6)	0.76(7)	17%	-0.02(1)	0.50(3)	0.86(4)	15%
P4	3.45(3)	1.65(3)	0.80(6)	17%	3.19(2)	1.99(2)	0.87(3)	77%	-0.11(9)	0	0.80(4)	6%
P5	3.49(1)	1.63(1)	1.08(2)	97%					-0.33(7)	0	0.80(5)	3%

[a] Ref. 4 , [b]isomer shift (mm/s) relative to $CaSnO_3$, [c]quadrupole splitting (mm/s), [d]linewidth (mm/s) and [e]relative area under resonance curve. Preparation methods: aqueous solution 15 minutes(**P1**) and 24 hours (**P2**), solid state (**P3**), dry-freezing (**P4**) and physical mixture (**P5**)

Mössbauer spectra of the inclusion compounds could not be fitted as a combination of a Sn(IV) site with only one Sn(II) site. A better fitting was obtained assuming one Sn(IV) site and two Sn(II) sites. The corresponding parameters are presented in Table 1.

Isomer shifts and quadrupole splittings for site 1 in all the products correspond to free SnF_2, whereas for site 2 they were attributed to the inclusion products. This means that a interaction takes place between guest (SnF_2) and the host molecules (α-cd), so that the electronic configuration of the guest molecule is affected by the cavity of cyclodextrin. The decrease in δ corresponds to a decrease in s electron density at the tin nucleus, and the increase in Δ corresponds to an increase of the electric-field gradient produced by an asymmetric distribution of charge about the nucleus. Since cyclodextrin has an electrostatic field in the cavity one can associate the variation observed in δ and Δ values to that field [8]. This interaction suggests that the guest molecules in the cavity are under negative pressure, resulting in a decrease in the overlap between inner s-shells of the tin atom and the valence orbitals of the ligands, and consequently a decrease in δ. We think that this electrostatic interaction also might be related to the increase in Δ values. Similar results have been observed for ferrocene-cyclodextrin and some ring-substituted ferrocene derivatives-cyclodextrin [8].

It has been observed in the reaction of SnF_2 with an aqueous solution of H_2O_2 that the products present Mössbauer parameters δ =3.00 to 3.21 mm/s and Δ = 1.76 to 2.04 mm/s, which are typical of products of hydrolysis of SnF_2, such as Sn_4OF_6 and Sn_2OF_2 [9]. Although our site 2 has δ and Δ values within these ranges the presence of any product of hydrolysis in our compounds is unlikely, since the literature compounds were obtained around 200 °C, and our products were prepared at room temperature.

Site 3, present in all the compounds, is characteristic of octahedrally oxygen coordinated Sn(IV) [9]. It can be seen from Table 1 that, except for free SnF_2 and compounds **P4** and **P5**, site 3 shows $\delta \cong 0$ and Δ=0.50-0.56 mm/s, which agree very well with previously reported data for SnO_2 [10,11].

The amount of Sn(II) from SnF_2 which oxidizes to SnO_2 depends on the method of preparation (Table 1). In the methods involving aqueous solutions SnO_2 relative area (RA) signals are 39% (**P1**) and 100% (**P2**) respectively. The latter corresponds to a complete oxidation which occurred when the reaction time changed from 15 minutes to 24 hours. Although the reaction in the solid state was performed under N_2, a SnO_2 RA signal of 15% (**P3**) was observed. We attributed this partial oxidation to the presence of water in the cyclodextrin cavity.

It should be mentioned that the relative intensity detected for SnO_2 in the Mössbauer spectra through the RA parameter does not correspond to the relative amount of SnO_2 in the sample, because RA also depends on the recoil- free fraction (f), which is not the same for SnO_2 and SnF_2. It is known that f(SnO_2) =0.44 and f(SnF_2)= 0.034 at room temperature, i.e., the recoil-free fraction of SnO_2 is almost 13 times as large as that of SnF_2 [12]. As a result, oxidation of 1% of the Sn(II) of SnF_2 to SnO_2 would give an SnO_2 signal of 13% of the total signal. Therefore, the amount of SnO_2 formed in the various preparative methods are much smaller than the RA values shown in Table 1.

On the other hand, it is quite reasonable to assume that the f-factor is the same for both Sn(II) sites. Therefore the RA parameter corresponds to the relative amount of site 2 in the product. From the RA values given in Table 1 one can conclude that the amount of inclusion compound (site 2) is large in the dry-freezing method (**P4**) and small in the solid state method (**P3**).

Although in the aqueous solution method (**P1 and P2**) the amount of inclusion compounds depends upon the time of reaction , **P1** showed to be a more efficient method of inclusion than the solid state process (**P4**). This result agrees with the importance of water for the inclusion phenomenon with cyclodextrin [13].

In order to study the influence of the size of the cavity of cyclodextrin in these processes, investigation of the inclusion of SnF_2 in β-cyclodextrin is in progress.

ACKNOWLEDGEMENTS

The authors would like to thank financial support by Brazilian agencies FAPEMIG, CNPq and FINEP.

References

[1] a. J. Szejtli: "Cyclodextrin Technology", Kluwer Academic Publishers (1989), b. Wolfram Saenger; Angew. Chem. Int. Ed. Engl. 19 (1980) 344

[2] R.D. Sinisterra, R. Najjar, O.L. Alves, P.S. Santos, C.A. Alves de Carvalho and A.L. Conde da Silva, J. Incl. Phenom., 22 (1995) 91

[3] W. Saenger: "Inclusion Compounds", Ed. J.L. Atwood, J.E.D. Davies and D.D. MacNicol, Academic Press, London, Vol. 2, Ch. 8 (1984)

[4] T. Birchall, P. A. W. Dean and R. J. Gillespie, J. Chem. Soc(A) (1971), 1777

[5] N. N. Greenwood and T. C. Gibb, "Mössbauer Spectroscopy", Chapman and Hall Ltd, London 1971, pg. 391

[6] George Corrales and Georges Denes, Hyp. Int., 90 (1994) 441

[7] L. Fournès, J. Grannec, Y. Potin and P. Hagenmuller, Solid State Comm., 59(12), (1976) 833

[8] Yonezo Maeda, Nobuhiro Ogawa and Yoshimasa Takashima, J. Chem. Soc. Dalton Trans., (1987) 627

[9] Georges Denes, Eva Laou and Abdualhafeed Muntasar, Hyp. Int., 90 (1994) 429

[10] F. Gauzzi, A. Maddalena, G. Principi and B. Verdini, "Applications of the Mössbauer Effect", ICAME'83, Vol.3 pg. 919, Ed. Yu. M. Kagan and I. S. Lyubutin.

[11] Joseph K. Lees and Paul A. Flinn, J. Chem. Phys., 48(02) (1968) 882

[12] Georges Denes, J. of Solid State Chem., 77 (1988) 54

[13] Akihiko Ueno, Fumio Moriwaki, Tetsuo Osa, Fumio Hamada and Koichi Murai, Tetrahedron Letters, Vol.26(7) (1985) 899

TIN-119 MÖSSBAUER STUDIES OF TIN-RUTHENIUM ORGANOHETEROBIMETALLICS

Edmilson M.Moura, Cleide M. Rodrigues, Daniela C.Santos, Helmuth G.L.Siebald[1]
Departamento de Química, ICEx, Universidade Federal de Minas Gerais, 31270-901 Belo Horizonte, MG, Brazil
A. Abras
Departamento de Física, ICEx, Universidade Federal de Minas Gerais, 30161-970 Belo Horizonte, MG, Brazil

Abstract

Two organoheterobimetallics of cyclopentadienyl (Cp) involving tin and ruthenium were prepared and studied by several physical methods, of which Tin-119 Mössbauer spectroscopy played a major role. The complexes were derived from $Ru(\eta^5\text{-}Cp)(PPh_3)_2Cl$ (PPh_3=triphenyl-phosphine) which reacted with NaN_3 and anhidrous $SnCl_2$ affording the final products:
$Ru(\eta^5\text{-}Cp)(PPh_3)_2N_3$ **1**, $Ru(\eta^5\text{-}Cp)(PPh_3)_2SnCl_3$ **2** and $Ru(\eta^5\text{-}Cp)(PPh_3)_2SnCl_2N_3$ **3**. Complexes **2** and **3** contain a direct Sn-Ru bond, as evidenced by IR and Mössbauer spectroscopy. The Mössbauer parameters obtained are typical of Sn(IV) species tetracoordinated. From the values of the isomer shift the electronegativities for the groups azido and the $[Ru(\eta^5\text{-}Cp)(PPh_3)_2]$ moiety were estimated.

1. Introduction

Organoheterobimetallics are very important in several fields such as catalysis and in vitro models for the study of molecular nitrogen fixation in agriculture[1]. Furthermore, the azide ion N_3^- is an entity utilized as an anion in the chemistry of metal complexes. Due to the existence of several filled, unfilled and delocalized orbitals, it is possible to transform the azido moiety in the metal coordination sphere by chemical reactions, thermal and/or electrolytic activation and photolysis[2-5]. In addition, azido complexes can lead to metal nitrides containing a metal-nitrogen triple bond [6]. Photolysis of azido complexes leads to the formation of azido radicals in the primary photochemical process [7] and these photoredox reactions have found commercial use in photoinitiation of radical polymerization [8]. The oxydation reaction in the presence of light and molecular dioxygen leads to coordinated NO_2^- and the reaction with $NOPF_6$ affords coordinated molecular nitrogen.
This work reports the preparation and characterization of Ru and Sn azido complexes for which the chemical comportment of the azido moiety in the metal coordination sphere of a transition metal (Ru) and a representative metal (Sn) is discussed.

2. Experimental

The final products are air stable but due to the air sensitivity of the solutions, all syntheses and other operations were carried out with rigorous exclusion of dioxygen by using Schlenck tubes, vacuum line, catheter-tubing and glove box techniques, in an atmosphere of dioxygen-free argon. Hydrated ruthenium trichloride (38% Ru; Johnson Matthey Ltd.) and sodium azide were used as obtained. Benzene was destilled over sodium benzophenone under argon and degassed before use. Ethanol and methanol were destilled over magnesium ethoxide and magnesium methoxide,respectively, and degassed. Dichloromethane, acetonitrile and n-hexane were distilled from activated 4Å molecular sieves and degassed.

---

[1]To whom correspondence should be addressed

2.1 Syntheses of the complexes

Ru(η^5-Cp)(PPh_3)$_2$Cl was prepared from literature methods [9]

Ru(η^5-Cp)(PPh_3)$_2N_3$ **1**.

An ethanol solution (100 mL) of Ru(η^5-Cp)(PPh_3)$_2$Cl (0.50 g, 0.69 mmol) was refluxed for 5 h with 0.67 g (10.33 mmol) of sodium azide. This solution was cooled to room temperature and the solvent was removed by filtration. The orange residue was washed with ethanol (2x10 mL). The solid was extracted with CH_2Cl_2 (4mL), the solution filtered off and precipited with 75 mL of n-hexane. The product is separated by filtration, washed with n-hexane (2x10 mL) and dried under argon. Yield: 93%, m.p. 127.5-129.5°C

Ru(η^5-Cp)(PPh_3)$_2SnCl_3$ **2**.

A solution of 0.200 g (0.27 mmol) of Ru(η^5-Cp)(PPh_3)$_2$Cl, 0.100 g of $SnCl_2$ (0.54 mmol), 30 mL of benzene and 120 mL of methanol, was refluxed for 40 min. This solution was cooled to room temperature, filtered off and the canary yellow residue washed with methanol (2x5 mL) and dried under argon. Yield: 94 %, m.p. 185.6-188°C.

Ru(η^5-Cp)(PPh_3)$_2SnCl_2N_3$ **3**.

A mixture of **1** (0.100 g, 0.14 mmol) in 15 mL of benzene and $SnCl_2$ (0.05 g, 0.27 mmol) in 60 mL of methanol was heated under reflux for 40 min. On cooling to room temperature, yellow crystals were formed. The precipitate was filtered off, washed with cold methanol and dried under vacuum. Yield: 80%, m.p. 187.2-190°C.

Several attempts to obtain single crystals for the complexes **2** and **3** were unsuccessful, and the products always precipitated as powders. Therefore crystal and molecular structures for the complexes were not possible to determine. All complexes were analysed for C, H and N and the analytical results led to the formulations presented in Table 1.

2.2 Physical Measurements

IR spectra were recorded as CsI discs in the range 4000-200 cm^{-1} in a Perkin Elmer FTIR 1800 spectrometer, and the most important absorption are included in Table 1. The ^{1}H NMR spectra were obtained at 400 MHz, in $CDCl_3$ solutions on a Bruker WM 400, using TMS as standard which results are given in Table 2. ^{119}Sn Mössbauer spectra were run in a standard equipment at liquid nitrogen temperature using a $CaSnO_3$ source kept at room temperature. All spectra were fitted by means of Lorentzian lines shapes. The isomer shift values reported in Table 2 are given with respect to this source.

3. Discussion of results

3.1 IR spectroscopy

Table 1 presents the most important IR absorptions for the complexes. The existence of Sn-Cl bonds was shown by the appearance of a band at 410 cm^{-1} for **2** and 395 cm^{-1} for **3**, which were assigned to the Sn-Cl vibration. The complexe **3** also presented evidence of Sn-N bonding, which originated the new band at 375 cm^{-1}. The presence of azido ligand in complexes **1** and **3** is also revealed by the strong band at 2066 and 2075 cm^{-1}, respectively, and by the weak bands in 675 cm^{-1} (**1**) and 665 cm^{-1} (**3**). The Ru-Sn band was not observable, insofar as the much greater reduced mass in this case lead to Ru-Sn absorption below the wave number range under study (4000-200 cm^{-1}). The IR evidence led us to conclude that the complexes contain Sn-Cl, Sn-N_3 and Ru-N_3 bonds.

Table 1 Empirical formulas, colour, main absorption in IR (cm^{-1}) and analytical data for the complexes

Complex (colour)	ν_{Sn-Cl}	ν_{NNN}	δ_{NNN}	ν_{M-N}	%C calc.	%C exp.	%H calc	%H exp.	%N calc.	%N Exp.
$SnCl_4$ [a]	408.2 369.1									
1 Ru(η^5-Cp)($PPh_3)_2$ N_3 (orange)		2066	675	390	67.20	66.87	4.81	4.95	5.73	5.88
2 Ru(η^5-Cp)($PPh_3)_2$ Sn Cl_3 (canary yellow)	410				53.77	54.31	3.85	3.96		
3 Ru(η^5-Cp)($PPh_3)_2SnCl_2N_3$ (yellow)	395	2075	665	375	53.39	53.82	3.82	3.69	4.56	4.69

[a] Ref. [10]

3.2 NMR spectroscopy

Table 2 shows the ^{1}H NMR data for the complexes, as well as their ^{119}Sn Mössbauer parameters. The chemical shifts relative to tetramethylsilane (TMS) showed values in the range one would expect for phenyl and Cp groups. Integration of the spectra comforms very well with the analytical results. Thus, complex **1** gave an area ratio of 6.6:1.1, complex **2** gave 6.59:1.08 and complex **3** gave 6.38:1.06. The sharpness of the peaks and the good area ratios point to the presence of discrete molecules in solution. If any self-association occurred among them this would certainly lead to more complex spectra or at least to line broadening, which was not observed.

3.3 Mössbauer spectroscopy

Tin-119 Mössbauer spectra at 85K for complexes **2** and **3** are shown in Fig. 1, and the corresponding parameters are included in Table 2.

Although these complexes formally derived from Sn(II) species $SnCl_2$, their isomer shift values indicate that tin is definitely four-covalent, which is typical of the compounds involving tin-metal bonds [11,12,13,14]. From Table 2 we can see that the difference between δ and Δ values for our complexes are negligible. This means that by replacing the ligand chloride by the ligand azido the s-character of Sn atom and the symmetry of the electron distribution around Sn nucleus are quite similar in both complexes. The δ values are consistent with $5sp^3$ hybridization of Sn atom in our complexes. Comparing δ=1.98-1.96 mms^{-1} with δ=4.07 mms^{-1} of $SnCl_2$ ($5sp^2$) it is seen that a reduction in δ is caused by a reduction in s-character of Sn atom, which goes from 33% for $SnCl_2$ to 25% for our complexes. The identical δ value obtained for complexes **2** and **3** suggests that the electronegativities of the azido and chlorine groups are about

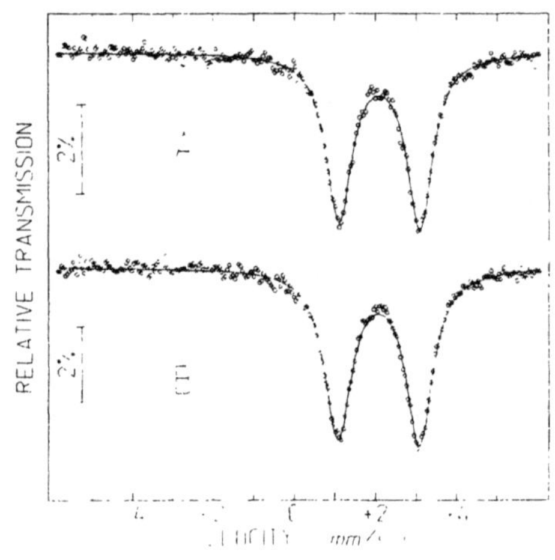

Figure 1. Mössbauer spectra at 85K for: Ru(η^5Cp)($PPh_3)_2SnCl_3$ (A) and Ru(η^5-Cp)($PPh_3)_2SnCl_2N_3$

the same. This result also indicates that Sn-N_3 and Sn-Cl π interactions are quite similar with predominating σ bonds. Furthermore, Sn-Ru is also a σ bond without backdonation via d orbital overlap from metal to ligand.

Table 2 ^{1}H NMR data and ^{119}Sn Mössbauer parameters at liquid nitrogen

Complex	^{1}H NMR[a]	δ (ppm)	δ	Δ	Γ
1 Ru(η^5-Cp)(PPh_3)$_2$$N_3$	7.40 (m)	4.22 (s)			
2 Ru(η^5-Cp)(PPh_3)$_2$$SnCl_3$	7.27 (m)	4.54 (s)	1.98(1)	1.96(1)	0.77(1)
3 Ru(η^5-Cp)(PPh_3)$_2$$SnCl_2N_3$	7.22 (m)	4.54 (s)	1.96(1)	1.99(1)	0.81(1)

[a] All NMR spectra were run in $CDCl_3$ solutions and δ(ppm) values are given relative to TMS taken as standard. The letters m and s in the first two columns refer to multiplet and singlet, respectively. δ: isomer shift (mms^{-1}) relative to $CaSnO_3$, Δ: quadrupole splitting (mms^{-1}), Γ: linewidth (mms^{-1}).

The existence of a quadrupole splitting in our complexes compared to Δ=0 for $SnCl_4$ indicates a severe break down in the symmetry of the electron distribution around Sn nucleus due to the electric-field gradient created by [Ru(η^5-Cp)(PPh_3)$_2$] moiety.

It has been shown for the species $SnCl_4$, $PhSnCl_3$, Ph_2SnCl_2 and Ph_3SnCl that there is a linear relation between the isomer shift (δ) and the sum of the Pauling electronegativities (χ), i.e., $\delta=-0.9437\Sigma\chi + 12.825$ [14]. Inserting $\chi_{Cl}=3.16$ [15] and δ=1.98 mms^{-1} for complex **2** into this expression we obtained χ=2.01 for the electronegativity of the [Ru(η^5-Cp)(PPh_3)$_2$] moiety. By the same calculation we get χ=3.18 for the electronegativity of the azido group, which is approximately the same of Cl, as expected. This result agrees with the fact that azido moiety is considered as a pseudo-halogen.

Acknowledgement The authors are grateful to finantial support from the Brazilian agencies FAPEMIG, CNPq, FINEP and PRPq-UFMG.

References

[1] P Smith, "Tin Chemicals", International Tin Research Institute, 1990
[2] A.P, Gaugham, K S. Bowman and Z. Dori, Inorg. Chem. 1972, 11, 601
[3] G. Stedman, J.Chem.Soc. 1960, 1702
[4] A.Haim and H.Taube, Inorg. Chem. 1963, 2, 1199
[5] H. Werner, W. Beck and Engelmann, Inorg. Chim. Acta 1969, 3, 331
[6] J. Chatt, C.D. Falk, G.J. Leigh and R.J. Paske, J. Chem. Soc. A 1969, 2288
[7] M.G. Evans, M. Santappa and N. Uri, J.Polym.Sci. 1951, 7, 243
[8] J.F. Endicott, M.Z. Hoffman and L.S. Beres, J.Phys. Chem. 1970, 74, 1021
[9] M.J. Bruce, C. Hameister, A.G. Swincer and R.C. Wallis, Inorganic Synthesis 1982,21, 78
[10] K. Nakamoto, Infrared and Raman Spectra of Inorganic and Coordination Compounds, 4th Ed. (Wiley, New York, 1986) p 130
[11] R.V. Parish, and P.J. Rowbotham, J. Chem. Soc. Dalton Trans. 1973, 37
[12] Roger J. Dickinson, R.V. Parish, P. J. Rowbotham, A.R. Manning and Paul Hackett, J. Chem. Soc. Dalton Trans. 1975, 424
[13] Geraldo M. de Lima, Carlos A.L. Filgueiras and A. Abras, Hyp. Int. 83 (1994) 183
[14] Eva V. Marques, Wellerson F. Ribeiro, Carlos A.L. Filgueiras and A. Abras, Hyp. Int. 96 (1995) 259.
[15] J.E. Huheey, Inorganic Chemistry-Principles of Structure and Reactivity, 3rd Ed. (Harper and Row, New York, 1983)

IV. Corrosion, Catalysis and Surfaces Phenomena

ELECTROCHROMISM IN IRON-OXIDE-BASED COATINGS OBTAINED BY SOL-GEL PROCESS

Y. ZENITAGOYA, J. DAVALOS, V. PEÑA** AND W. ESTRADA*

Facultad de Ciencias, Universidad Nacional de Ingeniería, Casilla 31-139, Lima, Perú.
* Instituto de Química-Física "Rocasolano", Madrid, España.
** Facultad de Física, Universidad Nacional Mayor de San Marcos, Lima, Perú.

Abstract

Iron-oxide-based coatings were produced by sol-gel process. Both glass and ITO-precoated-glass were used as substrates. An iron sol was prepared from an aqueous solution of Fe (III) chloride and by a dip-coating technique, followed by a heat treatment at 300 °C (and 500 °C) during 30 minutes, a film was grown over the substrate. By Mössbauer spectroscopy (CEMS and TMS method) and x-ray diffraction we found that films subjected to heat treatmebnt at 300 °C and 500 °C are amorphous, or eventually nanostructure, α/γ Fe_2O_3 and polycrystalline α-Fe_2O_3, respectively. Electrochromic behavior, in the range of 500-800 nm, was obtained only for superparamegnetic α/γ-Fe_2O_3 coatings. Electrochromism was not observed for polycrystalline α-Fe_2O_3.

Introduction

Electrochromism is an interesting phenomenon with potential application on non-emissive displays, smart windows, automotive mirrors, etc. An electrochromic material changes its optical transmittance, from a bleached to colored state, in a persistent and reversible way when cation insertion / extraction takes place under the application of an electric field. This behavior has been studied in numerous transition metal oxides, being the Wox and NiOx the most studied materials [1]. Coatings have been obtaining using many types of techniques, in which sol-gel method seems to be one of the most promising techniques, potentially useful for large area applications [2].

Sol-gel is an attractive method for thin film fabrication due to the possibility to get, in a simple way, a variety of structures and phases just controlling the xerogel characteristics. Intensive works have been performed in the last years in order to get electrochromic transition-metal-oxide coatings using the sol-gel technique [1,2]. Particularly electrochromic iron-oxide-coatings have been studied showing that this material is electrochromic iron-oxide-coatings have been studied showing that this material is electrochromic under certain conditions of sample preparation, in which the post-heat-treatment seems to be one of the critical parameter to be controlled [3]. However it is not yet clear the structural characteristics of the electrochromic iron-oxide-coatings, particularly it is controversial which phase is electrochromic: γ-Fe_2O_3 or α-Fe_2O_3 [3,4].

Experimental

Iron-oxide-based coatings were obtained by sol-gel process. Iron sol was prepared from an aqueous solution of $FeCl_3.6H_2O$ (0.37M) in which was added a controlled proportion of CH_3COOH (ac). Glass or glass/ITO was used as substrate; by dip-coating process the film was grown over the substrate using a rate of 6,5 cm/min and followed by a 30 minutes of post heat treatment. We have used two kind of temperatures for heat treatment: 300°C and 500 °C.

^{57}Fe Mössbauer spectra for film/substrate system were recorded, at room temperature (RT), in the modality of both CEMS and transmission (TMS) using a Mössbauer spectrometer with a ^{57}Co/Rh source and constant acceleration. A home-made parallel plate avalanche counter for CEMS was employed [5]. The spectra were computer fitted to Lorentzian lines using the usual constrains of equal area and width for two lines of doublets, and of equal width and areas in the ratio 3:x:1:x:3 for the six lines of sextets (x=2 for the transmission spectrum). The isomer shifts (IS) are referred to α-Fe.

X-ray diffractograms (XRD) of the films were recorded for film/substrate system using Cu-Kα radiation with a Philips PW 1730/10 diffractometer.

Electrochromism was studied in a three-electrode electrochemical cell using an aqueous solution of NaOH as electrolyte for cation insertion/extraction, SCE as reference electrode and Pt as counterelectrode. For optical measurement we used and OPTOMETRIC RS-325 spectrophotometer in the wavelength range of 300-800 nm..

Results and Discussions

The Mössbauer spectra for CEMS and TMS are shown in Figs. 1 and 2, respectively. The spectra of the film treated at 300 °C (figs.1a and 2a) show a broad doublet (Γ = 0.52 mm/s) whose parameters, IS = 0.37 ± 0.05 mm/s, QS = 0.86 ± 0.04 mm/s, are similars to the observed for following superparamagnetic (spm) products: oxyhydroxides, ferrihydrite, hematite, maghemite [6-8]. The XRD diffractogram did not show any peak for samples corresponding to figs.1a and 2a, indicating, in agreement with Mössbauer results, that these coatings are micro- or nanostructured..

The spectrum by CEMS of the film treated at 500 °C (fig 1b) shows a broad sextet (Γ=0.50 mm/s), whose parameters, IS=0.38±0.02 mm/s, QS=-0.220± 0.05 mm/s, H = 525 ± 8 kG, can be associated to polycrystaline hematite (α-Fe_2O_3). The spectrum by TMS of the same film (fig. 2b) present one sextet (Area ≈ 70%), associated to α-Fe_2O_3, and one doublet similar to the α/γ-Fe_2O_3, previously described. This product probably correspond to the internal part of the sample, near the interface film/substrate. The XRD data, in agreement with Mössbauer results, show peaks for (104), (012), (116) crystallographic planes corresponding to α-Fe_2O_3.

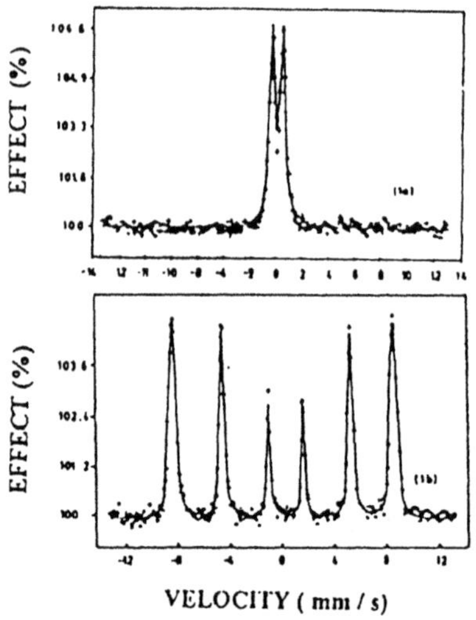

Fig. 1. CEMS at room temperature from iron oxide thin films deposited by dip-coating technique, (1a) heat treated at 300 °C and (1b) heat treated at 500 °C, during 30 minutes.

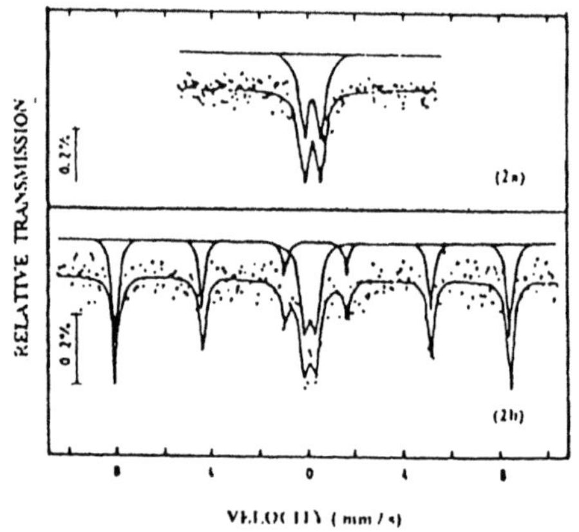

Fig. 2. TMS at room temperature from iron oxide thin films deposited by technique, (2a) heat treated at 300 °C and (2b) heat treated at 500 °C, during 30 minutes

Films subjected to heat treatment at 500 °C did not show any electrochromic behaviour, however those heated at 300 °C presented an pronounced electrochromism. In fact in the range of 500-800 nm optical transmittance can be recorded from bleached state (85% at $\lambda = 700$ nm) to colored state (40 % at $\lambda = 700$ nm).

At present stage we do not have yet a complete explanation of the electrochromic behaviour for iron-oxide-based coatings prepared by sol-gel and subjected to low heat treatment, but these first results are clear indications that the structural characteristic is critical, and probably amorphous or nano-structured coatings are needed for cation insertion/extraction into the material allowing that the electrochromism phenomenon to be present.

Conclusions

Electrochromic iron-oxide-based coatings can be obtained by sol-gel process. Films treated at 300 °C, constituted by micro or nano-crystalline α/γ-Fe_2O_3, present electrochromism, in contrast to those heated at 500 °C which are mainly constituted by large α-Fe_2O_3 particles.

Acknowledgments

We thanks to IPPS, Uppsala University, and Faculty of Science, UNI, Lima, for financial support. We are also grateful to Dra. Isabel Fonseca, "Rocasolano" Institute (Madrid), for technical assistance in XRD measurements.

References

[1]. C.G. Granqvist, "Handbook of Inorganic Electrochromic Materials", Elsevier Science (Amsterdam), 1995.

[2]. C.M. Lampert and C.G. Granqvist (eds.), "Large area Chromogenics: Materials Devices for Transmittance Control", SPIE Opt. Engr. Press, Bellingham (USA), 1990.

[3]. B. Orel, M. Macek and Svegl, K. Kalcher, Thin Solid Films, 246 (1994) 131.

[4]. L. Armelao, R. Bertoncelo, L. Crociani, G. Depaoli, G. Ganozzi, E. Tondello and M. Bettinelli, J.Mater.Chem., 5(1) (1995) 79.

[5]. J.R. Gancedo and M. Gracia, Hyp. Int., 29 (1987) 1097-1100.

[6]. W. Kundig, H. Bomwel, G. Constabaris and H. Linquist, Phys. Rev., 142 (1966) 327.

[7]. E. Murad and U. Schwertmann, Am. Miner., 65 (1980) 1044.

[8]. M.E. Brett, K.M. Parkin, M.J. Graham, J. Electrochem. Soc., 133 No 10 (1986) 2031.

STRUCTURAL AND ELECTRICAL CHARACTERIZATION OF TIN-OXIDE-BASED COATING OBTAINED BY SPRAY PYROLYSIS USING AN ALCOHOLIC SOLUTION WITH LOW AND HIGH CONTENT OF AMMONIUM FLUORIDE

H. BEDON, R. EKLUND, W. ESTRADA*

Facultad de Ciencias, Universidad Nacional de Ingeniería, Casilla 31-139, Lima, Perú.
*Royal Institute of Technology, Stockholm, Sweden.

Abstract

Fluorine doped SnO_2 at r = 0, 5, 9, 17, 23, 28, 33 and 38 % (r = F / (Sn+F), in weight), obtained by spray pyrolysis, were studied by Mössbauer spectroscopy (CEMS modality) and Hall effect measurements. When F-doping increases from r = 0 up to r = 23% there is an systematic increment in the Mössbauer parameters IS and QS. At the same time both the electrical conductivity and free carrier density increase. Further increments of r (> 23%) make IS and QS to decrease. Correlation between Mössbauer parameters and electrical properties indicates that fluorine doping (for r < 23%) in the films diminishes the oxygen vacancies. For r > 23% a new phase $SnOF_2$ and / or interstitial position of F is taking place.

Introduction

Tin-oxide-based coatings, doped with fluorine, have been extensively studied [1,2]. F-doped SnO_2 is known by its high electrical conductivity and spectral selectivity in the infrared region [1,2], which makes of tin-oxide-based coating an interesting material as heat mirror, transparent counter electrode, gas sensor, etc. [1,2,3]. For fabricating SnOx has been employed many techniques but there is not doubt that spray pyrolysis has been largely the most used one. To dope pyrolytic tin-oxide-based coatings with F usually is employed ammonium fluoride, but for many applications not more than 5 % in weight (F/F+Sn) is needed, therefore little attention has been paid on coatings produced using high content (more than 5 %) of ammonium fluoride in the starting solution for spraying, which seems to be interesting for gas sensing application [4]. In an earlier work [4] we studied the optical and morphology characterization of SnO_2 when high content of ammonium fluoride in the spraying solution was present. In this work we study the structural characterization of tin-oxide-based coatings doped at r = 0, 5, 9, 17, 23, 28, 33 and 38% (r = F/F+Sn, in weight) using Mössbauer spectroscopy

Experimental

The films were fabricated using the spray pyrolysis method described elsewhere [3,4]. The spraying solution was $SnCl_4 \cdot 5H_2O$ dissolved in ethanol (at 0.2 M) in which NH_4F, previously dissolved in destilled water, was added. With this mix was then prepared alcoholic solutions at r = 0, 5, 9, 17, 23, 28, 33 and 38 % .

We used a substrate temperature between 350°C and 360°C with a deposition time of 8 minutes. All coatings were deposited at the same spraying conditions, assuring that the relevant variation was only in the doping level.

Mössbauer spectra, CEMS modality, were obtained at room temperature. We employed a Ca^{119m} SnO_3 source; the velocity was calibrated by use of a ^{57}CoRh source and α -Fe absorber.

Electrical measurements were performed with the van der Paw technique [5]. A magnetic field from 0-0,5 T, with steps of 0,05 T, was employed for Hall effect measurements. We used samples of 1x1 cm^2 in size.

The thickness was determined by interference method using an spectrophotometer Optometric RS-375.

Result and Discussions

Fig. 1 and 2 shows the IS and QS vs. fluorine concentration, respectively (r). IS increases from -0.01 to 0.07 mm/s when r varies from 0 to around 20 %, and it decreases for higher r (see fig.1). QS shows also a similar behaviour. It increases from 0.47 to 0.60 mm/s for r = 0-23%, and start to decrease for higher values of r. The parameters were fitted and analyzed using a least square program. Those values of IS and QS indicates that Sn^{4+} predominates in the films [7,8,9], implying that coatings basically correspond to SnO_2 phase. Slight increasing of QS for r = 0-23% (fig.2) implies that there is a more asymmetry around Sn atom; this can be ascribed to: F takes substitutional oxygen positions and/or crystal lattice is distorted towards a less crystallinity. For r >23 %, QS starts to decrease with the increment of r; this indicates that a more symmetric distribution of F is taking place after this level of doping.

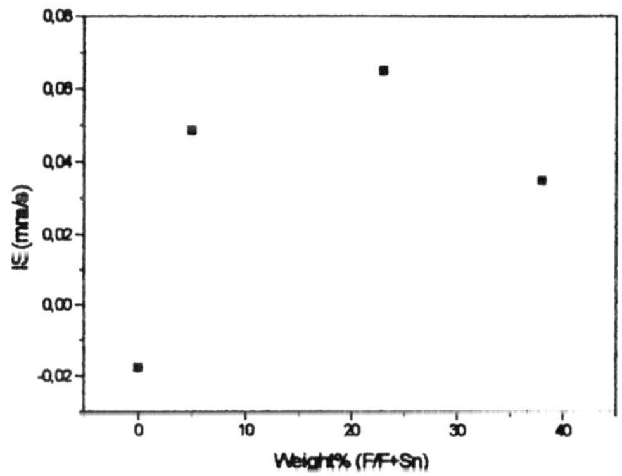

Fig. 1 The Isomer shift, IS, versus amount of fluor, F/ F+Sn, in the solution.

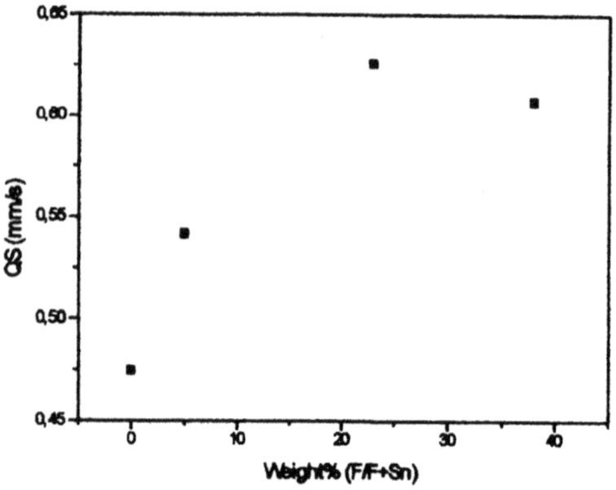

Fig. 2 The Quadrupole splitting, QS, versus amount of fluorine, F / F+Sn, in the solution.

Besides, when r = 0-23% the IS also increases. This is ascribed to the increment of free s-electrons density. This is in agreement to the substitutional argument for F, which liberate one electron to the conduction band. The decrease of electrical resistivity ρ and increment of n, as showed in fig. 3, reinforce this argument.

On the other hand, the IS-decreasing tendency for r >23% could be related to:
a. Better fluorine homogeneity around Sn, as proposed earlier.This could allow the $SnOF_2$ formation which binds the s-electron. This is also in agreement to the increment of electrical resistivity and the lowering of electron density, as it is showed in fig. 3.
b. Fluorine adopts interstitial positions in the lattice, and due to its electronegativity can localize electrons and so decreasing the free electron density, n.
However we should also mention that high ammonium chloride content in the spraying solution makes the coating surface to be covered by high density of pits, ~0,1μm-size, as we showed in an earlier work [4], contributing this also for resistivity increasing.

Fig.4 shows the variation of mobility, μ, according to the increment of doping: μ constantly rises with r. This result is somewhat unexpected, since one could expect that, after certain point, μ should decrease as the impurities act as scattering centers. However, the vacancies are probably stronger scattering centers, so assuming that fluorine starts filling them, the mobility could increase. Therefore this argument is in favor that vacancies in undoped films start to be filled by fluorine during doping process.

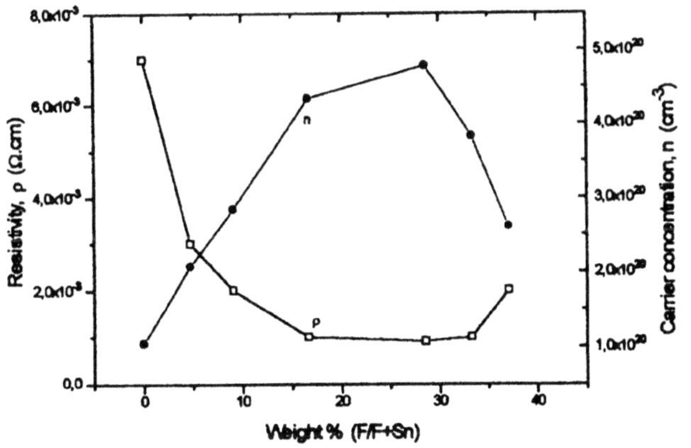

Fig. 3 The resistivity, ρ and the carrier concentration, n, versus F / F+Sn, in the spraying solution.

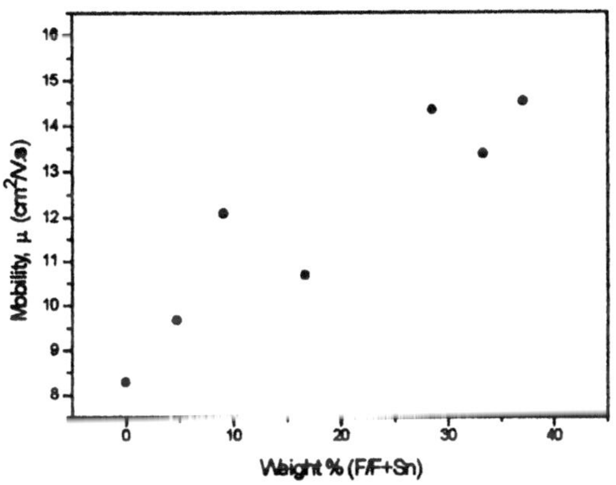

Fig. 4 The carrier mobility, μ, versus F / F+Sn, in the spraying solution.

Conclusions

Fluorine doping in tin-oxide-based coatings are probably substitutional in oxygen vacancies when percentage is low (r < 23%, F/F+Sn in weight). IS and QS increase and electrical resistivity decreases. Further increments in r produce the lowering in free carriers density and the increment of ρ. New phase of $SnOF_2$ and/or interstitial F could take place. This is in agreement to IS and QS decreasing. Doping process diminishes the vacancies concentration in the film.

Acknowledgments

We thanks to IPPS, Uppsala University, and Faculty of Science,UNI, for financial support.

References

[1] Bertil Stjerna, Ph.D. Thesis, Chalmers University, 1990 (unpublished)

[2] M.Miki-Yoshida and E.Andrade,Thin Solid Films, 224(1993)87

[3] M.R. Soares, P.H.Dionisio, I.J.R.Baumvol, W.H.Schreiner.,Thin Solid Films, 214(1992)6.

[4] D. Acosta, M. Zironi, E. Montoya and W. Estrada, accepted to be published in Thin Solid Film, 1996

[5] L.J Van der Paw, Philips Res. Rep. 13 (1958) 1.

[6] A.Tiburcio-Silver, A.Maldonado, A.Escobosa, E.Saucedo, J.M.Montoya and J.A.Moreno, Applied Surface Science 70/71 (1993)746.

[7] R.Swanepoel, J.Physics E:Instrumentation.,16(1983)1214.

[8] E.Leja, J.Korecki, K.Krop and K.Koll, Journal of Physique, 40 (1979)408

[9] B.Stjerna, C.G.Granqvist, A.Seidel, L.Häggström, J.Appl.Phys., 68(12)(1990)6241.

LOW TEMPERATURE AND RAPID SYNTHESIS OF Fe_2TiO_5 PHASES

Fabiana Gennari[a], Stella Duhalde[b]. and Gerardo Quintana[b]

[a] Consejo Nacional de Investigaciones Científicas y Técnicas, CONICET-CNEA, 8400 Bariloche, Argentina.
[b] Departamento de Física, Facultad de Ingeniería, UBA, Paseo Colón 850, 1063 Buenos Aires, Argentina.

Abstract

A novel method for low temperature and rapid synthesis of orthorhombic (pseudobrookite) and monoclinic Fe_2TiO_5 phases is presented. Amounts of Fe_2O_3 and TiO_2 oxides in ratio 7:3 wt % were encapsulated in Cl_2 atmosphere and heated for diferent period of time and at different temperatures between 700 and 950°C. Chlorine is found to depress the reported temperature for the formation of the Fe_2TiO_5 phases and substantially accelerates their synthesis. The obtained phases were characterised by x-ray diffraction (XRD), scanning electron microscopy (SEM) and Mössbauer spectroscopy (MS) at room temperature (RT) and at 25K. The transformation process from starting materials to final products and the role of chlorine are discussed in terms of the Mössbauer results.

INTRODUCTION

The pseudobrookite Fe_2TiO_5 has an orthorhombic structure that was first described by Pauling [1] from x-ray diffraction patterns of the natural single crystal. Fe and Ti atoms occupy the special positions 8f and 4c, respectively, of the orthorhombic space group. The unit cell contains four molecules and its dimensions are a=9.79, b=9.93 and c=3.72 A. A number of other phases have been shown to have the same structure, including $FeTi_2O_5$ (ferropseudobrookite), $MgTi_2O_5$ (karrooite), $(Mg,Fe)Ti_2O_5$ (armalcolite) and solid solutions in the Fe_2TiO_5-Ti_3O_5 series, etc. Considerable attention has been devoted to the investigation of the pseudobrookite because of its extraordinarily high degree of thermal expansion anisotropy and spin-glass like behaviour below 55K (transition temperature). This can be correlated with a metastable distribution of Fe^{3+} between the available 8f and 4c sites [2,3]. This distribution yields a large number of structurally inequivalent Fe^{3+} sites. Therefore, the low temperature Mössbauer pattern exhibits superposition of hyperfine splittings.

The monoclinic Fe_2TiO_5 transforms into the orthorhombic one above 1200K [4,5]. The structure of the monoclinic compound can be related to the rutile structure by means of crystallographic shear operation. Stacking faults are formed in the monoclinic phase because this structure is unstable with respect to the room temperature structure of orthorhombic Fe_2TiO_5. The monoclinic Fe_2TiO_5 phase was reported to be prepared using a high-temperature solution (HTS) method. The structure was determined by x ray analysis and Mössbauer spectroscopy, and was found to be isostructural with V_3O_5. Ti and Fe atoms occupy the positions 4a and 8f respectively, without discarding the existence of cation disorder. The Mössbauer spectrum consists of a symmetric doublet above the ordering temperature (T=350K), while the room-temperature one is composed of very broad outermost lines characteristic of a distribution of hyperfine field. At 78K the spectrum shows six well defined broadened absorption lines due to various nearest neighbour environments. Another report by

Hanzel et al [6], demonstrated that 4a sites are preferentially occupied by Fe^{3+}, while Fe^{3+} and Ti^{4+} are randomly distributed over 8f sites.

These results can be used to relate the site occupancy in both structures when the transformation is proceeding.

The aim of this report is to present a new route of synthesis of monoclinic and orthorhombic Fe_2TiO_5 and discuss the transformation process in terms of cation sites distribution obtained from Mössbauer spectra.

EXPERIMENTAL

The starting materials were (α-Fe_2O_3 (Spex Industries, Inc., USA) and anatase TiO_2 (Mallinckrolt Chemical Work) powders. The specific area were 3.0 m^2/g, and 9.4 m^2/g respectively, as measured by nitrogen adsorption at 77K (M 2205, Micromerities Instrument Corp., Norcross, GA). Chlorine (Indupa, Argentina) 99,8 % purity was used. TiO_2 powder had particles ranging from less than 1μm to about 60μm composed by small grains which had a mean size of 0.2μm. α-Fe_2O_3 powder had a particle distribution from about 200 to less than 10μm, which in turn were agglomerates composed of very small grains with a mean size of about 0.2μm. The samples were prepared by mechanical mixing of TiO_2 and Fe_2O_3 powders in a 7/3 ratio.

The new synthesis procedure allows to obtain large quantities of polycrystalline Fe_2TiO_5. It consists in the direct reaction of the powder mixture with a chlorine gas. The samples were placed within a spherical quartz capsule and held in vacuum for 1h until they were outgassed. Then, the chlorine was introduced to get a pressure of 0.1MPa at 950°C and finally the capsules were sealed. The encapsulated samples were heated at the reaction temperature for the selected time and quenched in air. After each thermal treatment the crystalline phases were studied by x-ray powder diffraction analysis (PW 1710, Philips Electronic Instruments) and by scanning electron microscopy.

Mössbauer measurements at RT and 25K were carried out with a constant acceleration spectrometer, using a ^{57}Co source in a Rhodium matrix. Spectra were fitted with the NORMOS program for crystalline sites and hyperfine fields distributions (HFD) when necessary. The Isomer shifts values are referred to iron metal. Absorbers were prepared with 50mg/cm^2.

RESULTS AND DISCUSSION

In previous studies [3], the orthorhombic Fe_2TiO_5 was prepared by heating in air a mixture of Fe_2O_3 and TiO_2 at 1000°C for more than 12hs. On the other hand, monoclinic Fe_2TiO_5 was synthesised at 1350°C using the high temperature solution method. This process requires the presence of alkaline earth oxide and fast cooling rate, which suppresses the synthesis of the orthorhombic phase [4]. On the contrary, in the present work, it is possible to obtain the monoclinic phase without the addition of basic oxides and simultaneously with the orthorhombic phase. The Fe_2TiO_5 formation mechanism involves chemical vapour transport using chlorine as the transport agent. This novel method requires lower temperatures and shorter heating times than those reported in literature.

The morphology associated with orthorhombic Fe_2TiO_5 formed in chlorine atmosphere at 850°C during 4hs is shown in Fig.1. It can be observed that some crystals consist of tablets/like material grown into the TiO_2 particle. The size of these tablets is about 8μm. These faceted grown crystals have the typical habit of pseudobrookite [7]. The small grains around

Figure 1: SEM image of Fe_2TiO_4 platelets.

pseudobrookite crystals correspond to remaining rutile phase. These crystals obtained at low temperature and short heating time suggest the supply of Ti and Fe from gaseous compounds.

The values of the hyperfine parameters obtained and the phases indexed by XRD are given in Table 1. The fitting, for the room temperature spectra was performed assuming two doublets for the 8f and 4c sites in the orthorhombic structure and an hyperfine magnetic field (HMF) distribution for the monoclinic one. The low temperature spectra were fitted assuming HMF distribution for both structures.

Table 1: Mössbauer parameters at RT, obtained considering HMF distribution for the monoclinic Fe_2TiO_5 (M) and crystalline sites for the pseudobrookite Fe_2TiO_5(PSB).H_m is the most probable HMF. In the last column A is anatase and R rutile. Errors in the parameters are 0.2T for H, and 0.02mm/s for QS and IS.

T (°C)	t (hs)	PSB						M			PSB/M+PSB	XRD
		QS_{8f} mm/s	IS_{8f} mm/s	QS_{4c} mm/s	IS_{4a} mm/s	A_{8f} (%)	A_{4c} (%)	H_m T	QS mm/s	IS mm/s	Relative Area (%)	Phases Indexed
750	20	0,76	0.35	0.45	0.35	38	62	32.0	-0.15	0.33	18	A,R,M
750	120	0.76	0.37	0.47	0.36	44	56	32.4	-0.17	0.36	28	R,M,PSB
800	4	0.75	0.39	0.44	0.39	43	57	31.0	-0.14	0.36	33	R,M,PSB
800	8	0.83	0.37	0.47	0.39	42	58	31.0	-0.13	0.47	41	R,M,PSB
800	20	0.92	0.38	0.54	0.38	43	57	-	-	-	100	R,PSB
800	120	0.92	0.39	0.53	0.39	44	56	-		-	100	R,PSB
850	4	0.92	0.38	0.52	0.38	53	47	-	-	-	100	R,PSB
900	4	0.92	0.38	0.53	0.38	53	47	-	-	-	100	R,PSB

The room temperature Mössbauer spectrum of the sample treated at 700°C for 120hs shows only the presence of α-Fe_2O_3 and an incipient doublet of very low intensity. But the corresponding spectrum of the sample heated during 8hs at 750°C consists of a doublet superimposed to a complex spectra with a relatively broad hyperfine field distribution, which evidences the formation of the pseudobrookite and the monoclinic phases. The spectra of the samples treated at 750°C during 120hs and at 800°C during 4 and 8 hs, show the same kind of pattern with a slight increment in the relative doublet area. Nevertheless, when the treatment is prolonged for 20hs at 800°C, only the presence of the orthorhombic phase is evidenced. At higher temperature and longer time treatments the relative occupation of the 8f and 4c sites by Fe^{3+} changes toward a state where all possible configurations are equal. This process occurs at high temperatures, when diffusion takes place. Some of the typical spectra are shown in Fig.2.

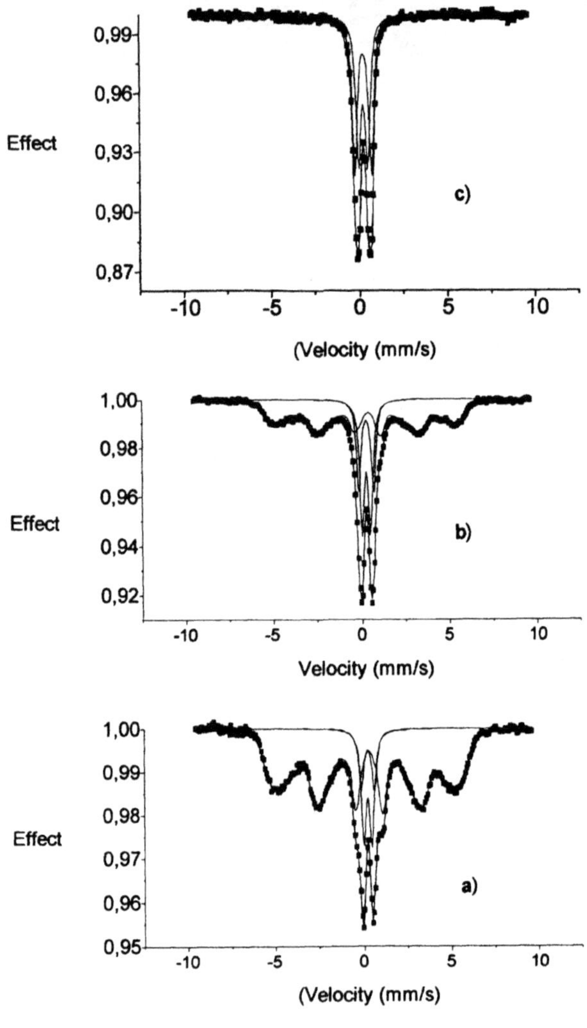

Figure 2 :Room temperature Mössbauer spectra as a function of the thermal treatment temperatute. a) 750°C, b) 800°C and c) 850°C.

Low temperature Mössbauer parameters obtained for the Monoclinic structure are:H_M , the maximum of the distribution, 48.8 T, QS = -0.14mm/s and IS= 0.49mm/s. The spectrum shows six well defined broadened absorption lines due to the superposition of several hyperfine sextet patterns belonging to various nearest neighbour environments. The PSB low temperature spectrum exhibits the superposition of a large number of hyperfine splittings. The obtained parameters, considering two HMF distribution related to the two available sites are: H_{M1}=38.5T, QS_1= 0.6mm/s, IS_1=0.49mm/s and H_{M2}=38T, QS_2= 0.06mm/s, IS_2=0.46mm/s. The relative intensities are HMF_1/HMF_2= 1.32.

No intermediate Cl-Fe compound is detected by MS. Nevertheless, the quadrupole splitting, for both orthorhombic and monoclinic phases obtained by heating the chemicals up to 800°C for no more than 8hs, differs from the values reported. Moreover, the rather broad linewidths of the doublet corresponding to the pseudobrookite structure in these samples, suggest that Cl could partially replace oxygen, leading to a variation in the Fe^{3+} environment in both phases. Mössbauer parameters corresponding to samples heated at 800°C for 20hs and at higher temperatures are in agreement with those reported elsewhere. Further x-ray diffraction refinement [8], and electric field gradient calculation will be carried out to corroborate this model.

References

[1] Pauling L. Z. Kristallogr. 73, 97-112, 1930.

[2] L.Gurewitz, U.Atzmony, Phys.Rev.B, 26, 6093-6098, 1982.

[3] J.M.Cruz, P.Morais, K.Skeff Neto, Phys.Lett.A, 116, 45-47, 1986.

[4] Drofenik et al. J. Solid. State Chem. 40, 47-51, 1981.

[5] Krasevec et al, J. Solid State Chem. 40, 52-58, 1981.

[6] Hanzel D. et al. Solid State Communications, 40, 899, 1981.

[7] E.S.Dana and D.H.Ford, Mineralogy, 4th, Part V (Wiley, London, 1932).

[8] F.Gennari, J.Andrade Gamboa and D.Pasquevich, unpublished results.

CARBIDE FORMATION BY ELECTROEROSION ON IRON SURFACES

E. D. Cabanillas
Departamento de Materiales, Centro Atómico Constituyentes,
Comisión Nacional de Energía Atómica, Av. del Libertador 8250,
1429 Buenos Aires, Argentina.

R. Viña, G. Punte, J. Desimoni, and R. C. Mercader
Departamento de Física, Universidad Nacional de La Plata,
C.C. 67, 1900 La Plata, Argentina.

Abstract

The formation of carbides produced by incorporation of carbon atoms of the dielectric as a function of the current and duration of the sparks used in the electroerosion of iron surfaces is investigated by CEMS and XRD. It is found that different types of carbides are produced depending both on the duration and the current of the sparks.

Introduction

The process of electroerosion is widely used for cutting metallic samples. In spite of the rather old development of the method [1-3], the formation of phases not present prior to the treatment were not observed until recently [4]. The nature of the processes involved is so rich in physical phenomena at each step and depends on so many parameters, like sparks transferred power and duration, nature of the dielectric, or nature of the sample material, that a complete understanding of the whole process is not possible at present and partial features have to be studied separately.

Conversion Electron Mössbauer Spectroscopy (CEMS) investigation of spark eroded surfaces of pure iron revealed that phases not present before the electroerosion treatment, α'- and γ-iron [4] are produced. In addition, Fe_3C and other iron carbides are produced at the surface of α-Fe treated samples. These carbides had not been reported in previous studies found in the literature [5] and had been only mentioned in ref.[4]. The present investigation aims at the identification of the carbides using CEMS and x-ray diffraction (XRD).

Experimental

Pure iron metal pieces 99.95at% purity (Goodfellow Fe 000450 with the composition: Mn: 3000; Si:1000; C<800; P<400; S<500 ppm) were spark planed in a commercial machine using the trade named oil "AWS 662 Esso" as dielectric, with a difference of potential of 80 V between electrodes. Spark currents were in the range of 3.3 to 25A, and its duration was performed for times between 2 and 3072μs.

The XRD characterization of selected samples was carried out on a Philips instrument using a conventional Bragg-Brentano para-focusing geometry and graphite monochromatized CuKα radiation. Data were recorded in the 20 to 125° 2θ range, in steps of 0.015°, with 8s counting time per step. Each set of data were analyzed by the Rietveld method [6]. The refinement was applied to the unit cell constants with the atomic coordinates of the structures of the chief crystalline phases taken from the literature.

CEMS spectra were taken in a conventional constant acceleration spectrometer of 512 channels with a 15mCi nominal activity ^{57}CoRh source. The electron detector was of the constant flow type used with a mixture of He (94%) and methane (6%). The data were fitted by a least-squares program that used Lorentzian line shapes with constraints to simulate the magnetic and quadrupole interactions found in the spectra.

Results

The XRD results obtained for samples treated with the shortest spark time (2μs), indicate that the phase formation is almost independent of the current intensity. Both the 3.3A and 25A samples data show α-Fe, ϵ-Fe_2C, and γ-Fe, austenite, in similar amounts, Fig. 1.

At 3.3A the emergence of other Fe carbides is observed with increasing spark times. At 3072μs the XRD pattern shows a preponderance of θ-Fe_3C, cementite, and small quantities of ϵ-Fe_2C, of γ-Fe, and of χ-Fe_5C_2, Hagg's carbide.

Data from samples treated with sparks of the largest duration (3072 μs) show that the phases formed change with current, e.g., the increase of the γ-Fe at the expense of carbides is observed.

The analysis of the evolution of phases as a function of spark time when the current is 25A, shows an increase of γ-Fe and a decrease of ϵ-Fe_2C up to 1024μs. Increasing amounts of θ-Fe_3C, that was absent at shorter times, are observed for 2048μs and 3072μs, while γ-Fe remains almost constant.

CEMS spectra of samples treated with 3.3A, and times of a) 2μs and b) 3072μs are shown in Fig. 2.

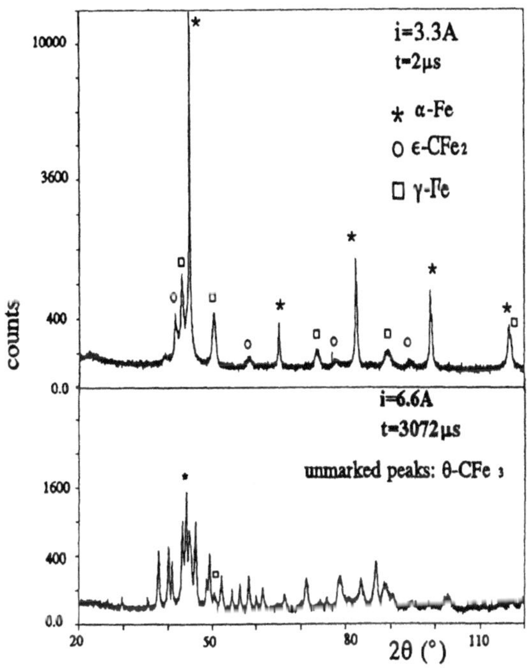

Fig. 1. X-ray diffraction data of spark-planed α-Fe samples at indicated discharge currents and times.

Discussion

Although it has not been performed to all samples, but those of extreme times and currents, the XRD data show at the same largest current a clear tendency toward the formation of a greater proportion of austenite and a smaller percentage of carbides with increasing spark times. We interpret this result as an indication that the cooling rate plays an important role in the carbide formation.

However, CEMS spectra are not conclusive. A more detailed estimation of the phase at which a certain hyperfine magnetic split signal belongs requires spectra of higher statistical soundness to be unambiguously assigned. Moreover, there are carbides with slightly different

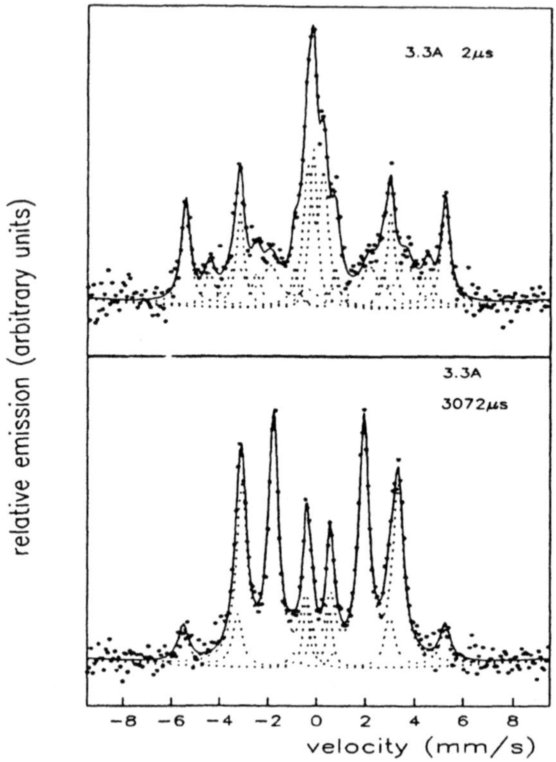

Fig. 2. Mössbauer spectra of spark-planed α-Fe samples with 3.3A at indicated times.

stoichiometries that yield very similar Mössbauer parameters.

Assuming that the carbides have formed at the outermost surface of the samples, we show in Fig. 3 the evolution of the sum of intensities of the magnetic signals belonging to all carbides and α' as a function of the spark power multiplied by the square root of time ($P\sqrt{t}$). In this figure two regimes of formation of carbides can be seen: one for $P\sqrt{t}$ up to $\approx 15Ws^{1/2}$ and a new one that starts at higher values. The quantity $P\sqrt{t}$ is directly related to the surface temperature according to the model of ref. [7].

Conclusions

These preliminary results of spark machined iron samples, show that a threshold $P\sqrt{t}$ value must be reached to ensure a high proportion of carbides to be found at the outermost surface. XRD results also support the surface carburization.

The present data are not enough to have a complete picture of the complex surface phenomena involved and more data are being collected currently to assess the effect of different spark rates and current intensities in the relative amounts of the observed carbide phases formed.

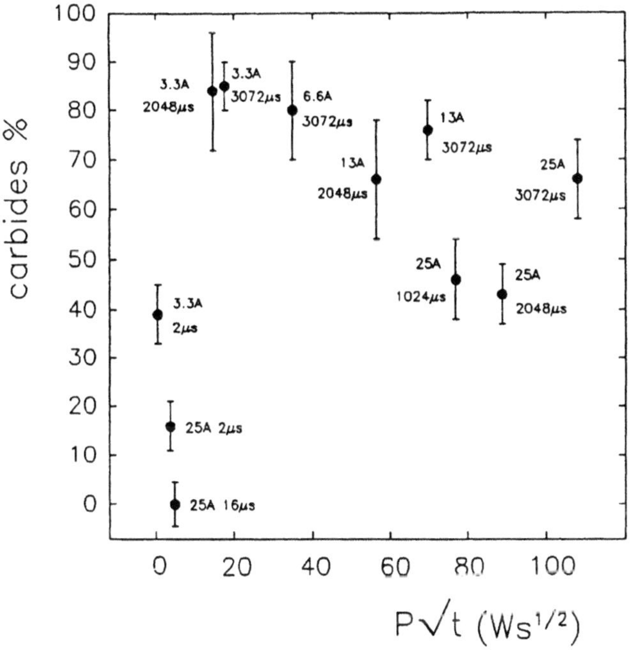

Fig. 3. Carbide percentages as a function of the spark power times the square root of discharge duration.

Acknowledgements

Authors acknowledge the financial support of Consejo Nacional de Investigaciones Científicas y Técnicas (CONICET), Argentina, through programs TENAES and PROFIMO. X-ray diffraction data were obtained at LANADI facilities. All authors are members of CONICET.

References

[1] B. R. Lazarenko and N. I. Lazarenko, Physics of the Spark Method of Machining Metals, Moscow, TsBTI MEP, 1946.

[2] B. R. Lazarenko and N. I. Lazarenko, Machine Tools Cutt. Tools, Moscow, 17 (1946) 8; and 18 (1947) 4.

[3] "Electrospark Machining of Metals", Volume 3, Edited by B. A. Krasyuk. Consultants Bureau, New York, 1965.

[4] "Phase transformations caused by spark planing of α-Fe surfaces", E. D. Cabanillas, J. Desimoni, G. Punte, and R. C. Mercader, J. Appl. Phys. 78 (1995) 2372.

[5] G. R. Wilms and J. B. Wade, Metallurgia 54 (1956) 263.

[6] A. Skthivel and R. A. Young, DBWS-9006PC, Program for Rietveld analysis of x-ray powder diffraction patterns, Georgia Institute of Technology, Atlanta (1991).

[7] S. Basu, and T. DebRoy, J. Appl. Phys. 72 (1992) 3317.

MÖSSBAUER STUDY OF THE COMPOSITION OF SURFACE LAYERS FORMED ON LOW ALLOYED STEELS

J.L. López, J. H. Dias Filho, S. L. Pereira and H.-D. Pfannes

Departamento de Física, Universidade Federal de Minas Gerais, C.P. 702, 30161-970, Belo Horizonte, Brazil

The composition of the barrier layer rust on a low alloyed steel was studied by transmission and conversion electron Mössbauer spectroscopy at 4.2, 110 and 300 K. The rust of two samples corroded for 72 hours in destilled water and aqueous NaCl-3% in weight-solution, respectively, was collected. A third sample was immersed for 72 hours in aqueous NaCl-10% in weight-solution and exposed to air for 3 months in a non-polluted rural region. All samples exhibited collapsed superparamagnetic spectra at 295 K and static spectra at 4.2 K, indicating the presence of α-, β- and γ- FeOOH. The relaxed spectra were simulated using a stochastic model and longitudinal relaxation and collective excitations.

1. Introduction

In the past decades great interest has been devoted to the study of the formation of the corrosion products in materials that were submitted to corrosion inhibiting processes [1-4]. In order to study corrosion processes of steel and quantify their inhibition it is important to know the phase composition of the corrosion products. The composition of the rust is determined by the nature of the particular environment as well as the total exposure time. In some environments, for example in destilled water saturated with air or in aqueous NaCl solution, an accelerated rusting process leads to fast corrosion of steel. We analyzed the phases of the corrosion products and evaluated their relative amounts in order to obtain information on the formation procedure of the layer. When rust is formed from alloyed steel immersed in water or NaCl solution the usual corrosion products identified by Mössbauer spectroscopy are α-,β-, γ-FeOOH and Fe_3O_4 [5- 8].

We observed that the main phase constituents of the corrosion products of our samples (low alloyed steel) formed in aqueous solution were : Goethite (α - FeOOH), Lepidocrocite (γ - FeOOH) and Akaganeite (β - FeOOH). We determined the composition and structure of the rust by X-ray diffraction and 57-Fe Mössbauer spectroscopy.

2. Experimental

Samples have been collected from low allowed steel plates exposed to corrosion in destilled water and aqueous solution in which the main corrosive agent was chloride. The rust was scraped from the surfaces of the test plates (sample I: steel plate treated in aqueous NaCl solution; sample II: corroded in destilled water; sample III: as I with subsequent exposure to air for 3 months). The Mössbauer source was 57-Co diffused in Rh. A conventional constant acceleration Mössbauer spectrometer in transmission geometry was used to collect transmission spectra at 300, 110 and 4.2K. A backscatter conversion electron Mössbauer spectroscopy (CEMS) type gas He/CH_4 flow proportional detector, developed in our laboratory, was used for recording CEMS spectra at room temperature. The static spectra were evaluated by fitting to a sum of Lorentzian lines using the Normos program [9], and the relaxed spectra were simulated using a stochastic model and longitudinal relaxation.

The samples were also analyzed by X-ray diffraction (Rigaku diffractometer, Cu-K_α radiation and graphite monochromator). Scans were realized in the range of 4°- 60° (2θ) at a speed of 1°/min, and the reflected intensities were recorded in a computer.

3. Results and Discussion

The X-ray powder diffractions (XRD) patterns of our samples showed broad unresolved reflections which indicate poor crystallinity. We have used the patterns to calculate the average particle size according to the expression $D=\lambda/(\beta \cos\theta)$, where λ is the wavelength, β is the linewidth at half height of the maximum intensity and θ is the diffraction angle. We found 50-64Å for well crystallized samples. Comparing with the ASTM diffraction file [10], our XRD results indicated the presence of the standard forms of rust (α-,β-,γ-FeOOH). Table1 lists the interplanar spacing (d) and intensities determined from the XRD measurements for the various components of the rust of samples I,II and III.

Table 1: Attributed phases, interplanar spacing d and relative intensities of the XRD-peaks of the corrosion products of the samples I, II and III.
(α,β,γ ≡ α-,β-,γ-FeOOH, resp.)

Phase	d / Å	Relative Intensities		
		I	II	III
γ	6.25	80	65	80
α	4.18	55	43	52
β	3.33	100	-	32
α	2.69	30	36	26
β	2.62	35	-	24
α,γ	2.46	95	100	100
γ	1.94	60	50	48
α	1.56	28	21	15

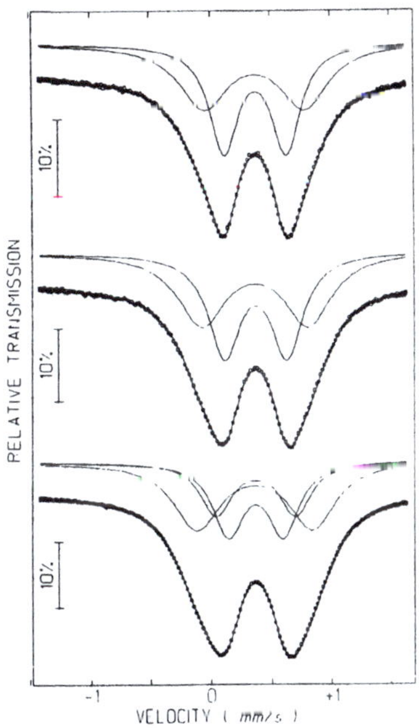

Fig.1. Mössbauer spectra at 300K of samples I (top), II (middle) and III (bottom). See Text.

The transmission spectra of the rust samples at room temperature of the deposited layer rust scraped after corroding in aqueous solution are shown in fig 1. They show only quadrupole split doublets with broadened lines, due to the presence of superparamagnetic ultrafine crystallites (< 100 Å) in acordance with the literature [11-15]. Fig.2 shows Mössbauer spectra at different temperatures of the samples. The average superparamagnetic Mössbauer blocking temperature can be estimated to approximately 110K. The spectra of the other samples at 110K are similar and can be fitted with sextets and a central doublet (γ-FeOOH) [16]. The spectra of all the samples were also recorded at 4.2K where no relaxation effects were found. From the values of the hyperfine parameters it is in principle possible to identify the different corrosion products [17]. The Mössbauer parameters obtained from the spectra are listed in table 2.

Table 2: Mössbauer parameters (transmission) of samples I (top), II (middle) and III (bottom). Quadrupole splitting (QS) and isomer shift (IS, rel.α-Fe) in mm/s, magnetic hyperfine field (BHF) in Tesla.

T/K	IS	QS	BHF	Phase
300	0.37	0.52	-	γ
	0.37	0.83	-	α,β
110	0.35	0.63	-	γ
	0.37	-0.15	44.2	β
	0.33	-0.10	47.6	β
	0.36	-0.18	49.5	α
4.2	0.37	0.03	44.1	γ
	0.35	-0.05	46.3	β
	0.36	-0.02	50.2	α
300	0.36	0.50	-	γ
	0.35	0.62	-	α
110	0.34	0.55	-	γ
	0.37	-0.15	49.3	α
4.2	0.39	0.03	44.3	γ
	0.38	-0.13	50.6	α
300	0.36	0.51	-	γ
	0.36	0.95	-	β
	0.37	0.62	-	α
110	0.34	0.55	-	γ
	0.35	-0.06	46.2	β
	0.34	-0.14	49.1	α
4.2	0.37	0.04	45.5	γ
	0.36	-0.24	46.3	β
	0.36	-0.16	50.8	α

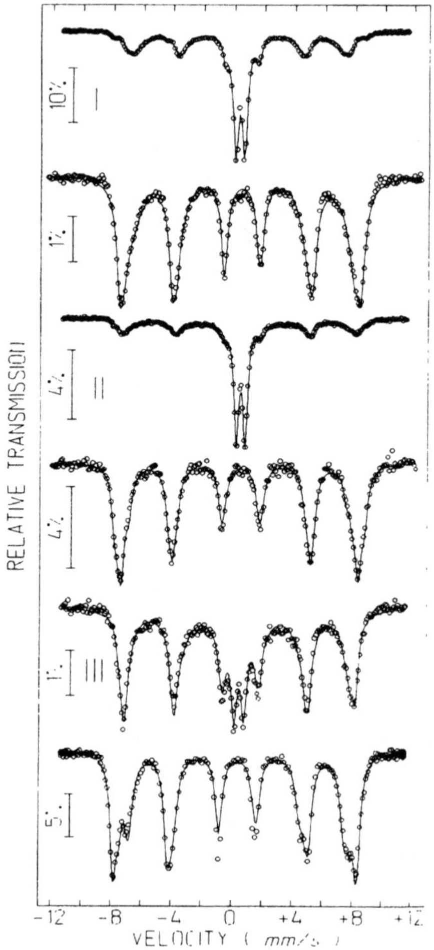

Fig.2. Mössbauer spectra of the samples I (top), II (middle) and III (bottom) at 110K and 4.2K.

Mössbauer CEMS spectra and simulations at room temperature are shown in fig. 3. The corresponding Mössbauer parameters are listed in table 3.

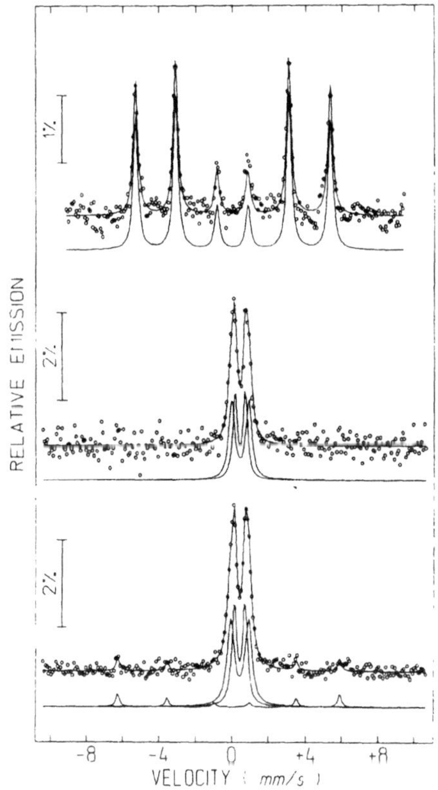

Fig. 3: CEMS spectra at 300K.
a) Non-corroded steel; b) sample I; sample;
c) as sample I with subsequent annealing at 100°C for 10 min.

Table 3: Mössbauer parameters of CEMS measurements. Quadrupole splitting QS and isomer shift IS (rel. α-Fe) in mm/s, magnetic hyperfine field BHF in Tesla.

Sample	IS	QS	BHF
Steel plate	0.00	0.02	33.04
Rust	0.39	0.73	-
	0.40	0.93	-
Rust+Heating	0.40	0.97	-
100C°/10 min	0.40	0.58	-
	-0.09	-0.18	37.8

We simulated Mössbauer spectra of microcrystalline α-,β- and γ-FeOOH using a relative reduction of about 15% in the magnetic hyperfine field near the blocking temperature caused by collective excitations. This reduction occurs because there exists a dependence of the effective magnetic hyperfine field on temperature given by $H_{eff}=H_o\langle\cos\theta\rangle_{av}$, with $\langle\cos\theta\rangle_{av}\approx 1-k_BT/2KV$ [18] (below blocking temperature). In samples with a broad particle size distribution, this leads to an asymmetric line shape. Our experimental spectra do not show this and therefore we took only an average particle size into account in our simulations. The spectra also don't show a simultaneous presence of static and relaxed patterns, as should be the case for a broad particle size distribution. The theoretical model used is based on a stochastic theory, using an internal field, longitudinal relaxation and collective excitations [18]. Our simulations were in good agreement with the experimental spectra (see fig.3).

4. Conclusions

The qualitative phase analysis, performed with the aid of Mössbauer spectroscopy, agrees for all samples with the XRD results. α-, β- and γ-FeOOH are the main components of the corrosion products present in the samples I, III and α-FeOOH,γ-FeOOH in the sample II. The particle size of the corrosion products studied were below 130Å, leading to completely collapsed spectra at room temperature.

References

[1] J. Dézsi, A. Vértes and L. Kiss, Journal of Radioanalytical Chemistry 2 (1969) 183.

[2] H. Jr. Leidheiser, G. W. Simmons and E. Kellerman, Croat. Chem. Acta 45 (1973) 257.

[3] M. J. Graham and M. Cohen, Corrosion 32 (1976) 432.

[4] T. Péev, M. K. Georgieva, S. Nagy and A. Vértes, Radioanal. Letters 33 (1978) 265.

[5] T. Misawa, K. Asami, K. Hashimoto and S. Shimodaira, Corrosion Sci. 14 (1974) 131.

[6] T. Kobayashi, K. Asami, Y. Izumi and R. Katano, Hyp. Int. 39 (1988) 419.

[7] K. Nomura, M. Tasaka and Y. Ujihira, Hyp. Int. 41 (1985) 681.

[8] A. K. Singh and T. Ericsson, Corros. Sci. 25 (1985) 931.

[9] R. A. Brand, J. Lauer and D. M. Herlach, Met. Phys. 12 (1983) 675.

[10]ASTM Podwer Diffraction File, ASTM STP 48-M2, ASTM, Philadelphia, Pennsylvannia, 1963.

[11]H. Kubsch and H. A. Schneider, Proc. Int. Conf. Mössbauer Spec. 1 B-13. Cracow (1975)

[12]T. Kobayashi, K. Fukumura, Y. Isozumi snd R. Katano, Hyp. Int. 57 (1990) 1923.

[13]C.Haces, N. R. Furet and L. Mulesshkova, Hyp. Int. 67 (1991) 587.

[14]S. Mϕrup, J. A. Dumesic and H Topsϕe, in: Application of Mössbauer Spectroscopy, Vol. II, ed. R. L. Cohen, (Academic Press, New York, 1980) p. 1-53.

[15]J. Teillet, J. M. Breton and D. Festy, Hyp. Int.93(1994)1801.

[16]E. De Grave, R. Persoons, D.G. Chambaere, R. E. Vandenberghe and L. H. Bowen, Phys. Chem. Minerals 13 (1986) 61.

[17]E. Murad and J. H Johnston in Mössbauer Spectroscopy Applied to Inorganic Chemistry, vol. 2, ed. Gary Long (Plenum Publishing Corporation, 1987) p. 507.

[18]S. Mϕrup, H. Topsϕe and J. Lipka, Journal de Physique, Colloque C6, suplément au n° 12, Tome 37 (1967) C6-287.

MÖSSBAUER STUDY ON THE COMBINED EFFECT OF SO_4^{2-} AND Cl^- IONS ON THE CORROSION OF LOW CARBON STEEL

A. L. Morales[1], D. Cartagena[1], J.L. Rendón[2], A. Valencia[2], C.A. Barrero[3], and C. Dauwe[3]

[1]Departamento de Física, Universidad de Antioquia, A.A. 1226, Medellín, Colombia
[2]Grupo de Corrosión y Protección, Departamento de Ingeniería de Materiales, Universidad de Antioquia, A.A.1226, Medellín, Colombia
[3]Department of Subatomic and Radiation Physics, University of Gent, Proeftuinstraat 86, B-900 Gent, Belgium

Abstract
We have found, by means of immersion-emersion tests, a range of combined concentrations of SO_4^{2-} and Cl^- ions which inhibits corrosion and produces a more protective rust layer. We also found that the corrosion rate, and the magnetite content of the rust follow a similar trend, being rather constant for the low concentrations of the pollutants and scaling up for the larger concentrations. The magnetite area ratios show a more stoichiometric compound for the lower concentrations, and also the corrosion potential evolves toward higher values indicating the formation of a more protective rust.

1. Introduction

The effect of pollutants on the atmospheric rust formation has been studied since the early days of corrosion studies[1,2], some decades later their combined action[3,4] was taken into account and it continuous to attract the interest of the scientific community[5,6], this has been so specially for industrial-marine atmospheres. Beneficial effects on corrosion rates in steel have been found in different contexts, e.g., Gaur et. al.[7] found that by adding 10% NaCl to sulfuric acid for pickling decreases the corrosion rate by a factor of two.
Golovina et. al.[8] have postulated the existence of a beneficial effect, i.e., a decrease of the corrosion rate, when the pollutant ions SO_4^{2-} and Cl^- are simultaneously present with a yet undetermined concentration, this finding was later confirmed by Corvo and Leon[9] in a laboratory simulation of rusting under atmospheric conditions. Based on this hypothesis Valencia et. al.[10] have performed immersion-emersion tests to borne out experimentally that this concentration exist and to find the appropriate concentrations of the contaminants, by following the evolution of the corrosion potential measured when the samples enter and leave the solution. Their results point to a concentration of ($SO_4^{2-} = 10^{-4}M + Cl^- = 1.0x10^{-3}M$) as a candidate for this useful effect, since in this case the corrosion potential was evolving all the time toward higher values and stayed very close to the one from the reference solution, containing only $SO_4^{2-} = 10^{-4}M$, which produce a more homogeneous, adherent and compact rust. We have extended this work further to explore concentrations around this value, to measure the rust features using Mössbauer spectroscopy, to obtain the corrosion rate, by a gravimetric method, and to measure the evolution of the corrosion potential as the experiment progresses.

2. Experimental

Our experimental set up, following Pourbaix's ideas[11], consisted in six vessels, each one supporting six 150x33x3 mm samples of AISI 1012 carbon steel with composition, C = 0.12, Si = 0.22, S = 0.016, Ni = 0.01, Mn = 0.49, Cu = 0.01, Cr = 0.07, P = 0.033, Mo = 0.003, Sn = 0.001 and Al = 0.003. The pollutant concentrations, of SO_4^{2-} and Cl^- ions, are shown in Table 1. The concentrations, reflecting typical atmospheric values, were chosen according to earlier work[10] which suggested a decrease in the corrosion rate for concentrations around ($SO_4^{2-} = 10^{-4}$M + $Cl^- = 1.0\times10^{-3}$M), so vessels 3, 4, 5 had concentrations close to this one, vessel 6 had a concentration not studied before and 1, 2 were taken as a reference to stress the combined effect of the contaminants. The solutions are aerated continuously and its pH was kept around 6.0 by periodically renewing them during the experiment. The samples rotated continuously with a period of about 43 minutes a third of it immersed in the solution and the rest outside under a temperature around 45 °C, for simulating the drying period. The experiment ran for 65 days, during this time the corrosion potential was measured twice a day when the specimen enters and leaves the solution, a saturated calomel electrode was used as reference.

Table 1
Pollutant concentrations for the different vessels

Vessel	1	2	3	4	5	6
Cl^-x10^{-3}M	0.0	1.0	1.0	1.5	2.0	2.5
SO_4^{2-}x10^{-4}M	1.0	0.0	1.0	1.0	1.0	1.0
Mean Corrosion Rate(μm/d)	0.027	0.044	0.052	0.041	0.141	0.118
Magnetite(%)	32.6	26.0	19.0	21.5	71.6	64.1
Area Ratio B/A	1.59	1.50	0.98	0.92	0.84	0.35

Four specimens from each container were used to measure the corrosion rate by means of a gravimetric test, according with ISO norm 156/4 N 153, which consisted in performing various cycles of etching, after the mechanical removal of the rust film, in the presence of an inhibitor to avoid etching of the steel substrate, and measuring the mass lost after each cycle. By extrapolating the values to cycle zero one can find the mass lost due to corrosion. The corrosion rate is obtained from the expression :

$$V_{CORR} = 10 \times m / (A \times \delta \times t) \,, \qquad (1)$$

where m is the mass lost in mg., A is the area exposed to corrosion in cm^2, δ is the material density in g cm^{-3}, and t is the exposure time in days. The results are shown in Table 1.
The rust obtained was classified in two types, one which was loosely bound and fell down by hitting the sample against a hard surface, the outer rust; and a second one which was scrapped from the surface, the inner rust. Both were analyzed by Mössbauer spectroscopy at room temperature and at 80 K.

3. Results and discussion

Table 2 shows our results for the Mössbauer spectra of the outer rust, where HF, IS, QS, AREA stand for the hyperfine magnetic field, isomer shift(with respect to α-Fe), quadrupole splitting, and relative areas respectively.. We are able to identify, using the fitting program MOSF developped by the Group in Gent, the rust components as magnetite and goethite from the sextets and the doublets are mixtures of lepidocrocite, akaganeite and superparagmagnetic goethite, for the systems containing Cl, and without akaganeite for reference system 1. Akaganeite was identified by means of the low temperature spectra.

Table 2

Analysis of Mössbauer spectra from outer rust(HF(KOe), IS(mm/s), QS(mm/s), AREA(%))

Vessel	OXIDE	HF	IS	QS	AREA	Vessel	OXIDE	HF	IS	QS	AREA
1	Fe_3O_4	489	0.27	-0.03	12.6	4	Fe_3O_4	489	0.28	-0.014	11.2
		460	0.68	0.004	20.0			460	0.65	-0.002	10.3
	α-FeOOH	313	0.31	-0.16	23.4		α-FeOOH	327	0.32	-0.1	35.6
			0.34	1.32	0.1				0.43	1.46	0.6
			0.36	0.84	26.7				0.36	0.84	16.6
			0.37	0.53	17.0				0.37	0.52	25.7
2	Fe_3O_4	489	0.27	-0.03	10.4	5	Fe_3O_4	490	0.29	-0.02	39.0
		459	0.65	-0.01	15.6			457	0.64	-0.002	32.6
	α-FeOOH	348	0.35	-0.003	35.3		α-FeOOH	327	-0.31	-0.52	16.2
			0.37	1.33	1.6				0.53	1.66	0.8
			0.35	0.88	13.8				0.38	1.0	2.3
			0.36	0.53	23.3				0.36	0.56	9.1
3	Fe_3O_4	488	0.27	-0.008	9.6	6	Fe_3O_4	492	0.28	0.005	47.6
		459	0.63	-0.008	9.4			458	0.58	0.06	16.5
	α-FeOOH	325	0.26	-0.19	35.5		α-FeOOH	307	0.32	-0.55	16.1
			0.48	1.65	0.4				0.43	0.56	0.2
			0.35	0.85	19.3				0.35	0.58	9.9
			0.36	0.52	25.9				0.31	0.55	9.5

By comparing the different spectra we notice that the most striking feature is the strong magnetite content for the higher concentrations of the Cl^- ion, i.e., of the order of $2.0x10^{-3}$M; while the other spectra show a lower content of magnetite, as can be seen from Table 1.

The area ratios(B/A) for tetrahedral(A), HF ~ 489 kOe, and octahedral(B), HF ~ 459 kOe, sites are shown at the bottom of Table 1. We notice that for the reference concentrations, with low corrosion rates, the area ratio is closer to the stoichiometric value of 1.8. This suggest the formation of larger and more crystalline magnetite particles which may grow as a consequence of the formation of a protective layer not allowing the diffusion of the contaminants and oxygen to the interior of the rust, a mechanism similar to the one propose by Davalos et. al.[12] for weathering steels. The protective layer may be less protective for the case of reference vessel 2 due to its higher corrosion rate. Our last conjecture is at odds with previous work[13] showing that the magnetite formed, under the action of Cl^-, in marine atmospheres is not protective. The area ratio is around one for vessels 3 and 4, suggesting a less protective rust but also showing corrosion rates not too far away from reference vessel 2. The relative area decreases further for the vessels with higher concentrations of Cl^-, indicating the formation of a porous rust and leading to higher corrosion rates. We can also notice that the addition of sulfate ions to vessels 3 and 4 decreases the amount of magnetite with respect to vessel 2.

The results for the spectra of the inner rust show a similar trend and its composition does not differ to much from the inner one, a fact found by other authors[14], e.g., the magnetite content of the inner rust increase by 3-5%.

The goethite content of the rust is an important parameter in the analysis since it has been propose by Miranda[15] that its presence favors the formation of a protective rust, our numbers in Table 2 seems to confirm this hypothesis since we have a larger amount of this oxide in the vessels with smaller corrosion rate, including the reference systems. The variation in the hyperfine parameters for goethite is due to its presence as particles of different sizes with poor crystallization, this shows up in the spectra as very broad sextets.

In Table 1, third row, we present our results for the mean corrosion rate in every vessel where we can notice the high values obtained for the larger concentrations of the Cl^- ion. Also, for comparison, we have included in Table 1 the percentage of magnetite in the rust and we stress the remarkable correlation between these two quantities since both take the highest values for vessels 5 and 6 and stay rather constant for the lower concentrations, vessels 3 and 4, both quantities are shown in Fig. 1 where we have normalized the values of the corrosion rate with respect to the magnetite content in vessel 5.

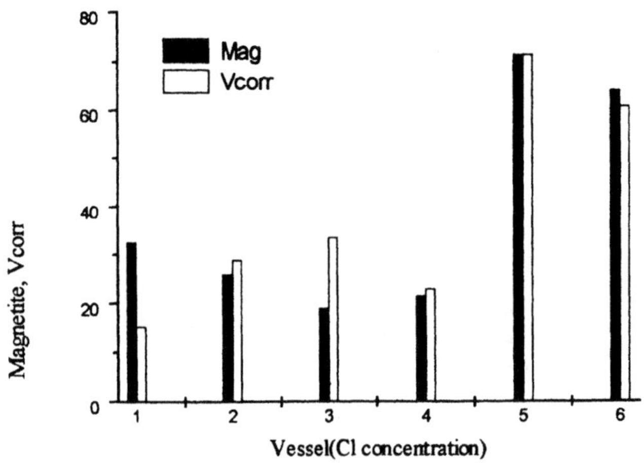

Fig. 1. Correlation between magnetite content and corrosion rate(see text).

In Fig. 2. we present the results for the corrosion potential, our results are in good agreement with those found by Valencia et. al.[10]. The corrosion potential is a consequence of the processes going on during the electrochemical reactions producing the rust; and of course, it will vary continuously as different processes are switch on and this happens particularly during the immersion an emersion of the samples. In previous studies[15,16] it has been used as a clue to the properties of the rust film grown on the steel surface, so for a potential which evolves toward higher values and at the same time the potential difference between entrance and exit of the samples is small, then the rust film is more protective. The value of the potential reflects the transformation processes occurring on the sample as has been established in the literature[17]. For vessel 1, we find a potential moving to higher values and then staying approximately constant for the last few days of the experiment; this is in agreement with the lowest corrosion rate found in this container; the rust formed is adherent and the sample shows sulfate nests typical of sulfate corrosion. Vessel 2, the other reference concentration, shows also an increasing potential, but with values lower than vessel 1, leading to a higher corrosion rate. We can see that for vessels 3 and 4, the lowest Cl^- concentrations, the corrosion potential evolves toward higher values but close to each other and to container 2, with corrosion rates not to far from each other. In contrast, vessels 5 and 6 show stronger variations and its potential(not shown in Fig. 2) evolve toward lower values, of the order of -0.3V, this may be the cause for the large amount of magnetite in these samples since it is formed around this potential. These containers showed the highest values of the corrosion rate; the magnetite produced was very far from stoichiometry and was also produced in the largest amounts. This behavior confirms further the conclusions above about the correlation between the corrosion rate and the magnetite content of the rust.

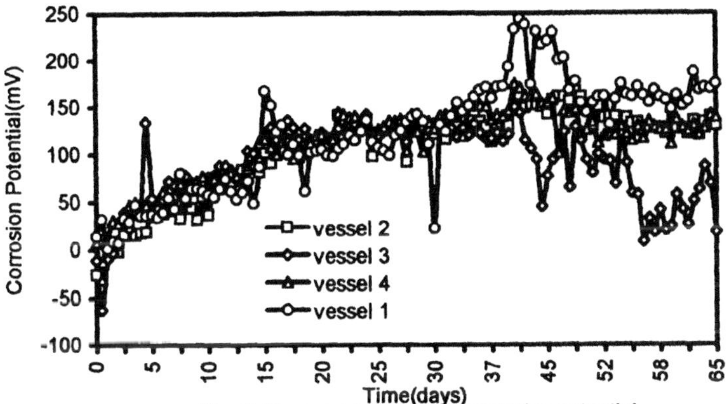

Fig. 2. Time evolution of the corrosion potential

According to the discussion above we have found that vessel 4 is the one showing the existence of the inhibitor effect from the combination of pollutants studied since its corrosion rate is smaller than reference vessel 2, its corrosion potential evolved toward higher values close to reference vessel 1, the stoichiometry of the magnetite found took a value intermediate for the batch considered here and the goethite content was among the highest.

We think that the arguments just given above can be fulfilled by other concentrations around the ones chosen for vessels 3 and 4, so we are planning more experiments to ascertain the validity of the present results.

Conclusions

In the present study we have found a range of concentrations where the combined effect of the pollutants can occur in a beneficial way. We have also found a correlation between the corrosion rate and the magnetite content of the rust, higher corrosion rates produce a higher quantity of poor stoichiometric magnetite. Our results indicate that the magnetite stoichiometry may be an important parameter to analyze the kinetics of the corrosion phenomena and also the protective nature of the rust. The corrosion potential evolves with time in a very different manner for the lower pollutant concentrations as compared to larger ones, confirming the hypothesis that evolution toward higher values leads to a more protective rust, i.e., lowering the corrosion rate. We find a limiting level for chloride concentrations, above which the magnetite production is enhanced dramatically and producing high corrosion rates; this may lead to the existence of competitive phenomena between chloride and sulfate ions as a mechanism inhibiting the magnetite formation and to lower corrosion rates. The idea that more goethite content leads to more protective rust is also in agreement with our results.

Acknowledgments

This work was partially supported by the Instituto Colombiano para el Desarrollo de la Ciencia y la Tecnología, Francisco José de Caldas, Colciencias.

References

[1] W.H.J. Vernon, Trans. Faraday Soc. 19 (1924) 839.
[2] U.R. Evans, Trans. Faraday Soc. 19 (1923) 201.
[3] Y.U.M. Pachenco, Zasch. Metall. 19 (1983) 103.
[4] J. Mertel, Rev. Ib. de C. y P. XVII (1986) 217
[5] M. Stratmann and J. Müller, Corros. Sci. 36 (1994) 327.
[6] C. Arroyave, F.A. Lopez and M. Morcillo, Corros. Sci. 37 (1995) 1751.
[7] B.Gaur, T.B. Singh and D.D.N. Singh, Corrosion 52 (1996) 154.
[8] G.V. Golovina, G.M. Floriantovich and I.M. Kolotyrkin, Zaschita Metallov 15 (1979).
[9] F. Corvo and Y. Leon, Rev. Iber. Corros. y Prot. XIX (1988) 291.
[10] A. Valencia, J. Arias and G. Villa, in: Proc. of the 5th Iberoamerican Congr. on Corrosion and Protection, Tenerife, Spain, 1995.
[11] M. Pourbaix, Rapport Technique CEBELCOR RT 160, Bruxelles, 1969
[12] J. Davalos, M. Gracia, J.F. Marco y J.R. Gancedo, Hyp. Int. 69 (1991) 871
[13] C. Arroyave, M.Sc. thesis, U. F. de Rio de Janeiro, Brazil, 1988.
[14] A.K. Sing, T. Ericsson and L. Häggström, Corros. Sci. 25 (1985) 931.
[15] L.M.R. Miranda, M.Sc. thesis, Bruxelles, 1980.
[16] M. Pourbaix, Corros. Sci. 30 (1990) 963.
[17] M. Stratmann and K. Hoffmann, Corros. Sci. 29 (1989) 1329.

MÖSSBAUER SPECTROSCOPIC INVESTIGATION OF CORROSION IN SCRUBBING AND CLEANING COMMERCIAL STEEL PRODUCTS

J. F. M. Borges[a], *M. I. da Costa Jr.*[a,1], *F. Locatelli*[a], *F. A. Molossi*[a], *J.I. Kunrath*[a], *G. Feldmann*[b], *R. Toukhvatoulline*[b], *R.J. Muenchen*[b], *G.E. Englert*[c]

[a] Instituto de Física, UFRGS - Cx.P. 15051, 91501-970 Porto Alegre, RS, Brasil

[b] DeFEM, UNIJUÍ - Cx.P. 560, 98700-000 Ijuí, RS, Brasil

[c] LAPEC/DeMet, UFRGS - 90035-190 Porto Alegre, RS, Brasil

Commercial steel wool, and steel scrubber have provided suitable samples for a study of corrosion and degradation of steel products used in every-day cleaning, and scrubbing tasks. The analysis made by x-ray powder diffraction, and Mössbauer spectroscopy shows that rust formed by corrosion of steel immersed in aqueous solution consists of iron hydroxides and oxides whose formation is described by the Hiller scheme. The reactive rust component is γ-FeOOH, and formation of more stable iron compounds shows that the original materials are degradable to the environment.

Key words: corrosion, commercial steel, Mössbauer spectroscopy

1. INTRODUCTION

Corrosion is the natural deterioration or destruction of a material as a result of its interaction with its environment. This term is applied mostly to metals and alloys, particularly to their reaction with oxygen (rusting). Corrosion of metals is generally confined to the surface. Generally, the corrosion processes are chemical or electrochemical, although physical and mechanical factors contribute to the corrosion.

Perhaps the most important chemical reaction of iron, from an economic standpoint at least, is the least desirable one: the reaction of iron, water, and oxygen to form hydrated iron oxide, or rust. The corrosion of iron has been studied carefully, and the formation of rust is known to be an electrochemical reaction. The mechanism of the corrosion process of iron in aqueous solution can be described by the Hiller scheme [1]:

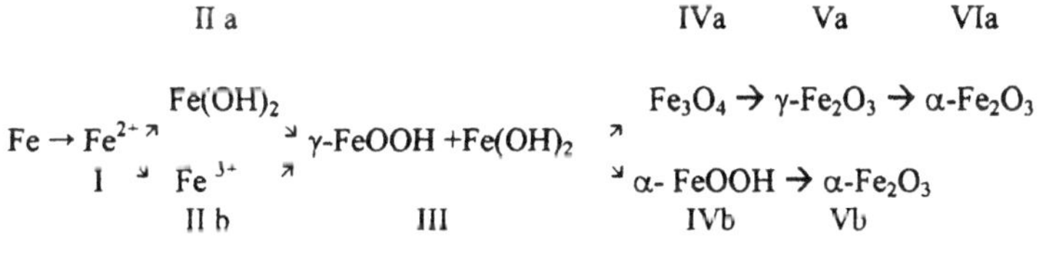

[1] Corresponding author

Rust formation must be the way of deterioration in watery conditions of some commercial steel products used in every-day scrubbing and cleaning tasks. Using appropriate analytical techniques such as x-ray diffraction, and Mössbauer spectroscopy we have investigated the degradation of such steel products into the environment.

2. EXPERIMENTAL PROCEDURES

Commercial steel wool of three different trademarks labeled A, B, C, and one commercial steel scrubber labeled D have provided suitable samples for such an investigation . The elemental composition of each material was determined by inductively coupled plasma atomic emission spectrometry (ICP-AES), which is the analytical technique with the largest number of applications in the analysis of iron alloys [2]. One steel wool pad of each trade-mark, and a whole steel scrubber were separately immersed in water filled plastic cups. After one month period, during which part of the immersed material corroded and precipitated, the remaining aqueous solutions with the precipitates were poured into appropriate paper filters. These were dried in open atmosphere at room temperature. The solid crust adhered to each filter was then removed by careful scraping, grounded into a mortar, and the resulting fine powder was used to make samples for x-ray diffraction analysis. Afterwards, 100 mg of each powder were separately mixed with sugar to provide uniform Mössbauer absorbers. The Mössbauer spectroscopy measurements were carried out using a constant acceleration spectrometer in transmission geometry with a ^{57}Co:Rh source. All measurements were done at room temperature. Labeling of rust samples is in accordance with that of the original materials. Isomer shift values are refered to alfa-iron at room temperature.

3. RESULTS AND DISCUSSION

Table 1 shows the percent atomic composition of the originally selected materials obtained with the ICP-AES technique.

Table 1.Relative % atomic composition of the original commercial steel.

steel	Fe	Si	Cr	Mn
A	98.48	0.263	0.192	0.791
B	98.98	0.191	0.192	0.641
C	98.91	0.206	0.185	0.696
D	98.09	0.319	0.462	1.395

One can see that steel wool samples A,B, and C have nearly the same proportions of Si, Cr, and Mn within the Fe matrix, while steel scrubber sample D has them in greater amounts. Silicon added to iron increases the hardness of steel; manganese is added to iron because it reduces iron oxide to form manganese oxide, and together with chromium, it improves the hardness, the durability and corrosion resistance of steel.

Due to the high iron content, the diffractograms display the same lines of bcc iron for all samples, and Mössbauer spectroscopic results are also alike.

The results of the x-ray diffraction analysis of the rust precipitates are shown in Figure 1. Most part of the rust is composed by lepidocrocite γ-FeOOH or Lp; magnetite Fe_3O_4 or Mt; and goethite α-FeOOH or Gt for samples A, B, and C. According to the Hiller scheme (phase II) the rust is dominated by the reactive compound γ-FeOOH that may be formed by deprotonation of the $Fe(OH)_2$ compound in the reaction [3]:

$$2Fe(OH)_2 \rightarrow 2\ \gamma\text{-}FeOOH + 2\ H^+ + 2\ e^-$$

As a subsequent reaction the formation of the spinel Fe_3O_4 is explained by the diffusion of Fe^{2+} ions into the lattice of γ-FeOOH, and a subsequent transformation of the lattice of lepidocrocite into the cubic lattice of the spinel. Another reaction that may take place is [3]:

$$2\ \gamma\text{-}FeOOH + Fe(OH)_2 \rightarrow Fe_3O_4 + 2\ H_2O$$

The diffractograms also show the presence of the more stable forms γ-Fe_2O_3, and α-FeOOH in rust samples A, B, and C indicating that Hiller scheme is an adequate description of the mechanism of the corrosion process in these cleaning commercial steel products.

The x-ray diffraction results for sample D can be seen in Figure 1D. Within the given statistics, one identifies γ-FeOOH, and α-FeOOH, which means that the corrosion process in this case is given by part of the Hiller scheme (phases III and IVb in the diagram). Again, the reactive rust component γ-FeOOH transforms into the more stable phase α-FeOOH, in the same span of time for the experiment, that is, one month period. The diffractogram displays very broad lines which can be attributed either to a chemical disorder or a small particle size distribution of the α-FeOOH crystallites. The later hypothesis is favored by Mössbauer spectroscopy results.

The spectrum at the bottom of Figure 2 consists of a single quadrupole doublet (QS = 0.55 mm/s; IS = 0.37 mm/s). It corresponds to γ-FeOOH, and to superparamagnetic α-FeOOH with an average particle size smaller than 10 nm [4]. The corresponding spectra for rust samples A, B, and C display a quadrupole doublet attributed to to γ-FeOOH, and superparamagnetic α-FeOOH superimposed by sextet line components assigned to Fe_3O_4 and γ-Fe_2O_3. The former spectral component is best described by a hyperfine field distribution due to a finely dispersed non-stoichiometric $Fe_{3-x}O_4$ that results from the phase transformation III to IVa in the Hiller scheme. Most probable hyperfine field values for the distribution are: 48.9 Tesla for tetrahedral sites, QS = 0.0 mm/s, and IS = 0.26 mm/s; 45.0 Tesla for octahedral sites, QS = 0.0 mm/s, and IS = 0.62 mm/s. No clear evidence of hematite α-Fe_2O_3 was found from the experimental results. The relative amounts of rust components can be calculated from the relative areas of the Mössbauer spectra (assuming equal recoiless absorption factors), and are shown in Table 2.

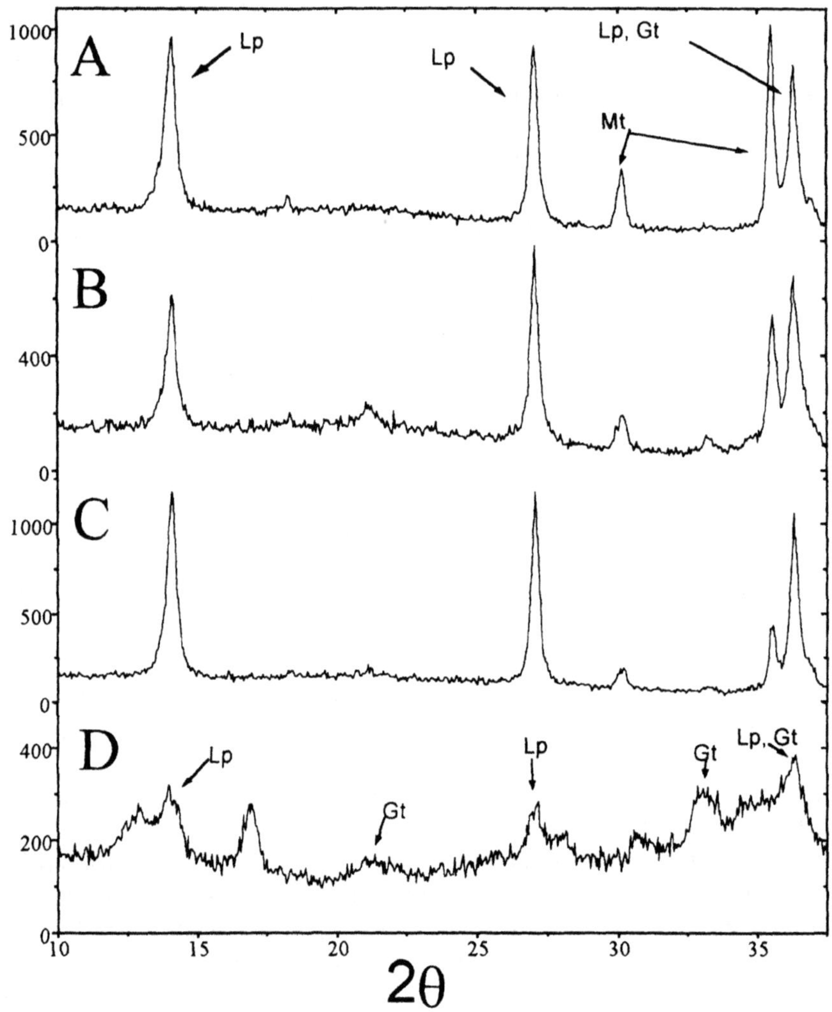

Figure 1. Selected portions of rust component diffractograms.

Table 2. Relative % amounts of the different phases of the Hiller scheme for rust materials calculated from Mössbauer spectral areas.

SAMPLE	III + IV	IVa + Va
A	60 %	40 %
B	65 %	35 %
C	75 %	25 %
D	100 %	-

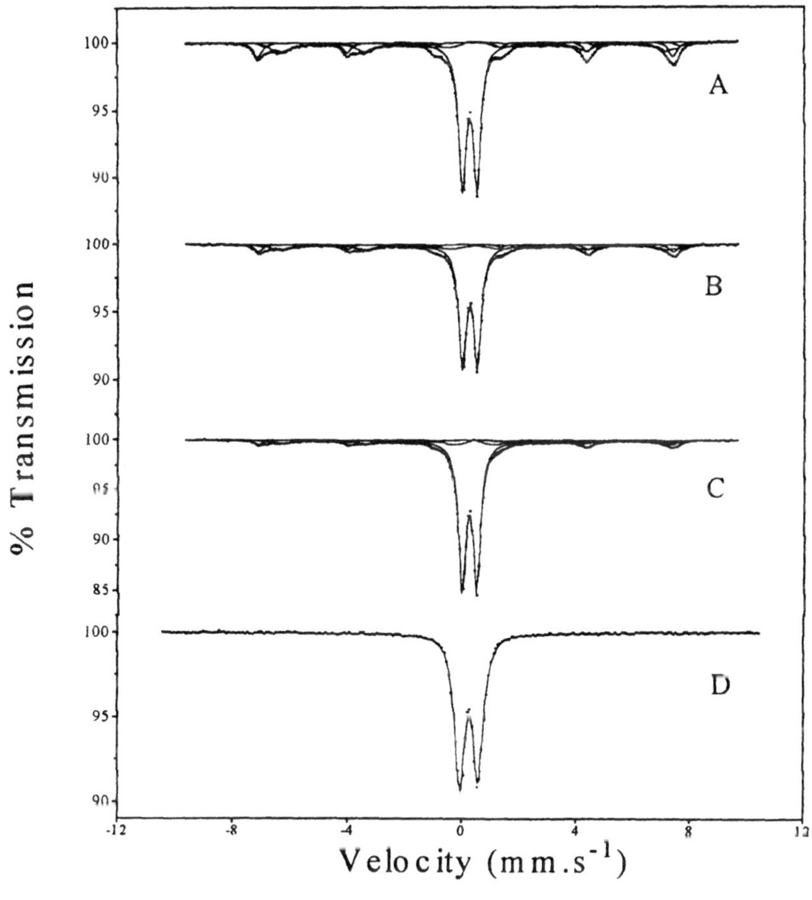

Figure 2. Mössbauer spectra of rust for each selected sample.

4. CONCLUSIONS

The analysis of the rust precipitates by x-ray diffraction, and Mössbauer spectroscopy techniques shows the presence of iron hydroxides and oxides. The reactive rust component is γ-FeOOH , and the formation of more stable rust compounds can be described by the Hiller scheme. Such stable compounds are commonly found among the soil constituents [4] which means that the original materials are perfectly degradable to the environment.

ACKNOWLEDGEMENTS

This work was supported by the Brazilian agencies CNPq, FINEP, and by PIBIC/CNPq-UFRGS.

REFERENCES

[1] J.E. Hiller, Werkstoffe und Korrosion 11 (1966) 943.
[2] J. Marshall, J. Carrol, J.S. Crighton and C.L.R. Barnard, J. Anal. Atom. Spectrom. 7 (1992) 349R.
[3] T. Peev, B. Mandjukova and I. Mandjukova, I., Corrosion 43 (1987) 739.
[4] E. Murad, Hyp. Inter. 47 (1989) 33.

SURFACE CARBON STEEL PRODUCTS AFTER LASER IRRADIATIONS UNDER DIFFERENT PROCESSING CONDITIONS

M. F. Creus and E. Gallego Lluesma
Laboratorio de Procesamiento Láser, CIOp
C.C. 124, 1900 La Plata, Argentina

J. Desimoni and R. C. Mercader
Departamento de Física, Facultad de Ciencias Exactas,
Universidad Nacional de La Plata, C.C. 67, 1900 La Plata, Argentina

A. C. Agudelo, R. Gancedo and J. F. Marco
Instituto de Química-Física "Rocasolano", CSIC, Serrano 119,
E-28006 Madrid, España.

This work deals with steel laser irradiations performed in air, without using anti-reflective coatings, under conditions suitable to obtain surface melting with an industrial CO2 laser. The results show that hematite, magnetite, wüstite, cementite and austenite can be formed on the steel surface as a result of the laser treatment. Qualitatively, all the treatments show to form reproducible surface compositions. However, under some processing parameters, the relative concentrations of the different phases can not be reproduced, probably due to the inhomogeneous spray of the molten metal produced by the shielding gas flow.

1. Introduction

The surface of metals is usually subjected to laser treatments to improve its physical-chemical properties. Although a lot of work has already been done to understand of the basic mechanisms that govern the whole process of the laser beam absorption and ensuing surface modifications, no model gives a thorough description of the critical dependence with sensitive parameters yet [1].

To study in depth the mechanisms of surface phase formation produced by laser melting, we have started a series of experiments with different carbon steels and laser parameters. Initially we have decided not to use anti-reflection coatings to avoid the extra difficulty that the intermixing of the substrate with the coating material would introduce in the interpretation of the results.

The aim of this work is to determine by CEMS the steel surface phase composition brought about by different laser irradiation treatments as well as to investigate the laser processing parameters which led to the formation of reproducible surface phase compositions.

2. Experimental

Sheets of 3.0 mm thickness of SAE1070 steel was used. Their chemical composition is: C 0.68 wt%, Mn 0.75 wt%, Si 0.21 wt%, S 0.012 wt%, P 0.015 wt%, and balance of Fe.

A sequence of abrasive treatments with 180-340-500-grit silicon carbide paper increased the fineness, removed oxide layers, and improved the surface finish. Afterwards the surface was cleaned using ethyl alcohol and tissue paper.

An industrial CO_2 laser (10.6μm) that produces a circular beam of mainly Gaussian power distribution was used to irradiate the surfaces at different scanning velocities and power densities. When the laser beam impinges on the steel surface part of it is reflected. The absorbed beam energy produces a melt pool which is used to sweep the area of the surface to be treated. To process uniformly the area without a coating, special cares have to be taken owing to the drastic change in the emissivity that occurs when the melting starts. Several tests were performed to ensure that once the melt pool starts it does not stop because of, e.g., too fast scanning velocities, too short beam pulses, or too low beam energies. The best results were obtained with a heat affected zone of constant depth, when the successive rectilinear tracks left by the melt pool scanning had an overlapping between 25 and 50% of the spot diameter. The surfaces of the irradiated samples were observed using an optical microscope without any further surface preparation.

Table I: Laser Processing Parameters

Parameters	Treatment A	Treatment B	Treatment C
Average power density (W/mm^2)	248	990	2830
Output power (W)	630	700	200
Scanning velocity (mm/s)	14.17	40.83	100.00
Overlapping (%)	30	42	50
Spot diameter (mm)	1.8	0.6	0.3
Frequency (Hz)	Continuous Wave	300	Continuous Wave
Interaction time (ms)	127	15	3
Air pressure (atm)	3.80	0.23	0.10

Table I shows the three laser operation conditions used in the present work. The instrument blows an air jet permanently on the surface that is impinged by the laser beam to protect the focussing lens from the particles ejected backwards by drawing the melt off the cut kerf ("shielding gas").

For each condition, a pair of samples were treated to check the equality of the each process. After the surface were treated, specimens of 40 mm × 35 mm were laser-cut to be examined by Conversion Electron Mössbauer Spectroscopy (CEMS). CEMS measurements were performed at room temperature using standard helium-methane flow and acetone gas electron proportional counters with constant acceleration drives. The data were collected in standard 512 multichannel scalers and analyzed by a least-square fitting program with constraints using Lorentzian line-shapes. The isomer shift are refered to α-Fe at room temperature.

3. Results and discussion

When steels of medium to high carbon content are ground abusively, the surface may be heated sufficiently to develop a thin rehardened martensitic surface layer accompanied by a tempered layer that is much deeper and of highly variable thickness [2]. CEMS results (not shown) obtained from the non-treated laser samples only showed the typical six-line ferrite pattern, indicating that our abrasion process did not produce any noticeable modification of the steel surface.

Figures 1 and 2 show CEMS spectra recorded from samples treated under the same processing parameter set (Treatment A). Table II collects the Mössbauer parameters obtained from the fit of the spectra and their assignment to different iron phases [3-5] and relative concentrations. It follows from this table that the surface phase composition of different sections of the material shows considerable variation.

Figures 3 and 4 show the CEMS spectra recorded from the samples treated using processing parameter set B. Table II demonstrates that the overall composition of the melted layers formed on the two samples under this treatment were similar.

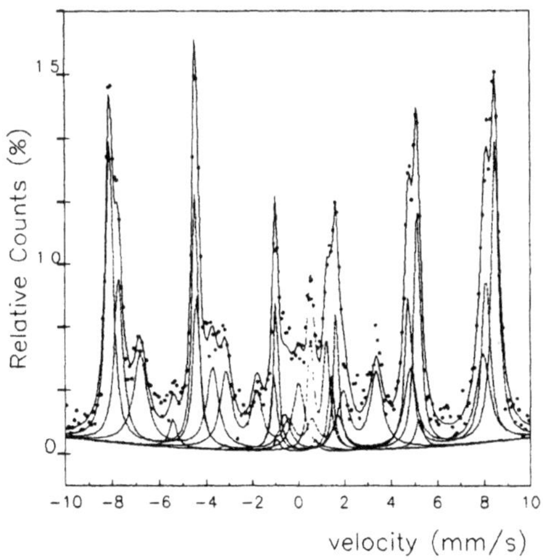

Fig.1. CEMS from sample 1 under treatment A.

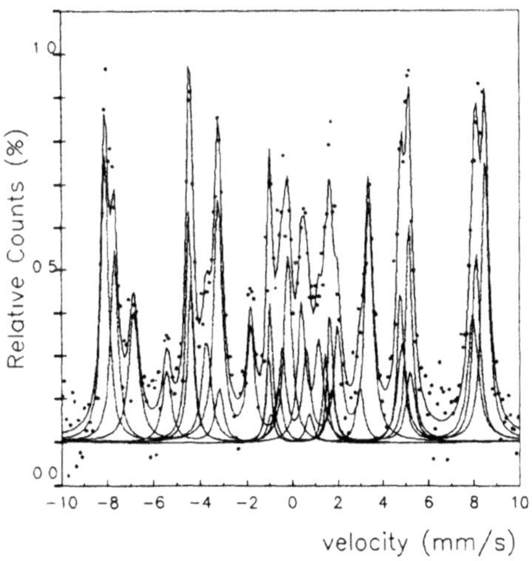

Fig. 2. CEMS from sample 2 under treatment A.

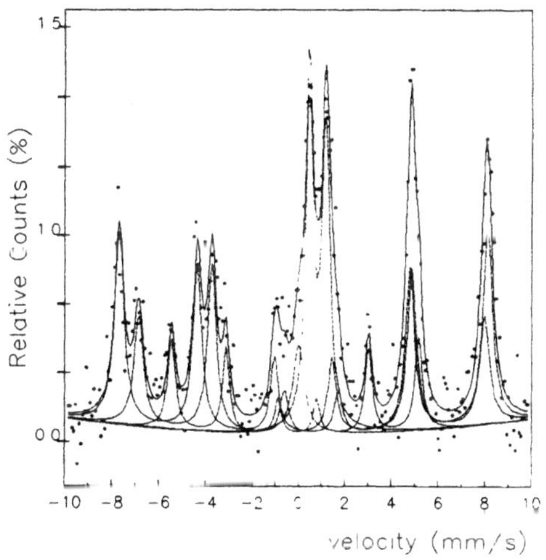

Fig. 3. CEMS from sample 1 under treatment B.

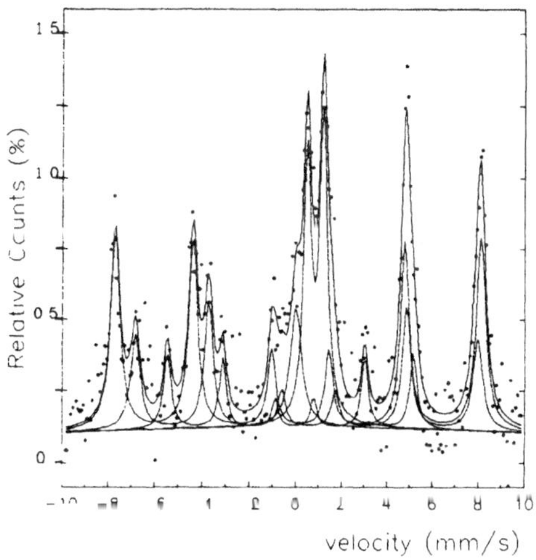

Fig. 4. CEMS from sample 2 under treatment B.

Table II: Surface species determined by CEMS

Sample	Phase	H (kG)	IS (mm/s)	2ε (mm/s)	Δ (mm/s)	Abundance (%)
A 1	Ferrite (α-Fe	330_4	0.00_8	0.00_6		2_2
	Austenite (γ-Fe)		0.106			3_2
	Carbide	202_2	0.21_3	0.00_2		15_3
	Wüstite ($Fe_{1-x}O$)		0.98_2		0.70_3	5_1
	Magnetite (Fe_3O_4)	490_1	0.28_1	0.00_1		22_3
		458_2	0.67_3	0.00_2		20_4
	Hematite (α-Fe_2O_3)	515_1	0.36_1	-0.20_6		33_3
A 2	Ferrite (α-Fe)	330_3	0.00_5	0.005		8_4
	Austenite (γ-Fe)		0.10_5			5_1
	Carbide	205_1	0.19_3	0.00_2		22_4
	Wüstite ($Fe_{1-x}O$)		0.88_3		0.76_6	6_2
	Magnetite (Fe_3O_4)	490_1	0.30_1	0.00_1		18_4
		460_2	0.66_3	0.00_2		17_4
	Hematite (α-Fe_2O_3)	515_1	0.37_1	-0.20_8		24_4
B 1	Ferrite (α-Fe)	330_2	0.00_2	0.00_2		13_4
	Austenite (γ-Fe)		0.10_8			5_2
	Wüstite ($Fe_{1-x}O$)		0.92_1		0.71_2	22_3
	Magnetite (Fe_3O_4)	490_1	0.30_1	0.00_1		35_5
		460_2	0.66_2	0.00_2		25_5
B 2	Ferrite (α-Fe)	330_2	0.00_3	0.00_3		12_4
	Austenite (γ-Fe)		0.10_6			7_2
	Wüstite ($Fe_{1-x}O$)		0.94_1		0.70_2	21_5
	Magnetite (Fe_3O_4)	490_1	0.30_2	0.00_2		39_7
		460_2	0.67_3	0.00_2		21_6
C 1	Ferrite (α-Fe)	330_2	0.00_2	0.002		10_3
	Austenite (γ-Fe)		0.10_6			7_2
	Wüstite ($Fe_{1-x}O$)		0.96_2		0.84_3	19_3
	Magnetite (Fe_3O_4)	489_1	0.32_2	0.00_1		33_5
		454_1	0.62_2	0.00_2		31_5

Isomer shift of ferrite at room temperature is taken as reference.
Estimated errors are shown as subindices.

Figure 5 displays the spectrum of one sample corresponding to treatment C. CEMS results (not shown) of a second sample treated under the same processing parameters present almost no variation.

While performing treatment A with a constant laser beam, small changes in the brightness of the laser spot -denoting variations in the surface emissivity- and presence of a reduced spray of small particles ejected from the melted zone have been observed. During the pulsed laser treatment B and the continuous laser method C this phenomenon was not appreciated. Similarly to the results presented in ref. [6], our CEMS results show the presence of oxides that probably form a layer covering the resolidified droplets ejected from the melted zone.

The present results show differences in the relative concentrations of the surface phases formed after treatment A. This lack of reproducible surface phase compositions after treatment A could be interpreted as a consecuence of the way that the gas jet blows on the surface. As already mentioned, successive rectilinear resolidified troughs were overlapped to cover the surface of the samples. The shielding gas flow not only cools the surface, but also blows part of the melted material away from the melted zone in a manner that mainly depends on the pressure of the shielding gas. The higher the pressure, the more inhomogeneous the spray becomes. A dissimilar distribution of small resolidified droplets on the laser track induces a change in the way that the beam power is absorbed by the surface (see Fig. 6). When the nozzle moves along and across the surface, it finds regions where the beam is unevenly reflected and absorbed. When the next scanning begins overlapped with the previous trough with different density of solidified droplets, the energy deposited per unit time is different. Therefore, as a consequence of the melted material dragging and the cooling by the gas, different quantities of material are removed from the melted zone to one side of the laser track. Since the resolidification velocity depends mainly on the ratio volume-surface of the droplets, the thin layers of surface iron oxides change their relative proportions along the laser melted track.

The pulsed beam surface treatment B and continuous beam surface method C, however, have demonstrated to yield more reproducible results. The main difference must be due to the fact that the blowing shielding gas does not have the same effect on the cooling of the surface, as with the treatment A. Since there are not resolidified droplets on the laser troughs -heat radiation-conduction flows are uniform- (see Figs. 7 and 8) the cooling velocity can be considered constant along them. This last one can be almost totally suppressed using low shielding gas pressures and reducing the laser spot diameter on the surface for the same averange power density.

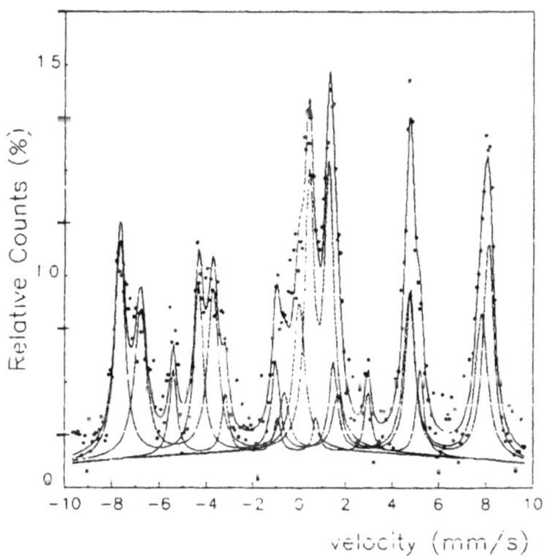

Fig.5 CEMS from sample 1 under treatment C

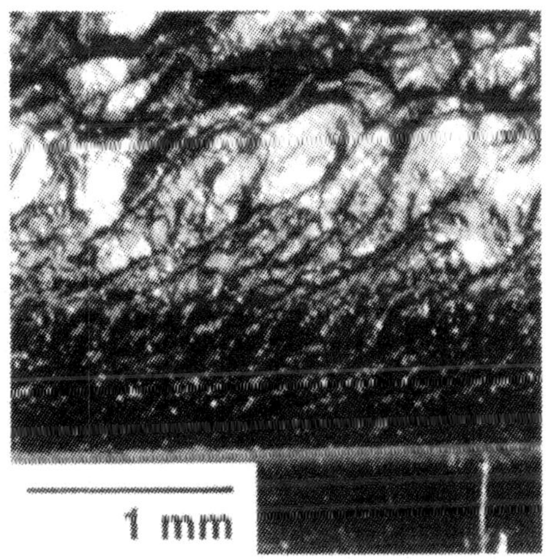

Fig.6 Enlarged picture of the sample surface under treatment A

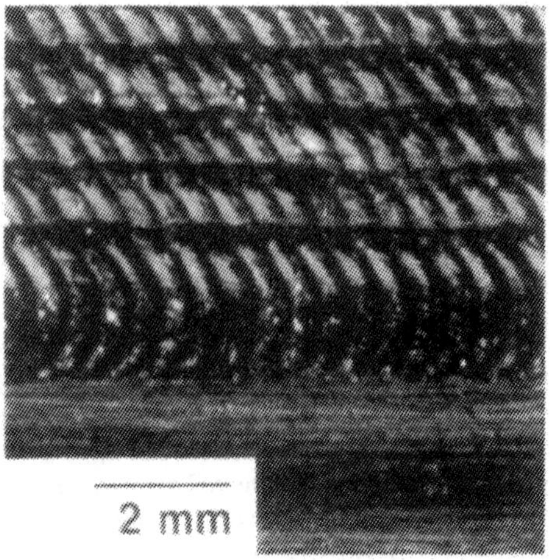

Fig.7 Enlarged picture of the sample surface under treatment B

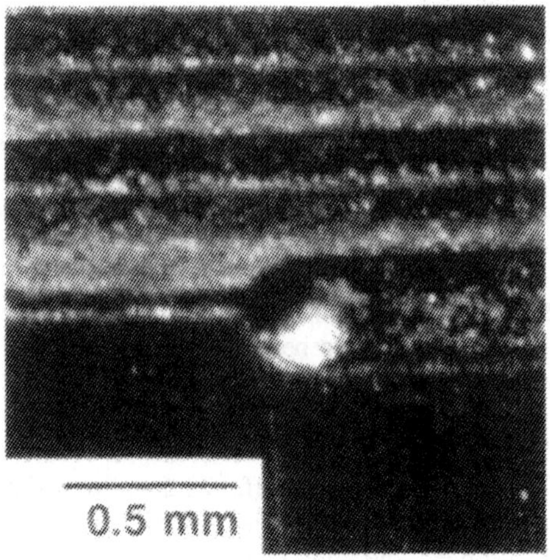

Fig.8 Enlarged picture of the sample surface under treatment C

4. Conclusions

In medium-carbon steel surfaces melted without anti-reflective coating by laser treatments in air, the thin surface layer formed as a result of the treatment is mainly constituted by iron oxides. Two of the three laser processing parameter sets investigated in this paper have shown to form reproducible surface phase compositions from both the qualitative and quantitative points of view. The lack of reproducibility of the treatment A could be attributed to the blowing effect of the high pressure of the shielding gas and the inhomogeneous spray of molten metal that it produces.

Acknowledgements

Partial economic support by PID-BID-CONICET 1116/91, PID-CONICET 3071800/91/92, PID-BID-CONICET 516/93, TENAES CONICET Argentina, and COLCIENCIAS, Colombia, is acknowledged. M.F.C. is fellow of Comisión Nacional de Ciencia y Técnica (CONICET), Argentina. E. G. L. is member of Comisión de Investigaciones Científicas de la provincia de Buenos Aires, Argentina. J. D. and R.C.M. are member of Carrera del Investigador Científico, CONICET, Argentina.

References

[1] W. M. Steen, Laser Material Processing, ed. Springer-Verlag , London (1991), 49.
[2] Metals Handbook, Vol. 9, ed. American Society for Metals (1985), 38.
[3] E. Murad and J.H. Johnston, in: Mössbauer Spectroscopy Applied to Inorganic Chemistry, Vol. 2, ed. Gary J. Long, Plenum Publishing Corporation (1987), 507.
[4] R. N. Greenwood and N. N. Gibb, Mössbauer Spectroscopy, Chapman and Hall, London (1971), 241.
[5] M. Ron in Applications of Mossbauer Spectroscopy, vol. II, ed. R.L. Cohen, Academic Press, New York (1980), 329.
[6] A. Ivarson, J. Powell, J. Kamalu, and C. Magnusson, J. Mat. Proc. Tech, 40, (1994), 359.

V. Magnetism and Magnetic Materials

MECHANISM OF BARIUM HEXAFERRITE FORMATION SEEN BY MOSSBAUER SPECTROSCOPY

Silvia Jacobo [a], Stella Duhalde [b], Celia Saragovi [c] and Miguel Angel Blesa [d]

[a] Departamento de Química, Facultad de Ingeniería, UBA, [b] Departamento de Física, Facultad de Ingeniería, UBA, Paseo Colón 850, 1063 Buenos Aires, Argentina.
[c] Departamento de Física, Comisión Nacional de Energía Atómica (CNEA), [d] Departamento de Química de Reactores, CNEA. Av. del Libertador 8250, 1429 Buenos Aires and INQUIMAE, FCEN, UBA, Buenos Aires, Argentina.

Abstract

Previous studies in our laboratory demonstrated the factibility of barium hexaferrite synthesis by a wet method that implies the reaction of sodium ferrate (VI) and barium chloride in alkaline media. Upon thermal treatment, the precursor yields the magnetic mixed oxide at a temperature as low as 670^{0}C.
The present communication reports the results of a Mössbauer study of the evolution of the precursor upon heating up to 850^{0}C. Mössbauer spectra obtained at RT and 15K demonstrate the gradual transformation of superparamagnetic particles into well formed barium hexaferrite. The results are in agreement with x-ray diffraction data.

1. Introduction

Hexagonal hard ferrites represented by $BaFe_{12}0_{19}$ are widely used in permanent magnets and recording media because of their large coercive force and fine grain structure.

Adelsköld [1] showed that this compound is isomorphous with the mineral magnetoplumbite. The oxygen ions form a hexagonal closed-packed lattice with substitutional Ba ions. The ferric ions are found in the interstices of the oxygen lattice. The structure can be considered to consist of two structural subunits: an "S", or spinel block, and a hexagonal "R"block. The 12 Fe^{3+} ions per unit formula are distributed amongst five sublattices: 6 in the octahedral sites **k** (spin up), 1 in the octahedral **a** site (spin up), 2 in the octahedral f_{VI} sites (spin down), 2 in the tetrahedral f_{IV} sites (spin down), and 1 in the pseudo-hexahedral site **e** (spin up). Therefore, the ^{57}Fe Mössbauer spectrum of $BaFe_{12}O_{19}$ at room temperature consists of five magnetically split and overlapping subspectra [2-4].

Due to its considerable commercial and technical importance, various preparation methods have been developed. The most common is the solid state reaction; firing mixtures of iron oxide and $BaCO_3$ at high temperatures produces multi-domain particles of low coercivity. Other methods, as the chemical precipitation of precursors from liquid solutions, are very suitable in the sense that they produce an intimate mixture of homogeneous composition [5]. Small crystalline hexaferrite particles can easily be obtained by heat treatment at relatively low temperatures.

An alternative synthetic procedure that involves coprecipitation of a precursor from a strongly alkaline solution in the presence of an oxidant, has been described by us earlier [6].

In this paper the evolution of the precursor and of the products obtained by this new method during the thermal treatment is analysed. The samples were characterized by Mossbauer spectroscopy (MS) and x-ray diffraction (XRD) and complemented with differential thermal analysis (DTA) and scanning electron microscopy (SEM).

2. Experimental

A ferrate solution obtained by adding $Fe(NO_3)_3.9H_2O$, to a 3M solution of NaOH , was mixed with $BaCl_2.2H_2O$. After standing for 24 hr, the slurry was treated at 80°C, then filtered and thoroughly washed. The residual solid was left overnight at 110°C in an oven. This precursor was subsequently heat treated during six hours at several different temperatures (570, 590, 620, 670, 710, 760 and 800°C).

XRD analysis was carried out with a Rigaku θ, 2θ diffractometer with Cu-K_α radiation.

Transmission Mössbauer spectroscopy was performed at room-temperature (RT) and when necessary, also at 15K. The obtained spectra were fitted assuming crystalline sites and hyperfine field distributions (HFD). In this case, both the parameter corresponding to the maximum of the distribution (H_M) and the average value $\langle H_a \rangle$ of the hyperfine field are quoted. The isomer shifts values are referred to iron metal.

3. Results and Discussion.

Some characteristic Mössbauer spectra and x-ray diffractograms of samples subjected to the different heat treatments are shown in Fig.1 and Fig.2 respectively.

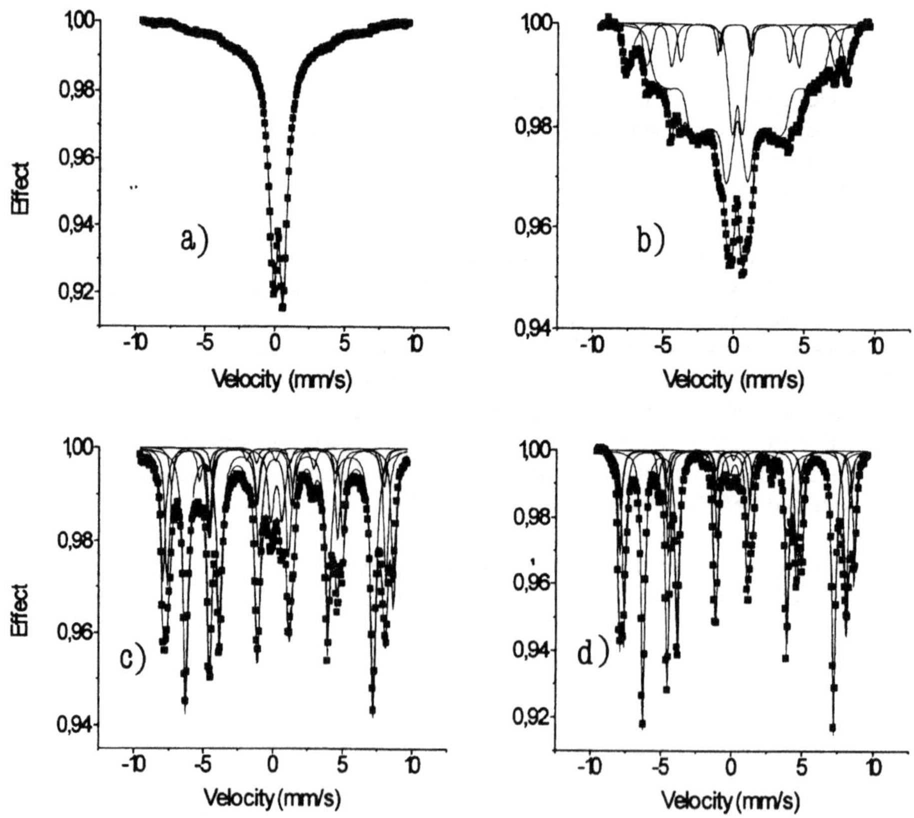

Figure 1: Room temperature Mössbauer spectra corresponding to different thermal treatment temperatures. a) precursor, 110°C, b) 570°C, c) 620°C, d) 800°C.

The obtained x-ray and Mössbauer patterns of the samples treated up to 620°C show a poor crystallization. The RT Mössbauer spectrum of the precursor (Fig.1a), indicates the presence of a distribution of small sized particles, which probably corresponds to ferrihydrites without discarding the presence of very small goethite. The obtained HFD values are: H_M= 3.18 T, $\langle H_a \rangle$= 13.8 T, with the same QS and IS values for all the elemental sextets (QS= 0.02mm/s and IS= 0.37mm/s). When lowering the temperature down to 15K the distribution of magnetic field is more clearly defined and its average value $\langle H_a \rangle$ increases, (H_M= 42.64 T, $\langle H_a \rangle$= 32.90 T). In coincidence, the XRD pattern shows a broad halo superimposed to the $Ba(OH)_2.2H_2O$ peaks (Fig.2a).

For the sample heated at 570°C (Fig.1b.), the RT MS spectrum shows the appearance of the ferrite more intense 12k and $4f_{iv,vi}$ sites signals superimposed to a broad HFD and a central doublet. The spectrum was fitted with a doublet (QS= 0.70mm/s, IS= 0.36mm/s) and three HFD (H_{M1}= 31.0 T, $\langle H_{a1} \rangle$= 22.7 T, H_{M2}= 41.4 T, $\langle H_{a2} \rangle$= 41.1 T, H_{M3}= 48.5 T, $\langle H_{a3} \rangle$= 47.9 T). In all these cases the HFD were fitted using only one QS and IS values for each distribution. The HFD_1 represents the 71% of the total area, while the narrow HFD_2 and HFD_3 are assigned to the above mentioned ferrite sites. It should be mentioned that the central doublet, which is not present in the precursor low temperature spectrum , remains without an appreciable change in its parameters upon lowering the temperature down to 15K. In coincidence, the XRD pattern shows the ferrite most intense reflections , some reflections from the precursor and extra peaks that could not be accurately indexed (Fig 2b).

In the case of the sample heated at 590°C, the RT MS spectrum shows the appearance of the five sextets characteristic of the barium hexaferrite besides the central doublet and a less intense HFD.

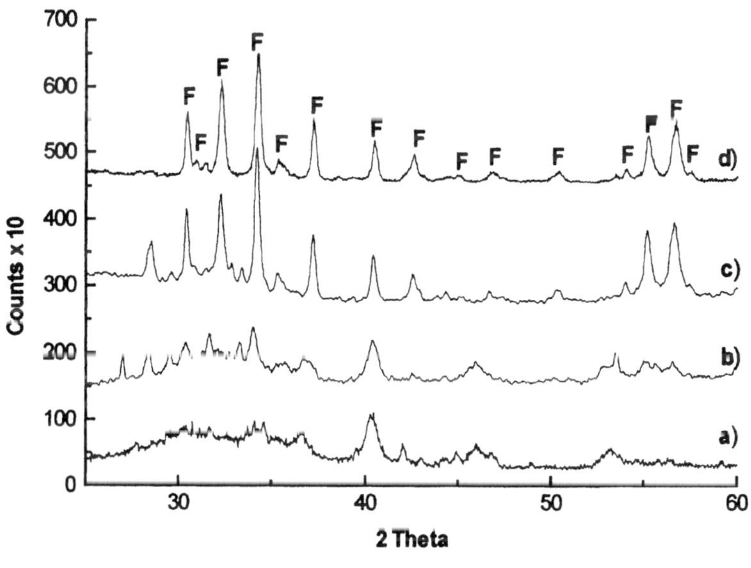

Figure 2: x-ray diffraction patterns corresponding to: a) precursor, b)sample treated at 570°C , c) sample treated at 620°C and d)710°C. Marks on the spectra : F, ferrite reflections.

Coincidentally, the XRD pattern shows the characteristic reflections of the ferrite phase and the same extra peaks found in the former sample, but with lower intensities.

The samples treated at the higher temperatures, above 620°C, show MS spectra and x-ray patterns similar to the ones displayed in Fig 1c, 1d and 2c, 2d respectively. In all the cases, the MS fitted parameters are fairly the same and indicate a well crystallized $BaFe_{12}O_{19}$ ferrite [3,4]. The obtained parameters at RT from these samples are displayed in Table 1. The errors in the parameters are 0.2T, 0.02mm/s and 0.03mm/s for H, QS and IS respectively.

Table 1: Obtained Mössbauer parameters at RT

Ferrite sites	H(T)	QS(mm/s)	IS(mm/s)
12k	41.9	0.42	0.35
2a	50.2	0.08	0.25
2e	40.1	2.12	0.26
4fiv	48.8	0.18	0.24
4fvi	51.6	0.15	0.37

The central doublet intensity decreases with the increasing treatment temperature, and remains when the MS spectrum is obtained at 15K in coincidence with the behaviour already mentioned. In agreement with the MS results, the XRD patterns of these crystallized samples indicate the presence of the barium hexaferrite as a majority phase and almost no traces of extra peaks. The presence of other phases in minor quantities is common to all preparative procedures; they are usually removed later by leaching in HCl

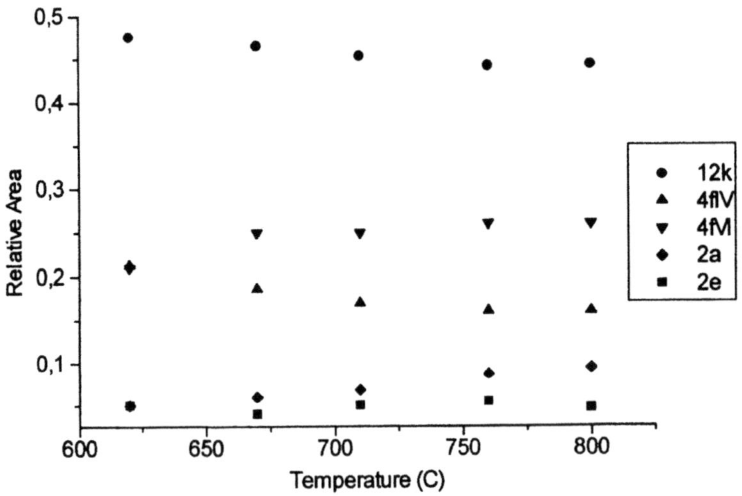

Figure 3: relative intensities of the five ferrite sites normalizated to the ferrite area and as a function of the thermal treatment temperature.

[7]. From the subspectra areas the relative proportion of the five ferrite sites were calculated under the assumption of the equality of the f factors. In Fig. 3, these values are

displayed as a function of the thermal treatment temperature only for samples treated above 620°C.

The above mentioned results suggest that the precursor (110°C) was a mixture of crystalline barium hydroxide and amorphous or small-sized iron hydroxide. At temperatures as low as 570°C the ferrite starts to crystallize and at 670°C is completely formed. Further thermal treatment changes the relative populations of the Fe sites in the crystalline barium hexaferrite. Examination of the powders by means of SEM showed that the precursor particle sizes were about 200 nm in diameter and that particles with sizes less than 500 nm in diameter were obtained by firing at about 800°C.

From the XRD and Mössbauer data it is shown that this new route of synthesis produces the $BaFe_{12}O_{19}$ ferrite at the lowest reported temperature. It should be noticed that the MS results indicate a hexaferrite formation temperature lower than that found by DTA, which showed an exothermic peak centred at 730°C (heating rate = 6K/min). This fact is not surprising since, as it is well known, the DTA data are not adequate to measure transition temperatures with precision, because they are very largely affected by factors such as sample mass, heating rate, instrumental features, and gaseous atmosphere [8].

Further work is in progress to determine the nature of the doublet as a superparamagnetic or an intermediate phase and the influence of the sites occupancy on the magnetic properties.

Acknowledgements

Partially supported by grant EX-036 (UBA). MAB is a member of CONICET.

References

[1] Adelsköld, V, Mineralogi och Geologi, (1938), 12A, 29,1.
[2] Hollatz, R, Sturhahn, W, Frohn, J, Hyp. Int. (1988), 41 479.
[3] Thompson, G and Evans, B, J,Magn.Magn Mat., (1991), 95 L142.
[4] Ylli, F and Gjonaj, Z., Italian Phys.Soc.Conference Proceedings, (1996), Vol.50, 259, ICAME'95, ed. I.Ortalli. Rimini, Italy
[5] Ross W., J. Am. Chem. Society (1980), 63, 11/12, 601.
[6] Jacobo S.E., Domingo-Pascual C., Rodriguez Clemente R., Blesa M.A., J. of Mat. Science, (1996), in press
[7] Ross W., Haak H., Voigt C. and Hempel K.A. , Colloque C_1, supplément au n°4, tome 38, (1977), C_1- 35.
[8] Galwey, Dollimore, Comprehensive Chemical Kinetics, Vol 22 , de. C.H.Bamford and C.F.H.Tipper. Elsevier Scientific Publishing Company. (1980).

MÖSSBAUER STUDY OF R_2Fe_{17} COMPOUNDS CONTAINING INTERSTITIAL FLUORINE

F.A.BATISTA, J.D.ARDISSON, A.I.C.PERSIANO and L.O.LADEIRA,*

Departamento de Física-ICEx-UFMG, CP 702, 30161-970 Belo Horizonte-MG, BRAZIL

*Centro de Desenvolvimento da Tecnologia Nuclear-CNEN, Belo Horizonte-MG, BRAZIL

ABSTRACT

In this work we present some unprecedented results obtained by using an original process, developed by our research group, to add halogen atoms interstitially to R_2Fe_{17} compounds in order to produce a new class of material for hard magnetic applications. The preliminary data, for intermediate times of reaction, analyzed by Mössbauer spectroscopy, X-ray diffraction and vibrating sample magnetometry show that the incorporation of about 1 halogen atom per formula unit (f.u.) expands the cell-volume of the 2:17 compounds by about 1-2 % producing an increase in the Curie temperature of the material.

1. Introduction

The increasing dependence between the industrial activities and the modern technology requires a continuous research on new materials. This is particularly important in the production of permanent magnets. Figure 1 shows the evolution of commercial hard magnetic materials in the last 100 years considering the energy product - $(BH)_{max}$ - as the parameter of improvement. We note a huge increase of this parameter from the beginning of the seventies, powered up by the discovery of the *rare earth - 3d transition metals*-based compounds (R-M systems). The high coercivity of this class of permanent magnets is due to the occurrence of uniaxial magnetocrystalline anisotropy associated with the presence of the R element.

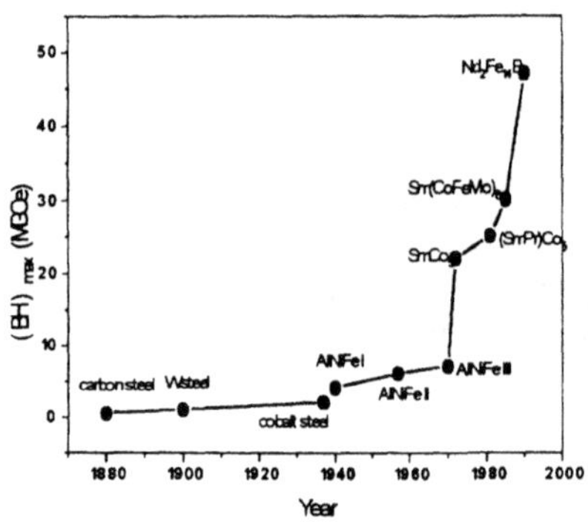

Fig.1 - Evolution of the $(BH)_{max}$ parameter of some leading commercial permanent magnets in the last 100 years.

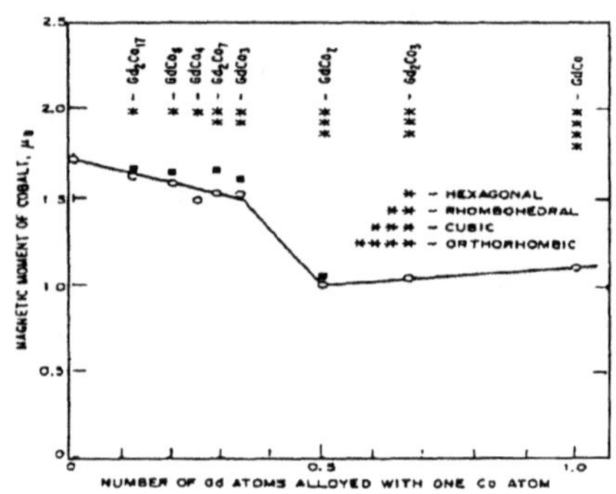

Fig.2 - Dependence of the magnetic moment of Co in Gd-Co compounds on the number of cobalt atoms in the alloy.

Figure 2 shows the increase of the magnetic moment of cobalt with its concentration in the system Gd-Co. This Figure demonstrates that the 3d-3d interaction plays the major role on the definition of the magnetic moment of the alloy. The R_2M_{17} stoichiometry has the largest concentration of 3d elements of all R-M compounds. Considering that cobalt is expensive and a strategic material, the use of M=Fe makes the R_2Fe_{17} compounds a promising alternative. These compounds present high magnetic saturation but low Curie temperatures (T_c close to room temperature), and planar anisotropy, which are not favorable to produce high quality permanent magnets. At the end of the eighties three different interstitial additions to R_2Fe_{17} compounds, namely H, C and N [3-7], have succeeded to improve the magnetic properties of the material, with increases in T_c of about 200 - 400°C. Moreover the addition of N to the alloy containing Sm induces uniaxial magnetocrystalline anisotropy in the material.

In this work we follow Qi-Nian et al [8] which consider Cl a promising addition to 2:17 alloys and Fähle and Beuerle [9] that propose that the addition of 3 fluorine atoms to Y_2Fe_{17} could be the best dopant to this system in order to produce large magnetic moments, high T_c and favorable magnetocrystalline anisotropy. We present and discuss some unprecedented experimental results due to the incorporation of a new addition of fluorine to R_2Fe_{17} alloys (R = Pr, Nd, Sm) by using a special technique developed by ourselves.

2. Experimental

The R_2Fe_{17} alloys (R=Pr, Nd or Sm) were prepared by arc melting the primary components with purity better than 99.9% followed by a homogenization treatment at 1100°C for about 10 days. The material was ground and sieved to particle-sizes between 20 and 60 μm. The gas addition was performed through two different ways: method **(a)** - by an indirect reaction at 320°C by exposing the material to the gas products of the thermal decomposition of polytetrafluoroethylene for varying times; method **(b)** - by the direct reaction [10] with pure fluorine gas in a Sieverts apparatus. The fluorine content in the material was determined by the selective-ion electrode analysis technique.The magnetic and structural features of the alloys, at room temperature, were investigated by ^{57}Fe Mössbauer spectroscopy (constant acceleration; ^{57}Co in Rh source) and X-ray diffraction analysis, respectively.

3. Results and discussion.

We developed two methods for adding fluorine atoms to R_2Fe_{17} compounds [10]. The efficacy of this methodology and certification that exclusively fluorine atoms have been incorporated to the material was confirmed by comparing the results obtained from the indirect reaction technique (decomposition of PTFE), with those produced by using a controlled reaction employing the highly oxidant and hazardous halogen elements with 99.9% purity. Figure 3 shows the Mössbauer spectra of samples produced by the two methods used to add fluorine atoms to R_2Fe_{17}. There was a suspicion that method "a" could provoke the incorporation of the 2 primary elements of PTFE, C and F, to the alloy. The Mössbauer spectra of Fig. 3 show essentially the same features in the two cases which confirms that both methods add only fluorine atoms to the material.

The X -Ray diffraction analysis indicates that the addition of the halogen atoms preserves the crystal structure of the parent alloy R_2Fe_{17}. Figure 4 shows in detail the shift to low angles of the (300) line in Pr_2Fe_{17}, before and after the additions of F or Cl. The reduction of the intensity of line (300) upon the gas addition was also observed in the other lines of the X-ray diffractiograms, simultaneously to the emergence of a broad peak corresponding to the (101) line of α-Fe.

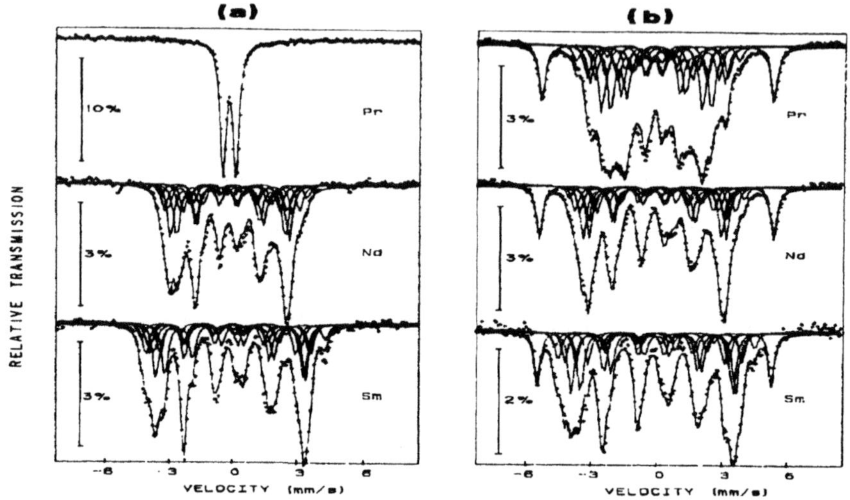

Figure 5 - ^{57}Fe Mössbauer spectra of R_2Fe_{17} (R=Pr, Nd or Sm), measured at RT, (a) before and (b) after fluorination at 320 C for 90 minutes. The continuous lines are the best fits to the data points; the dashed lines represent the component subspectra; the outermost absorption peaks shown in (b) correspond to lines 1 and 6 of alpha-iron.

parameter phase	$\langle B_{hf}\rangle$ Tesla	$\Delta\langle IS\rangle$ mm/sec	a Å	c Å	$\langle \Delta V/V\rangle$ %	T_c ^oC
Pr_2Fe_{17}	-		8 58	12.46		10
Pr_2Fe_{17} + fluorine	15 4	+0.08	8.61	12,47	0.9	42
Nd_2Fe_{17}	17.1		8.54	12.46		57
Nd_2Fe_{17} + fluorine	20.4	+0.04	8.60	12.46	0.6	65
Sm_2Fe_{17}	21.6		8.55	12 44		116
Sm_2Fe_{17} + fluorine	24.0	+0.02	8.60	12.47	1.3	155

Table 1 - Weighted average hyperfine fields $\langle B_{hf}\rangle$ of the 2:17 phases, changes in the weighted average isomer shifts $\Delta\langle IS\rangle$, lattice parameters a and c and average relative volume expansions $\langle \Delta V/V\rangle$ of the phases due to fluorination - measured at room temperature - and Curie temperatures T_C. The typical errors in the six parameters are 0.2 Tesla, 0.01 mm/s, 0.01 Å, 0.01 Å, 0.3% and 2°C respectively.

Recent results obtained in samples submitted to much longer fluorination treatments (~20 days at the same conditions as referred earlier) show that the Pr based alloy reaches another state of stability. The material keeps the same crystallographic structure of the original 2:17 phase and exhibits cell volume expansions of about 6%, similar to that observed after the addition of other gases such as N to the same base alloy [5]. Fig. 6 shows the MS of Pr_2Fe_{17} fluorinated measured at room temperature. It's interesting to notice that, besides the large increase in the hyperfine field, this spectrum is much narrow than that of the material treated for 90 minutes and a 4 subspectra fit was thoroughly accepted. The reduction of the spectral complexity could be interpreted as a consequence of a possible spin reorientation from basal to axial. The magnetization curves for this compound confirm an improvement in both the coercivity and remanence producing $(BH)_{max}$ values of about one order of magnitude larger in comparison with the binary alloy.

This occurrence, together with the observation of a refinement of the powder particles after fluorination confirms that, similarly to other gas additions to 2:17 compounds, the incorporation of the halogen atoms induces the micro-fragmentation of this material; futhermore the rise of a broad line of α-Fe indicates that the process involves some disproporcionation of the material within a sub-microscope scale, also observed by MS but not by SEM.

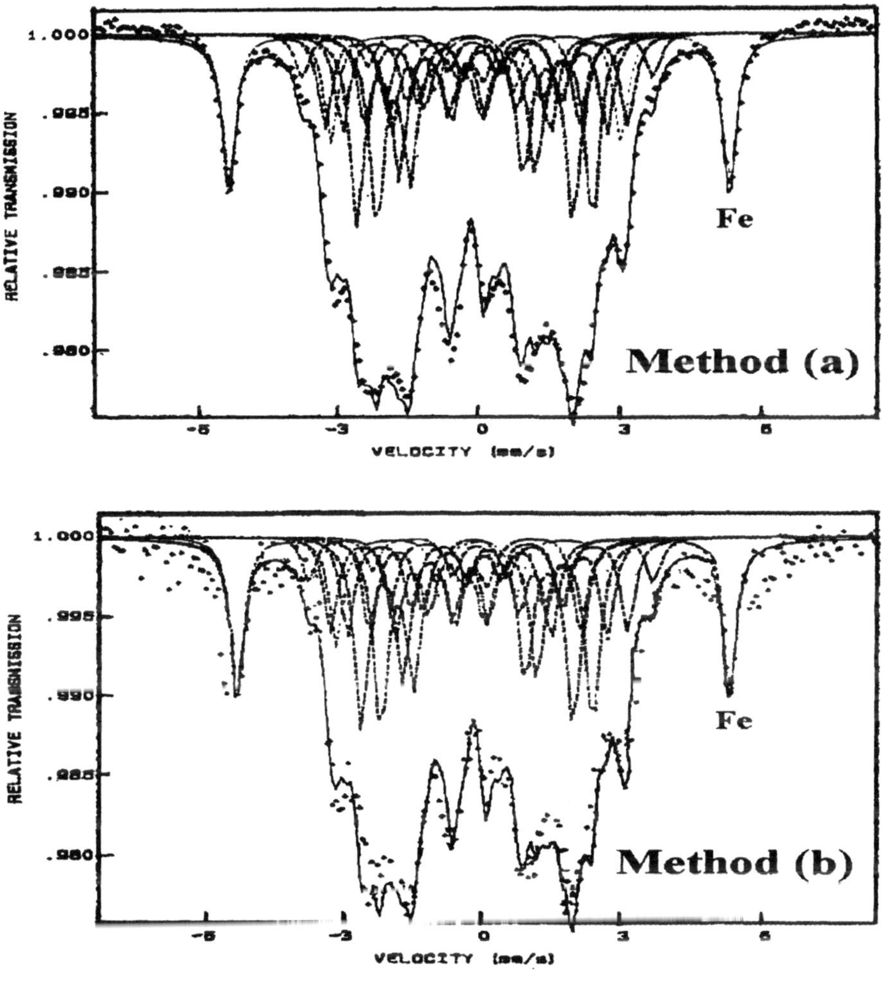

Figure 3 - Mössbauer spectra of Pr_2Fe_{17} alloys submitted to the two different fluorination processes: **(a)** decomposition of polytetrafluorethilene, and **(b)** by direct reaction with pure fluorine gas. The continuous lines are the best fits to the data points; the dashed lines represent the component subspectra.

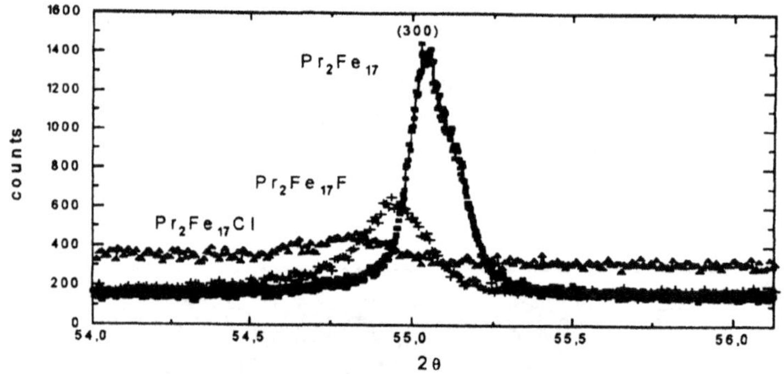

Figure 4 - X-ray (300) lines from the diffractograms of the Pr_2Fe_{17} alloy before (squares) and after the additions of F (crosses) or Cl (triangles).

Figure 5 shows the effects on the Mössbauer spectra obtained by using the same process to add fluorine to Pr, Nd and Sm based alloys. The room temperature Mössbauer spectra of the Pr_2Fe_{17} treated at 320 C for different intermediate times, up to 90 minutes, change from a doublet, characteristic of the paramagnetic binary compound, to a well defined set of 7 sextets (relative areas proportional to 2:1:2:2:4:2:4) + alpha-iron. This indicates that, despite the occurrence of a small disproportionation of the 2:17 alloy, the incorporation of about 1 F atom per f.u. produces the referred lattice expansions and a consequent increase in the Curie temperature (Tc) of the material, without any change in the planar easy-axis. In all cases the process improves the magnetic properties of the material measured at room temperature.

Table 1 displays some information concerning hyperfine parameters, structural data and Curie temperatures before and after fluorination of the same alloys as referred above. Note that in all cases the lattice parameters, T_c and the hyperfine fields increase as a consequence of the gas addition. Like the other gas additions to the 2:17 alloys, these changes are most probably provoked by the presence of interstitial fluorine atoms within the structure. The moderate changes in all spectra suggest that the alloys have absorbed a relatively small number of fluorine atoms per f.u. when comparing this with the addition of other gases as reported in the literature [3-5] (about 3 atoms per f.u.). Indeed, measurements of the sample's mass gain due to the reaction with F and the selective-ion electrode analysis suggest the absorption of about 1 F atom per f.u. The changes in the isomer shifts $\Delta\langle IS\rangle$ upon fluorination, toward positive values can be attributed to both the lattice expansion as well as to charge transfers. The value of the parameter $\Delta\langle IS\rangle/\Delta\langle \ln V\rangle$ due to fluorination of the three R-based alloys studied here gives an average of (5 ± 3) mm/s - is significantly larger than the typical values of alpha-iron due exclusively to volume expansion ($\Delta IS/\Delta \ln V$ = 1.31 mm/s) [11]) as well as due to nitrogenation of the same base alloys (1.6-2) mm/s [12,13]) which demonstrates that a significant charge transfer from the iron atoms to fluorine takes place, as it should be expected for a halogen element.

The presence of this oxidant element within the R_2Fe_{17} structure represents an important stabilization factor to prevent the deterioration of the material when exposed to the ordinary atmosphere. In fact, we observed that, contrarily to the pure 2:17 alloys, the fluorinated material keep their chemical integrity even after one year laid in air. Moreover the highly polarized atomic bond between the halogen and the neighboring atoms could induce, in appropriate conditions of fluorine occupancy and/or the R element used, a spin reorientation leading an uniaxial anisotropy of the compound

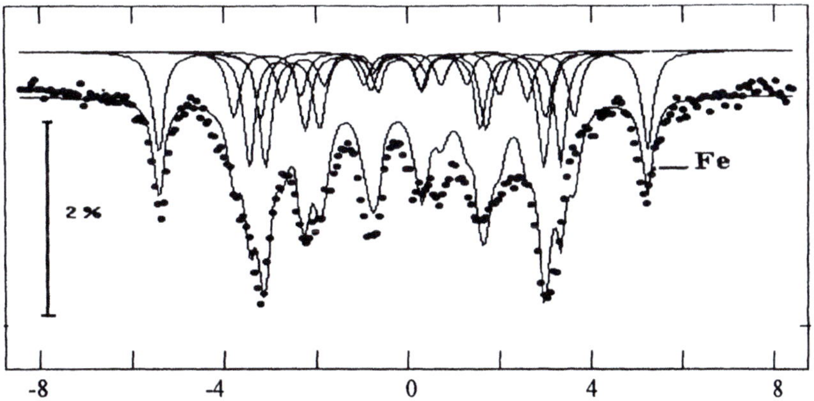

Figure 6 - Mössbauer spectrum of Pr_2Fe_{17} fluorinated for ~20 days at 320^{0}C. The dashed lines represent the 4 subspectra fit.

Conclusions

We have developed two methods to introduce atoms of fluorine insterstitially in R_2Fe_{17} alloys (R=Pr, Nd or Sm) keeping their chemical integrity. Some improvements in their magnetic properties like an increase in T_c and $(BH)_{max}$ as well as the stabilization of the alloy against oxidation were achieved. Further studies are required to stablish the suitability of fluorinated 2:17 compounds as permanent magnets.

Acknowledgments

The authors would like to thank the CDTN for the technical support and the Brazilian Agencies CAPES, CNPq, FINEP and FAPEMIG for the financial support

References

[1] B.D.Cullity, "Introduction to Magnetic Materials" Addusib-Wesley Publishing Company, Reading, Massachusetts, London, (1972).
[2] E.A.Nesbit and J.H.Wernik, "Rare Earth Permanent Magnets", Acad. Press,N.Y.,USA, (1973)
[3] Xiang-Zhong Wang, K. Donnelly, J.M.D. Coey, B. Chevalier, J. Etourneau and T. Berlureau, *J. Mater. Sci.* 23 (1988) 329.
[4] Xia-Ping Zhong, R.J. Radwanski , F.R. de Boer, T.H. Jacobs and K.H.J. Buschow, *J. Magn. Magn. Mater.* 86 (1990) 333
[5] J.M.D. Coey and Hong Sun *J. Magn. Magn. Mater* 87 (1990) L251.
[6] A.I.C. Persiano, J.D. Ardisson, F.A. Batista, C.C. Colucci and S. Gama, *J. Magn. Magn. Mater* 136 (1994) 149.
[7] J.D.Ardisson, PhD thesis, Depart. Physic, Federal University of Minas Gerais - Brazil (1995)
[8] Qi-Nian Qi, Hong Sun and J.M.D.Coey, *Hyperfine Interactions* 68 (1991), 27.
[9] M. Fähnle and T. Beuerle, *Phys. Stat. Sol. (b)*, 177- 2 (1993) K95
[10] J.D .Ardisson, A.I.C.Persiano, L.O.Ladeira and F.A.Batista, process patent requisition (1996)
[11] D.L.Williamson, Mössbauer Isomer Shifts, ed G.K.Shenoy and F.E. Wagner (North Holland, Amsterdam, 1978) p.317
[12] Bo-Ping Hu, H Shuo Li, Hong Sun and J.M.D.Coey, J.Phys.: Condens. Matter 3 (1991) 3983.
13] G.J.Long, O.A.Prigle, F.Gradjean, K.H.J.Buschow, J.Appl. Phys. 72 (10) (1992) 4845.

MÖSSBAUER STUDIES OF NICKEL AND ZINC FERRITES

J. H. Dias Filho[1], H.-D. Pfannes[1], S. L. Pereira[1], J. L. López[1], P. C. Morais[2], A. L. Tronconi[2] and F. A. Tourinho[3], A. A. Mendes Filho[4] and L.C.B. de Miranda Pinto[4]

[1] *Departamento de Física, Universidade Federal de Minas Gerais, C.P. 702, 30161-970, Belo Horizonte, Brazil*
[2] *Departamento de Física, Universidade de Brasília, 70910-900, Brasília, Brazil*
[3] *Departamento de Química, Universidade de Brasília, 70910-900, Brasília, Brazil*
[4] *Setor de Materiais Óticos e Eletrônicos*
Fundação Centro Tecnológico de Minas Gerais - CETEC
Av. José Cândido da Silveira, 2000, 31170-000, Belo Horizonte, MG, Brazil

Ferromagnetic Ni ferrites have been synthesized by the coprecipitation and condensation method from the chemical reaction among aqueous solutions of metal mixtures, both in alkaline medium. Zn ferrites were synthesized by a hydrothermal process. Powder samples with an average particle diameter of about 10nm were studied by Mössbauer spectroscopy in the range of 4.2-300 K. The Zn-ferrite samples do not exhibit magnetic spectra even at low temperatures. The Ni-ferrite samples are monodisperse to a good approximation, presenting completely collapsed spectra at room temperature and static spectra at low temperatures. Values of the saturation magnetization of the samples measured with a vibrating sample magnetometer will be presented as well as simulations of the relaxed spectra for Ni samples, obtained by using a stochastic model, longitudinal relaxation and collective excitations.

1. Introduction

Ferrites are complex magnetic oxides containing the ferric oxide (Fe_2O_3) as their basic component. In a natural manner the crystallographies of these materials fall into three types: (I) the cubic ferrites of the spinel type, (II) the cubic ferrites of the garnet type and (III) the hexagonal ferrites [1]. The subject of this present work, Nickel and Zinc ferrites, are of the spinel type and derive their formula MFe_2O_4 (where M represents a divalent ion of metal) from that of the spinel $MgAl_2O_4$.

These type of ferrites are also called ferrospinels because they have the same crystal structure as the spinel. The structure is based on a face-centered cubic lattice of the oxygen ions. There are eight formula units per unit cell. Each cell contains 32 O^{-2} anions, 16 Fe^{3+} and 8 M^{2+} cations. The lattice constant is of the order of 8.5 Å. There are 64 tetrahedral, or *A*- sites and 32 octahedral, or *B*- sites per unit cell, surrounded by four and six oxygen ions at equal distance, respectively. In $ZnFe_2O_4$ the 16 Fe^{III} ions occupy one-half of the *B*- sites, and the 8 M^{II} ions occupy one-eighth of the *A*- sites. This is the normal spinel structure. However, $NiFe_2O_4$ crystallizes in an inverse spinel structure, in which the M^{II} ions occupy *B*- sites and the Fe^{III} ions are evenly divided between *A*- and *B*- sites. The spin coupling scheme in the normal spinel structure indicates an antiparallel spin alignment of the Fe^{III} ions a zero magnetic moment of the Mn^{II} ions and, according to this, Zn-ferrite should be an antiferromagnet.

Solid solutions obtained when up to 50% Zn ferrite is added to others of the inverse structure like Ni- [2], Mn-, Fe and Co-ferrites [1] show an enhancement of the saturation magnetization.

Very small magnetic particles are of great interest because their properties are significantly different from those of the respective bulk materials. Particles with dimensions below 100 Å often exhibit superparamagnetic relaxation [3], i. e., fluctuations of the magnetization direction between the easy directions. They are commonly found in natural samples of soils, rocks, corrosion products and live organisms. Particles powder ferrites of natural and synthetic materials, have applications in, for example, magnetic devices.

2. Theory

A superparamagnetic crystallite normally has a size of a few nanometers and a resulting spin typically of the order of 10^4 (order of the number of atoms). Transitions between any two states should be considered for calculation of the Mössbauer spectrum and the inversion of an extremely large matrix of rank $2\cdot 10^4$ would be needed. However, using the *thermally averaged states* we can consider only transitions between two states (states separated by the energy barrier) [3]. For such a system, the theoretical Mössbauer spectrum can be calculated assuming an internal magnetic field [4], longitudinal relaxation and a stochastic model [5], using the superoperator formalism [6]. If an external magnetic field is applied on the system (powder sample), we must perform an integration over all possible orientations between the gamma direction and the quantization direction.

In an uniaxial crystallite with volume V and anisotropy constant K (magnetic anisotropy energy per volume), the energy barrier separating the two energy minima (with no external magnetic field) may be written in the form:

$$E(\theta) = KV \sin^2 \theta, \qquad (1)$$

where θ is the angle between the magnetization vector and the easy direction. At a temperature in which the thermal energy k_BT is of the order of the energy barrier (blocking temperature), transitions between the two minima (θ=0 and θ=π) can occur. The superparamagnetic relaxation is described in terms of these transitions. Collective excitations (small oscillations of the magnetization vector around the easy direction) cause a reduction of the hyperfine field. Near the blocking temperature, the reduction is of the order of 15%. The effective magnetic field can be calculated by [3]

$$H_{eff} = H(V = \infty) \cdot \langle \cos(\theta) \rangle_T ,$$

where

$$\langle \cos(\theta) \rangle_T = \frac{\int_0^{\pi/2} \exp\{-E(\theta)/k_BT\} \cos\theta \sin\theta d\theta}{\int_0^{\pi/2} \exp\{-E(\theta)/k_BT\} \sin\theta d\theta}. \qquad (2)$$

The superparamagnetic relaxation time τ depends on KV/k_BT and for $KV/k_BT>1$, [3,7]

$$\tau = \frac{M_s \pi^{1/2}}{2K\gamma_0} \left(\frac{KV}{k_BT}\right)^{-1/2} \exp\left(\frac{KV}{k_BT}\right), \qquad (3)$$

where M_s is the saturation magnetization and γ_0 is the gyromagnetic ratio.

A broad particle size distribution combined with collective excitations effects cause asymmetrical broadening of the Mössbauer lines.

3. Experimental

The ferrites $NiFe_2O_4$ and $ZnFe_2O_4$ were synthesized by coprecipitation and hydrothermal methods, respectively. Mössbauer spectra of the samples as powder were recorded using a constant acceleration spectrometer in the 4.2-300 K temperature range with a ^{57}Co in Rh source. The average particle size was determined by means of X-ray diffraction (Rigaku diffractometer, Cu-K_α radiation and graphite monochromator). Scans were realized in the range of 4°-160° (2θ) at a speed of 4°/min. Hysteresis loops and the saturation magnetization were determined in the 4.2-200 K range with a vibrating sample magnetometer. The magnetization values are presented in table I for a sample of 20 mg.Figure 1 shows the hysteresis loop at 4.2 K for both ferrites.

Table I. Saturation magnetization of the samples

	Saturation Magnetization (emu)	
T(K)	$ZnFe_2O$	$NiFe_2O$
4.2	0.03(7)	0.02(3)
20	0.03(6)	0.02(3)
50	0.03(5)	0.02(3)
100	0.03(4)	0.02(3)
125	0 03(2)	0.02(3)
200	0 02(8)	0.02(3)

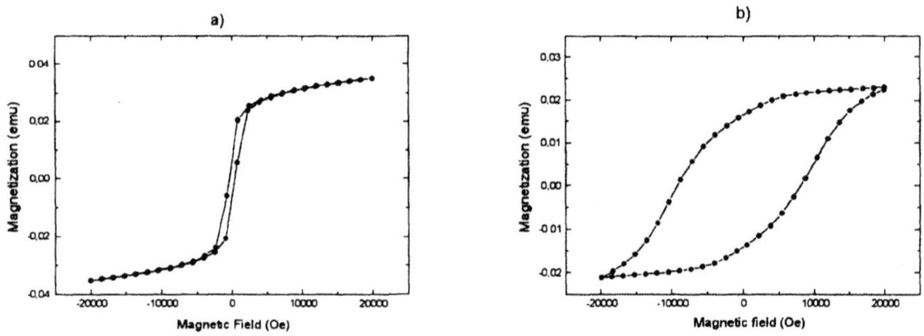

Fig. 1 a) hysteresis loop of the Zn ferrite
b) hysteresis loop of the Ni ferrite

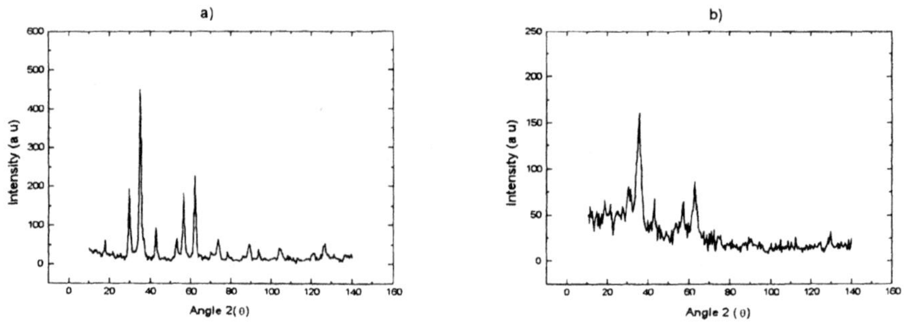

Fig. 2. a) X-ray diffraction pattern of the $ZnFe_2O_4$ sample
b) X-ray diffraction pattern of the $NiFe_2O_4$ sample

4. Results and Conclusions

Mössbauer spectra of Ni-ferrite show a magnetic splitting at 4.2 K and relaxed spectra, with broad lines at 80 K. Using the values for the anisotropy constant from the literature ($K=6.7 \cdot 10^3$ J/m^3) [1], the values for the saturation magnetization listed in table I, and a gaussian particle size theoretical distribution we have simulated Mössbauer spectra for the Ni-ferrites for several temperatures. Fig.3 shows the experimental and simulated spectra. The blocking temperature for this sample was estimated to 97 K.

The average particle size was determined by means of X-ray diffraction. We found 114 Å for $NiFe_2O_4$ and 112 Å for $ZnFe_2O_4$.The X-ray diffraction patterns of the samples are shown in fig.2a,2b.

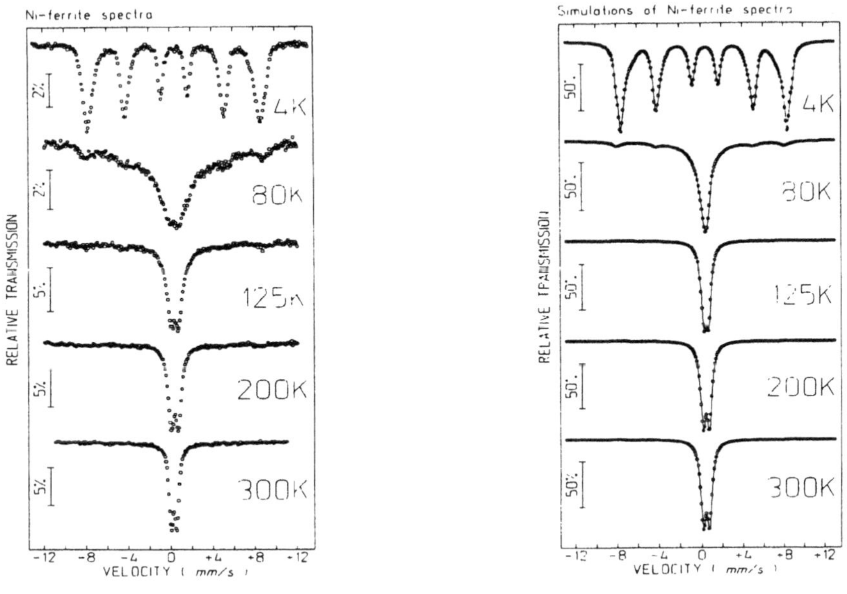

Fig. 3 Experimental and simulated Mössbauer spectra of the $NiFe_2O_4$ sample

Mössbauer measurements of the Zn-ferrite at 4.2K show a magnetic splitting with a large hyperfine field. Based on the magnetometer measurements, fig. 2.b, we can say that this sample is ferrimagnetic at 4.2 K and exhibits relaxed with broad lines at 10 K. We have also considered two subspectra, a gaussian particle size distribution and values of the anisotropy constant from the literature ($6 \cdot 10^3$ J/m^3) [8] to simulate Mössbauer spectra for this ferrite. Fig. 4 shows the experimental and simulated Mössbauer spectra. The blocking temperature for this sample was estimated to 35 K.

It's necessary to have a very good experimental particle size distribution,, with a great number of points, in order to get an also good theoretical distribution and good simulations.

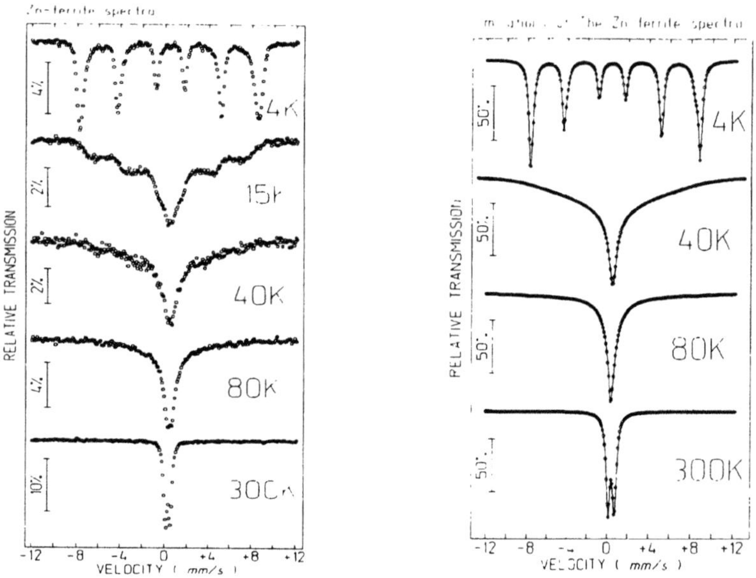

Fig. 4 Experimental and simulated Mössbauer spectra of the $ZnFe_2O_4$ sample

Acknowledgment

The support of the Brazilian research agencies CNPq and FAPEMIG is gratefully acknowledged.

References

[1] Chih-Wen chen, in: *Magnetism and Metallurgy of soft Magnetic Materials*, Dover Publications Inc, (New York, 1986) p. 208.
[2] J. M. Daniels and A. Rosencwaig, Canadian Journal of Physics 48, n° 4, (1970) 381.
[3] S. Mϕrup, H. Topsϕe and J. Lipka, Journal de Physique, Colloque C6, supplément au n° 12, Tome 37, (1976) C6-287.
[4] Wickman, H., Mössbauer Effect Methodology, vol. 2 (1966) 822.
[5] M. J. Clauser and M. Blume, Physical Review B, 3, n° 3, (1971) 583.
[6] A. Putzka , J. Phys. C 21 (1988) 3517.
[7] A. Aharoni, Phys. Rev. 135, (1964) A447.
[8] R. S. Tebble, D. J. Craik: *Magnetic Materials*, Wiley-Interscience, (London, 1969) p. 254.

MÖSSBAUER ANALYSIS OF THE HDDR $Nd_{14}Fe_{79}B_7$ MAGNETIC MATERIALS

M. B. Fernández van Raap, C. E. Rodríguez Torres, F. H. Sánchez and L. Mendoza Zélis
Departamento de Física, Universidad Nacional de La Plata, c.c. 67 (1900) La Plata, Argentina

F. Actis and G. Cohen
Centro de Investigación de Materiales y Metrología, Instituto Nacional de Tecnología Industrial, c.c. central 884, 500 Cordoba, Argentina

H. Nagel
Nagel&Krönert Engenieering GmbH, Südfeld 2, 59174 Kamen, Germany.

The Nd-Fe-B magnet powders with high coercivity (≈ 1.5 T) were produced by the hydrogenation, disproportionation, desorption, recombination (HDDR) process in a laboratory plant. Mössbauer Effect (ME) characterization of the process has been performed on several fully treated samples prepared by changing the control parameters, i.e. heating rates, disproportionation and recombination times and temperatures, for which significant changes are found in the microstructure and magnetic properties. For these materials no change is observed in the ME spectra comparing with the as cast material indicating that the main feature of the HDDR is at the microstructural scale. The hydrogenated and disproportionated material were also characterized, for the latter, besides the α-Fe, Fe_2B and $Nd_{1+\varepsilon}Fe_4B_4$ phases, α-Fe with Nd dissolved (1 %) is detected.

INTRODUCTION

The hydrogenation, disproportionation, desorption, recombination (HDDR) process[1], [2]. is known as a method of producing highly coercive Nd-Fe-B powder through hydrogen induced structural changes. The high coercivity of the magnet powder is considered to be due to the small crystallite size of recombined $Nd_2Fe_{14}B$.

The purpose of the present work is the characterization at the nanoscopic scale, as seen by ME, of the high coercive Nd-Fe-B powders produced by the HDDR processing route in a laboratory plant, where scaling up is being performed. This characterization should reveal whether magnetic properties of HDDR treated materials are also driven by the presence of point defects, highly disperse secondary phases, etc.

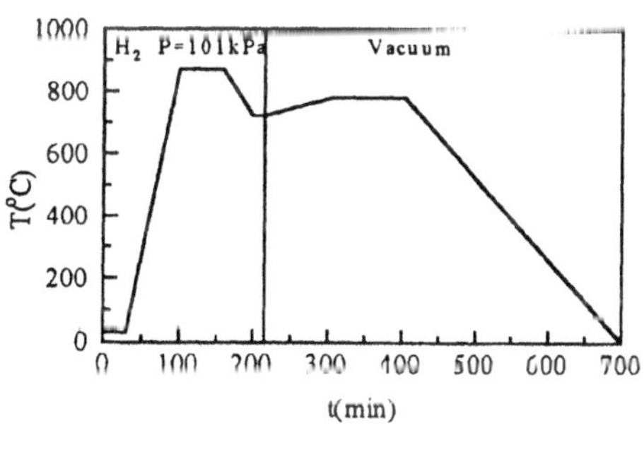

Figure 1: Diagram of a typical experimental HDDR cycle.

EXPERIMENTAL PROCEDURE

Commercially available ternary $Nd_{14}Fe_{79}B_7$ alloy was subjected to several HDDR processes in a rotatory kiln. Sample mass was typically 0.5 kg. A complete typical experimental cycle is shown in fig. 1. For the hydrogenation and disproportionation steps the hydrogen pressure was held at 101 kPa. Hydrogenation was carried out at 30°C during 40 minutes. The recombination times were stopped when the H_2 pressure was at 10 Pa in the samples fully treated

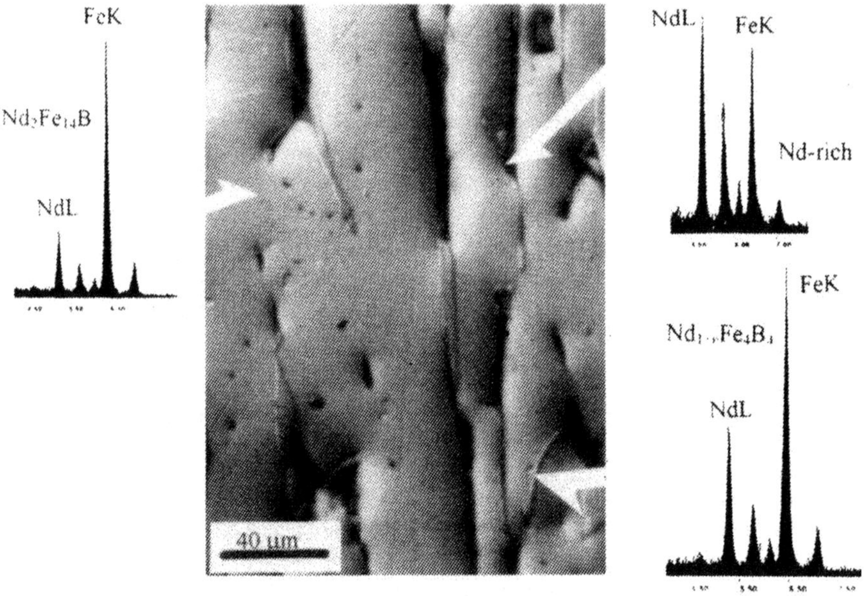

Figure 2: Optical Micrograph and EDAX of the as cast $Nd_{14}Fe_{79}B_7$ alloy.

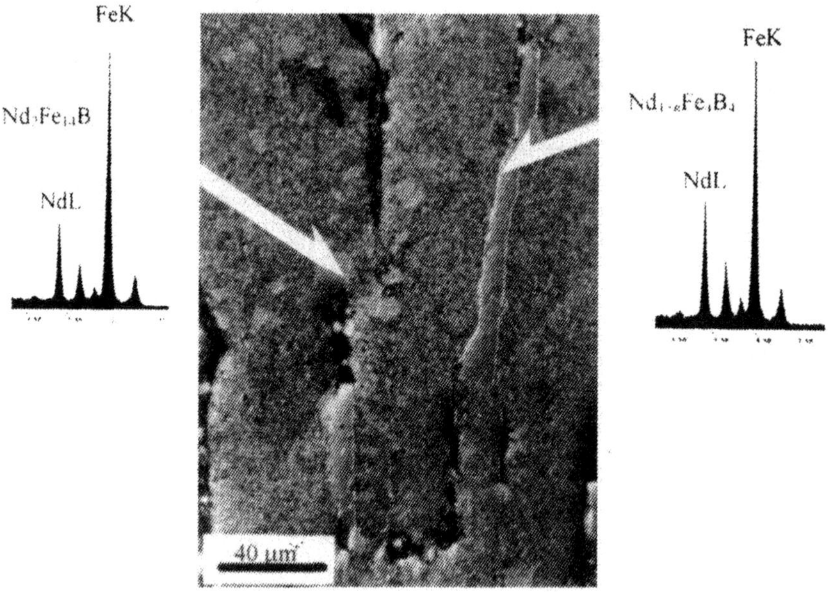

Figure 3: Optical micrograph and EDAX of the HDDR fully treated powder

For details see reference[4]. Ingot and powder microstructures were characterized with optical microscopy and EDAX microanalysis.
For the characterization of the intermediate steps same thermal batches were interrupted after hydrogenation and after disproportionation at 870°C during 10, 30, 45 and 120 minutes.
XRD experiments were carried out with CuKα radiation in a Philips PW1710 diffractometer. The ^{57}Fe Mössbauer measurements were performed under transmission geometry with a standard constant acceleration spectrometer having a ^{57}Co*Rh* radioactive source. Quoted isomer shifts (δ_{IS}) are referred to α-Fe at room temperature (RT). Measurements were performed at RT.
Demagnetization curves were obtained from samples separated in size fractions, after press in a die to 100-500 MPa and pulse magnetize on a 6 T magnetic field.

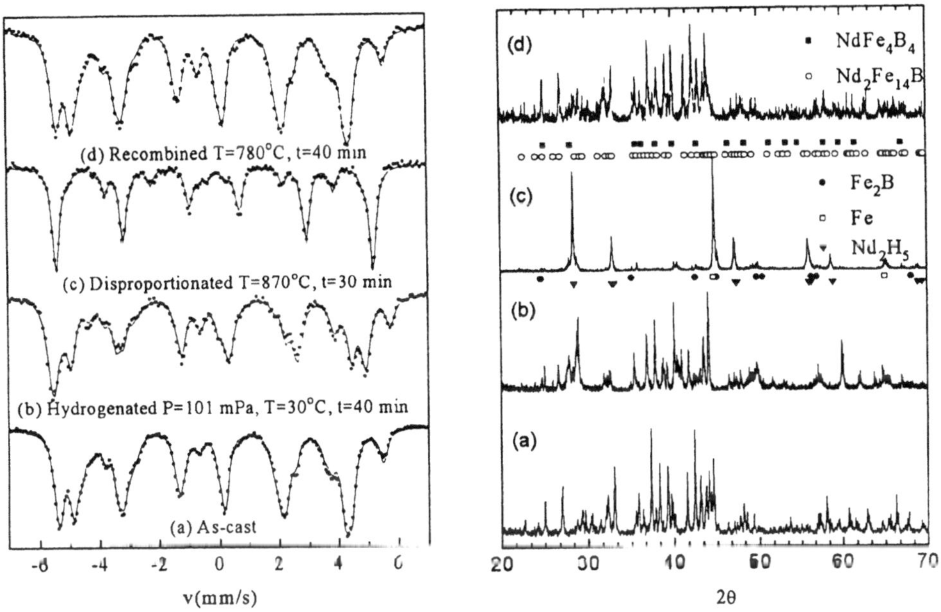

Figure 4: ME and XRD patterns of as-cast, hydrogenated, disproportionated and recombined $Nd_{14}Fe_{79}B_7$ alloy.

RESULTS AND DISCUSSION

The characterization of the as cast alloy as seen by optical microscopy, EDAX, ME and XRD, microanalysis is shown in fig.2 and fig.4.a. The alloy consists of the three main phases $Nd_2Fe_{14}B$ matrix (σ) phase, $Nd_{1+\varepsilon}Fe_4B_4$ (η) phase and eutectic Nd-rich at the grain boundaries. The $Nd_2Fe_{14}B$ crystal structure is tetragonal with six nonequivalent iron sites. The assignment of the six spectral components is still an open question. At least seven alternative ways for fitting the Mössbauer spectrum have been published[5]. The ME spectrum (fig. 4.a) was fitted with the superposition of six magnetic sextets whose relative intensities were constrained to be as it is required from the occupation of crystallographic nonequivalent iron sites in the $Nd_2Fe_{14}B$ structure (See table I). F. Grandjean et al.[5] assignment was chosen. A central paramagnetic doublet was also needed in accordance with the presence of η phase The Nd-rich phase could have iron dissolved, if this were the case, a singlet with a positive δ_{IS} would be expected [6]. No singlet could be fitted, so the amount of iron in the Nd-rich phase is too small to be detected and/or the singlet can not be resolved in the too complicated spectrum.

A complete HDDR process sequence as seen by ME and XRD is shown in fig. 4 with (b) after hydrogenation at 30°C during 40 minutes, (c) after disproportionation at 870°C during 30 minutes and (d) after recombination at 780°C ,during 251 minutes. For (b) the ME spectrum shows an increase in the average hyperfine field from 29.1 T to 30 9 T, as it is expected for the formation of a stable $Nd_2Fe_{14}BH_x$ hydride [7].

Table I: ^{57}Fe Mössbauer parameters derived from spectra a-d in fig.4.

Site	H_{hf} (T)	Δ_{QS} (mm/s)	δ_{IS} (mm/s)	f	Spectrum
$16k_1$	28.2	0.242	-0.062	0.285	a
	29.2	0.249	0.006	0.259	b
	28.6	0.229	-0.075	0.259	d
$16k_2$	30.4	0.130	-0.122	0.271	a
	32.4	0.134	0.028	0.271	b
	30.5	0.138	-0.120	0.271	d
$8j_1$	26 0	0.101	-0.089	0.136	a
	28.5	0.557	-0.001	0.136	b
	26.2	0.136	-0.108	0.136	d
$8j_2$	34.1	0.632	0.088	0.136	a
	35 4	0.597	0.126	0.136	b
	34 3	0.649	0.093	0.136	d
4c	24.1	0.165	-0.046	0.068	a
	25.4	0.291	-0.015	0.068	b
	24.7	-0.02	-0.001	0.068	d
4e	29.0	0.164	-0.046	0.068	a
	32.4	0.128	-0.095	0.068	b
	28.7	0.026	-0.000	0.068	d
doublet		0.649	0.036	0.037	a
($Nd_{1+\varepsilon}Fe_4B_4$)		0.594	0.060	0.063	b
		0.594	0.041	0.062	d
α-Fe	33.0	0.002	0.004	0.750	c
α-Fe(Nd)	30.8	0.016	0.040	0.066	c
Fe_2B	23 8	-0.034	0.127	0.151	c
$Nd_{1+\varepsilon}Fe_4B_4$		0.605	0.066	0.033	c

An increase in the lattice parameter of about 2 % is deduced from the XRD pattern in agreement with a large absorption of hydrogen, values as large as x≅2.5 have been reported for hydrogen treatment at room temperature [9]. No change is observed for the $Nd_{1+\varepsilon}Fe_4B_4$ subspectrum in the hydrogenated sample. The ME spectrum obtained from the disproporcionated powder fig. 4.c was fitted with four hyperfine interactions, listed in table I, 33 T and 23.8 T hyperfine field interactions are characteristic of α-Fe and Fe_2B phases respectively. This is in agreement with the disproportionation reaction of $Nd_2Fe_{14}BH_x$+(2+z) H_2 into 12 α-Fe+Fe_2B+2 $NdH_{2+z+x/2}$, where z is function of pressure and temperature [8]. For P=101 kPa and T≥630°C the known approximated composition is $NdH_{2\,7}$. The hyperfine field of 30 7 T may be an indication of small amount (7%) of

Fe atoms in α-Fe with Nd nearest neighbors This would correspond to a random solid solution with about 1 at % of Nd. Although a value of 6 is expected for ratio of iron atoms in α-Fe to iron atoms in Fe_2B for this fit, in which no restriction was imposed for the relative intensities, the measured value is 5 4 This may be a consequence of slightly different Debye Waller factors for the different phases

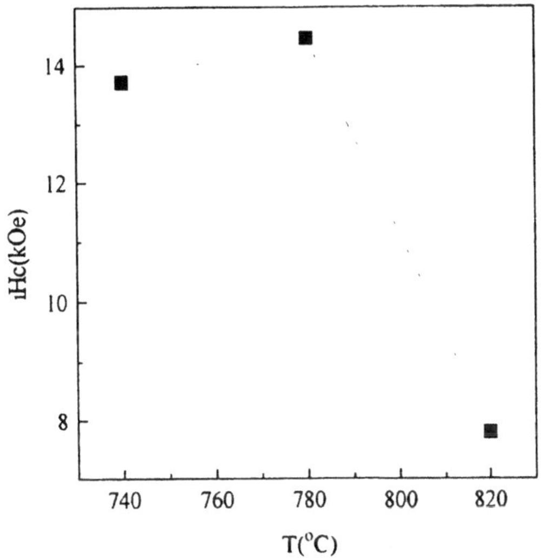

Figure 5: Variation of the intrinsic coercivity with recombination temperature

For this material a columnar type structure consistent of Nd hydride cylinders embedded in α-Fe and 100 nm average grains of Fe_2B has been reported[3] The XRD pattern was indexed with the reflections of α-Fe, Fe_2B phases and Nd_2H_5 hydride (see fig 4)

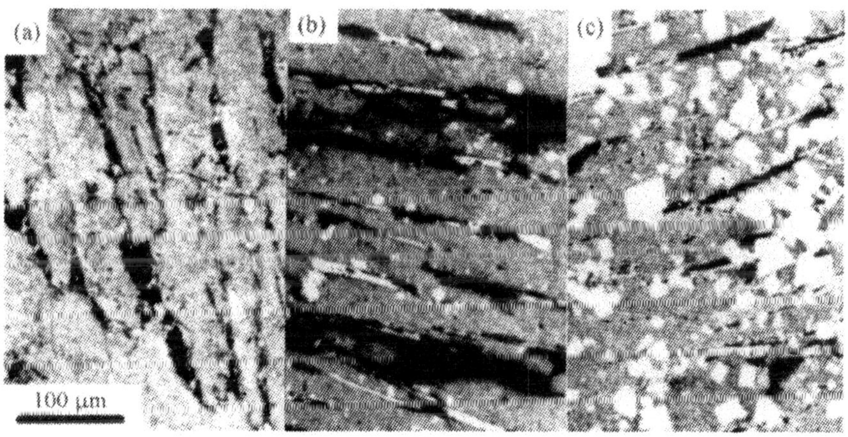

Figure 6: Optical Micrographs of fully treated samples recombined at 740 (a), 780 (b) and 820°C (c)

Finally, the ME spectra of several fully treated samples has been recorded, see for instance spectrum (d) in fig. 4. The corresponding optical micrograph is shown in fig. 3, by the HDDR process the $Nd_2Fe_{14}B$ matrix has been replaced by a fine grained phase that shows a higher Nd content, also large faceted grains appear whose composition is the same of the ingot matrix. Drastic changes in the microstructure[4] and in the magnetic properties of the alloy are found by changing the heating rate, recombination time and/or the temperature. For instance, by increasing the recombination temperature from 780°C to 820°C an abnormal grain growth is observed correlated with a dramatic decrease of the coercive field (See fig. 5). However, no change is observed in the Mössbauer spectra either for this modification or others tried in the HDDR process parameters[4]. Therefore the change of the magnetic properties can not be ascribed to the existence of highly disperse secondary phases nor to high concentration of points defects in the HDDR samples. Optical micrographs of fully treated powders for recombination temperatures of 740°C, 780°C, 820 °C are shown in fig. 6.

SUMMARY AND CONCLUSIONS

Distinctive ME spectra were recorded from the as cast, hydrogenated, disproportionated and recombined samples.

No change at the short range order size is observed on the fully HDDR treated powders compared to the as cast material. The main change introduced by the HDDR route is a microstructural feature and consists in the conversion of the coarse grained $Nd_2Fe_{14}B$ phase into a submicron size material.

Finally, it is worth of mention that the ME appears to be an appropriated technique to study the disproportionation and recombination kinetics, since the α-Fe signal can be well resolved from the $Nd_2Fe_{14}B$ one. This work is in progress.

ACKNOWLEDGMENTS

The authors gratefully acknowledge the partial financial support from CONICET (through Programa TENAES) and CONICOR of Republica Argentina, from the Secretariat of Science of the European Community, and G. T. Z.

REFERENCES

[1]I.R. Harris, *Proccedigs of the 12th International Workshop on Rare-Earth Magnets and their Applications*, Canberra, Australia, July 1992, pg.327.

[2]T.Takeshita and R. Nakayama, *Proceedings of the 12th International Workshop on Rare-Earth Magnets and their Applications*, Canberra, Australia, July 1992, pg.670.

[3] M.Matzinger, J. Fidler, O. Gutfleisch, and I.R. Harris. IEEE TRANSACTIONS ON MAGNETICS **31** (1995) 3635.

[4] F.A.Actis, H. Nagel, G. B. Cohen, O. M. Ragg and I. R. Harris. *Proc. 14th Int. Workshop on Rare-Earth Magnets and Their Applications*, Sao Paulo Brasil (1996) pg.174.

[5] O.A.Pringle and Gary J.Long in *Mössbauer Spectroscopy Applied to Inorganic Chemistry* vol 3 (Eds. G.J.Long and F. Grandjean) Plenum Press (1989).

[6] Shuming Pam et al. Proc. of the 5th Int. Symp. on Magn. Anisot. and Coercivity. Bad Soden-FRG (1987) 439.

[7]J.M.Cadogan and J.M.Coey, Appl. Phys. Lett. **48** (1986) 442.

[8]O.Gutfleisch, M. Matzinger, J. Fidler and Y. R. Harris. J. of Magn. Mag. Mat. **147** (1995) 320.

[9] R.Nakayama and T. Takeshita,J Appl. Phys. **74** (1993) 2719

STUDY OF THE MAGNETIC BEHAVIOUR OF THE $Fe_pMn_xAl_q$ ALLOYS IN ITS DISORDERED PHASE, AT ROOM TEMPERATURE, USING MÖSSBAUER SPECTROSCOPY.

A. J. Sánchez Asseff, G. A. Pérez Alcázar, A. Bohórquez.
Universidad del Valle, Dpto. de Física, A. A. 25360. Cali

We present an experimental study of the magnetic behaviour of the disordered quenched stainless steel alloys of $Fe_pMn_xAl_q$ in its fcc phase, enhancing the range covered in preceding investigations in order to have a better estimation of the composition regions for which the material, at room temperature, changes from a antiferromagnetic to a paramagnetic behaviour. These changes are detected with certain reliability by extrapolating the alloy's mean hyperfine magnetic field results derived from the corresponding Mössbauer spectra. The "critical compositions" were found to be close to the reported starting border of the fcc-bcc structural phase transition, being a little upper for compositions minor than 60 at.% Fe and a little lower thereafter, but falling apart after 70 at.% Fe.
This work was sponsored by Universidad del Valle and Colciencias.

1. Introduction

There are two published reports on the magnetic properties of the $\mathbf{Fe_pMn_xAl_q}$ alloys in its disordered fcc antiferromagnetic phase: an experimental study at room temperature by Mössbauer Spectroscopy and X-ray diffraction [1] and a theoretical proposal of a diluted Ising model with two random competitive bonds [2] to interpret that experimental information base. With this work we have obtained a more detailed estimation of the mean hyperfine magnetic field (MHF) of the stable compositions in this phase since the number of samples under investigation was greatly increased.

2. Experimental procedure

The ternary alloy compositions were obtained by mixing and compacting 99.9% purity powders of Fe, Mn and Al. The mixtures were melted under an argon atmosphere in an arc furnace. Then, the alloys were vacuum encapsulated, annealed for a week at 1000°C and quenched in iced water. With this preparation method, and using X-ray diffraction, Chakrabarti [3] obtained a structural phase diagram, part of which is shown in figure 1. Black dots in the figure represent our alloy set. As it can be seen, some of the prepared alloys fall into the structural phase coexisting regions.

By weighting the samples before and after melting them we have determined that the material loses are beneath the 3% of the initial weight; when this is the case, by chemical analysis we have found out that the loses of Mn are lower than 1.5%, of Fe lower than 1% and of Al lower than 0.5%. Alloys with loses greater than 3% are rejected.

We have classified the $Fe_xMn_pAl_q$ samples, from the composition point of view, in terms of series of constant at.% Fe (x=cte). Each series starts with 0 at.% Al (q=0), i.e., with a binary alloy of Fe-Mn. Next come the ternary alloys, obtained by progressive addition of Al in 2.5 at.% steps. There are six series, with a 5 at.% Fe difference between them; the first one is for x=40 and the last one for x=65.

From the magnetic point of view, each series starts with an antiferromagnetic binary alloy, having known from literature that, at room temperature (RT), its MHF is saturated [4] and that when the at. % Al increases, the lattice parameter also increases while the MHF decreases [1].

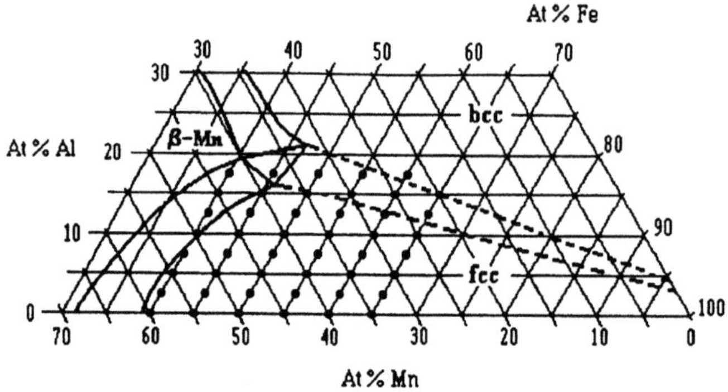

Fig. 1: Section of Chakrabarti's structural diagram [3].
Black dots represent studied alloy compositions.

3. Experimental results and discussion

Mössbauer spectra of alloy powder samples at RT were taken for all the samples. Most of the spectra, particularly those which belong to alloys far from, or not too close to, the structural phase transition border, where better fitted with a magnetic field distribution. For those for which this procedure was not proper we tried the fitting either with two magnetic distributions or with a single paramagnetic line together with a magnetic distribution, keeping the best adjusting curves. The two magnetic distribution fitting seemed to be a good election for the cases where the antiferromagnetic phase turned to be weak but its presence was still detectable, and the presence of the ferromagnetic phase was clearly appearing. The paramagnetic line along with the magnetic distribution was a suitable choice for the cases where the transition to a paramagnetic behaviour had already being reached and the ferromagnetic phase was also present.

As an example we have chosen the 55 at. % Fe series; its adjusting results are resumed in Table 1, and also in figure 2. Both presentations are ordered by the increase in the %Al composition: horizontally for the table and vertically for the figure (%Mn composition is implicit). In the figure, data and their adjusted curves are shown at the left, while at the right we have their magnetic field distributions.

Table 1. Adjusting parameters for the alloys of Fe-55% series.

Atomic % Al	0,0	2,5	5,0	7,5	10,0	12,5	15,0	17,5
Magnetic Distribution								
δ_{ISH} (mm/s)	-0,242	-0,246	-0,255	-0,218	-0,189	-0,154	-0,106	-0,093
ε (mm/s)	-0,006	-0,012	-0,003	-0,007	-0,013	0,024	-0,001	-0,019
H_{hf} (T)	2,804	2,808	2,681	2,687	2,054	0,939	28,148	23,751
σ_H (T)	1,015	1,114	1,057	1,144	1,119	0,551	3,830	4,412
Paramagnetic Site								
Relative area (%)	-	-	-	-	-	-	47,3	25,6
δ_{ISX} (mm/s)	-	-	-	-	-	-	-0,161	-0,163
FWHM (mm/s)	-	-	-	-	-	-	0,609	0,435

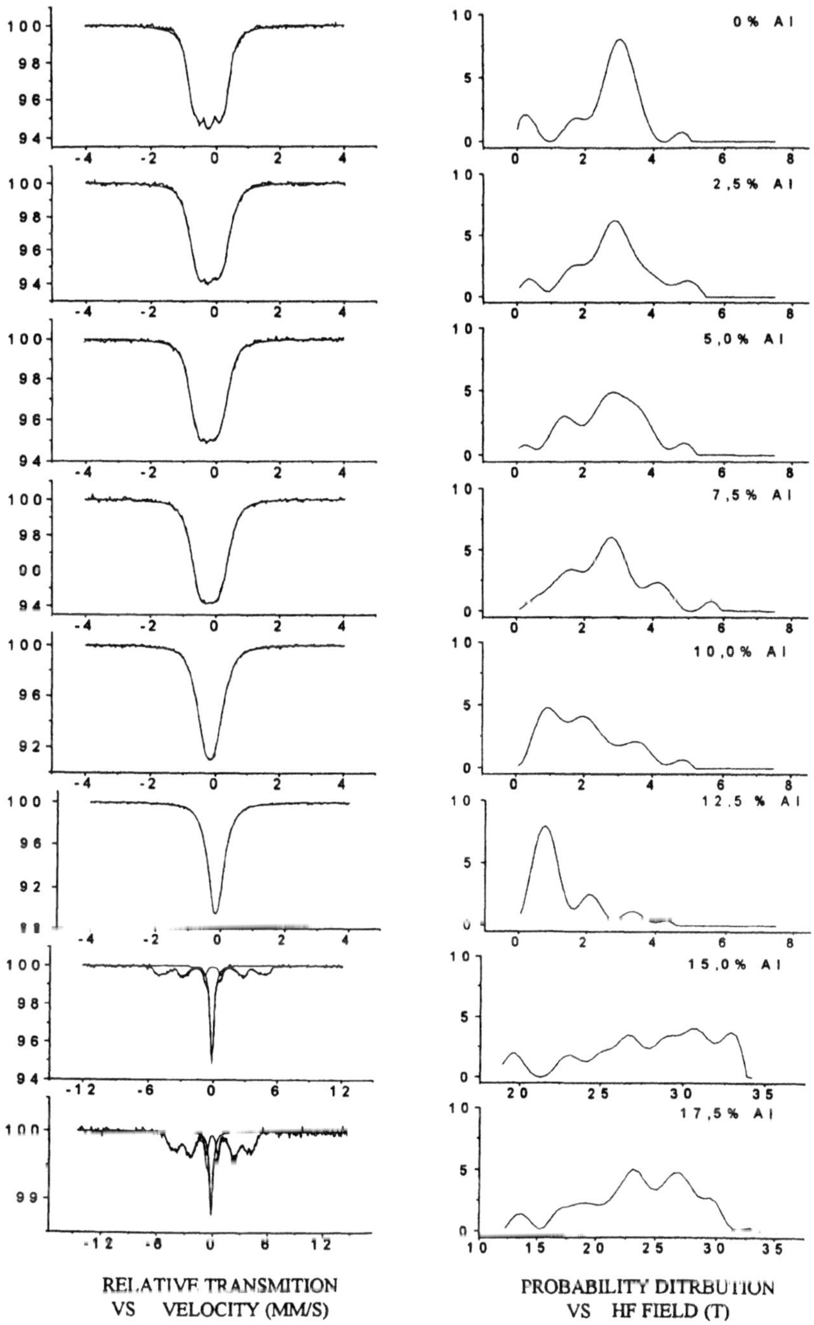

Figure 2: 55 at.% Fe Series: Data, their adjusting curves and the corresponding HF distribution.

In this illustrating case, the first six spectra were adjusted with one magnetic field distribution and the last two required the use of a paramagnetic site along with a magnetic distribution - rather than two magnetic distributions - to get a more appropriated adjustment; their Mössbauer spectra clearly show the presence of the bcc ferromagnetic phase which forces a broader field distribution, for much higher fields, in the adjusting procedure. These results are somehow related to the fact that these two alloys, of the 55 at.% Fe series, belong to the structural phase coexisting region, as it can be corroborated by looking at the diagram in figure 1.

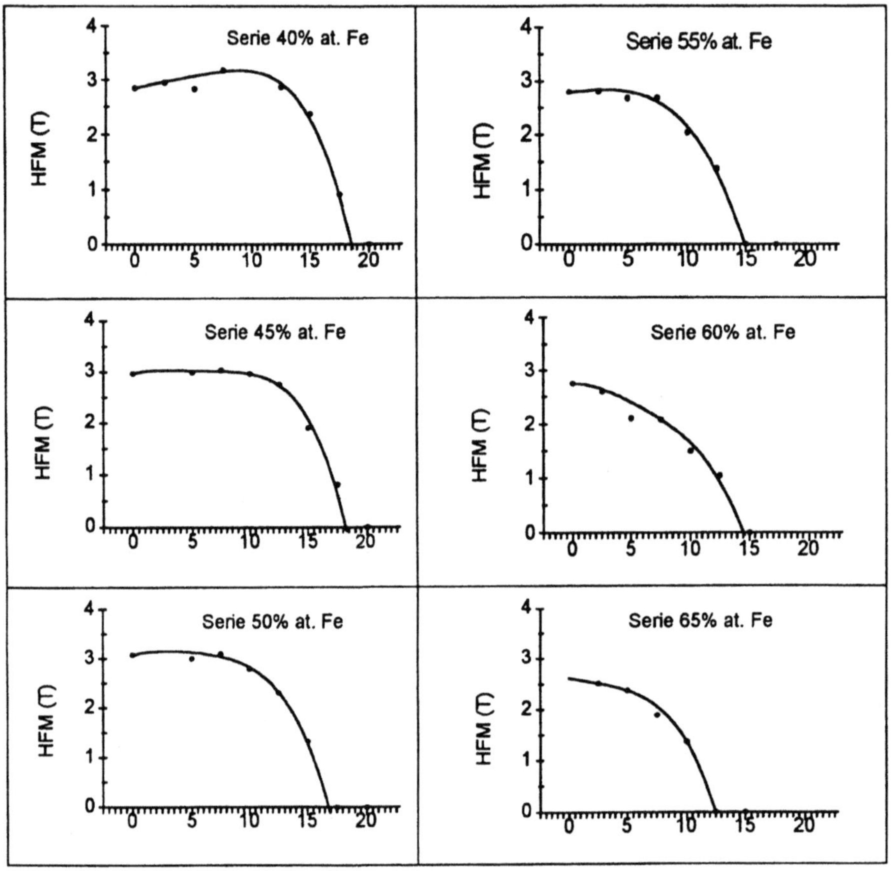

Figure 3: Mean hyperfine magnetic field (MHF) in Teslas vs. %Al composition for the six Fe at. % series. Fitting curves help extrpolation to determine the "critical compositions".

In a like manner, this same behaviour was also encountered in the other studied series. The adjusting procedure led us to presume that the antiferromagnetic MHF practically disappears at some, usually intermediate, alloy compositions, that can be determine by extrapolation, as follows. In figure 3 we may observe the values of the mean hyperfine field (MHF), in Teslas, plotted against at.% Al compositions, for the six Fe series. All the plotted curves show a MHF decreasing behaviour which may be extrapolated in order to determine the crossing points, i.e. the compositions, for which it may have

become zero. The compositions so encountered may be called "critical", since after those compositions no antiferromagnetic behaviour is detected; their values have been reported in Table 2.
As it was mentioned at the beginning, we decided to present our data and its analysis from the constant **at.%Fe** point of view. When a similar procedure is followed from the constant **at.%Al** point of view, additional extrapolation curves may be drawn, as it was already done in [1], and some complementary "critical compositions" may be obtained. These other values are also included in the refered Table.

Table 2: Encountered "critical compositions" from extrapolation of the MHF behaviour.

at. % Fe series		
Fe	**Mn**	**Al**
40,0	41,4	18,6
45,0	36,9	18,1
50,0	33,2	16,8
55,0	30,1	14,9
60,0	25,6	14,4
65,0	23,0	12,0

at. %Al series		
Fe	**Mn**	**Al**
78,8	13,7	7,5
69	21,0	10,0
64,5	23,0	12,5
55,0	30,0	15,0

These critical compositions have been placed on the phase diagram of figure 4. As can be observed, the points fall close to the border of the coexisting region proposed by Chakrabarti, where the structural phase fcc-bcc transition starts.

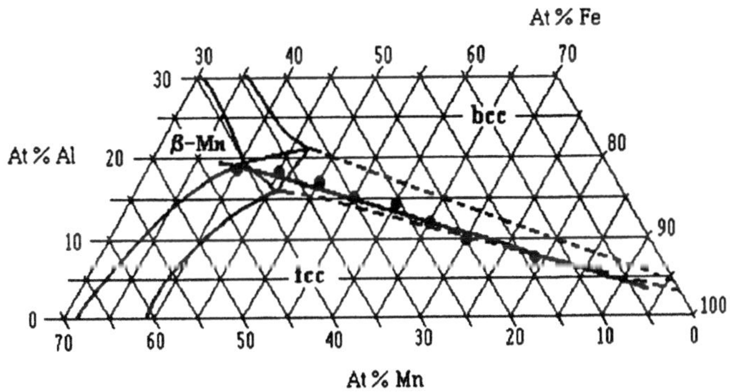

Fig. 4. Fe-Mn-Al critical compositions placed on a Chakrabarti's diagram. The curve is the result of interpolating data in order to stablish an antiferro-paramagnetic transition border.

4- Final considerations

As it was said, the "critical compositions", from the magnetic point of view, have some direct relation with the structural phase transition, since they are close to Chakrabarti's reported starting border of the fcc-bcc coexisting region. A transition line between the points has been drawn, being a little upper for Fe compositions minor than 80 at. % and a little lower afterwards. This last predicted tendency ought to be studied closely in order to confirm its validity since we suspect that in this region, where the austenitic

phase becomes unstable, and it starts to have martensitic characteristics, and its structure tends to be hcp, the research has to be managed with a wider scope and a more circumspect treatment.
With this study we may assure that the austenitic quenched stainless steels made out of the ternary stable alloys is practically antiferromagnetic in the whole studied fcc structural phase. Data reported conform an extended experimental base for estimating the exchange interaction constants from the determination of the adjusting parameters using a diluted and random bond Ising model with two competitive interactions (Mn-Mn, Mn-Fe) as it has been already done with the previous experimental work [2]. Now, we have a more extensive experimental base to define those magnitudes in order to be able to carry out a theoretical study for the just established antiferro-paramagnetic transition line.

Acknowledgments

The authors wish to thank the support given by Universidad del Valle and Colciencias to accomplish this work and to present it at the LACAME'96.

References

[1] G.A. Pérez Alcazar, E Galvão da Silva and C. Paduani. Hyp. Int. **66**, 221-230, (1991).
[2] A. Osorio, L. Zamora, and G. A. Pérez Alcázar. Phys. Rev. B, **53**(13), 8176(4)(1996).
[3] D.J. Chakrabarti, Metall. Trans. B **8**, 121 (1977).
[4] C. Paduani, J. A. Plaska and E. Glavão da Silva. Phys. Rev B **44,** 9715, (1991).

MOSSBAUER INVESTIGATION OF R AND Zn SITES IN Fe-DOPED R_2BaZnO_5 (R=RARE EARTH)

S. J. Stewart, G. F. Goya and R. C. Mercader

Departamento de Física, Universidad Nacional de La Plata,
C.C. 67, 1900 La Plata, Argentina

Abstract

Mössbauer measurements have been performed in 1.0at% ^{57}Fe-doped R_2BaZnO_5 oxides (R=Y, Sm, Eu, Gd, Dy, Ho) of the orthorhombic symmetry, space group *Pnma*-type structure. $Gd_2BaZn_{0.99}Fe_{0.01}O_5$ was also investigated by Electron Paramagnetic Resonance giving a g=6 value signal. The Mössbauer results indicate that Fe is in all cases in the Fe^{3+} valence stated, with $\Delta \approx 0.8$mm/s characteristic of Fe substituting Zn in the structure. This replacement was successfully achieved in different proportions for all compounds of the series.

Introduction

The diversity of magnetic structures and low dimensional magnetic orderings in some members of the family of R_2BaMO_5 ternary oxides (R=rare earth or yttrium and M=3d transition metal) has made it the target of numerous investigations. Currently there is a lack of information on the type of parameters that control these properties. In the past through magnetic measurements and doping with Mössbauer probes we have been able to advance in the knowledge of interesting features of these oxides[1-5]. However, most of these studies were carried out on systems that had localized magnetic moments at the transition metal site.

To study in detail the nature of the interactions of each magnetic sublattice, we have started to substitute a non-magnetic transition metal, namely Zn, for iron in a series of different oxides of the orthorhombic structural type with the *Pnma* space group with different R atoms. In this paper we report the first Mössbauer and x-ray studies of this series. EPR was also performed for one member of the series (R=Gd).

Experimental

Samples of R_2BaZnO_5 (R=Eu, Sm, Gd, Dy, and Ho) were prepared by standard solid state reaction from R_2O_3(99.999%), CO_3Ba(99.99%), α-$^{57}Fe_2O_3$ (enriched to $\approx$95%), and ZnO(99.999%) powders mixed in stoichiometric amounts. The homogenized mixture was first fired in air at 500°C, then reground and reheated three times at 900°C for 24 hours. Powder x-ray diffraction characterization was performed using Cu-Kα radiation. The data were refined by Rietveld profile analysis using DBWS 9006PC[6]. Mössbauer measurements were taken at room temperature in transmission geometry with a nominal 10mCi ^{57}Co source in a Rh matrix, with a standard 512 channels multiscaler. A non-linear least-squares program was used to fit the spectra to Lorentzian line-shapes. In some cases it was necessary to use a shape-independent distribution fitting program[7]. The isomer shifts are referred to α-Fe at 300 K. EPR data for $Gd_2BaZn_{0.99}Fe_{0.01}O_5$ were obtained with a Bruker ESP300 spectrometer in X-band ($\approx$9.4GHz), between 2 and 300K.

Results

Table I shows the hyperfine parameters after the fittings performed in the way described. Fig.1 shows spectra for R=Eu, and Gd, and Fig. 2 for R=Sm, and Y, that are representative of the different quadrupole signals found. The isomer shift values of all but Sm compounds analyzed belong to Fe^{3+}. The quadrupole splittings can are $\Delta \approx 0.8$ mm/s. The Sm compound displays also a signal of $\Delta = 1.37 \pm 0.02$ with $\delta = 0.13 \pm 0.02$mm/s.

X-ray data showed that the samples were monophasic within the sensitivity of powder diffraction experiments.

An EPR signal of g=6 was found for the first time in this type of doped compounds.

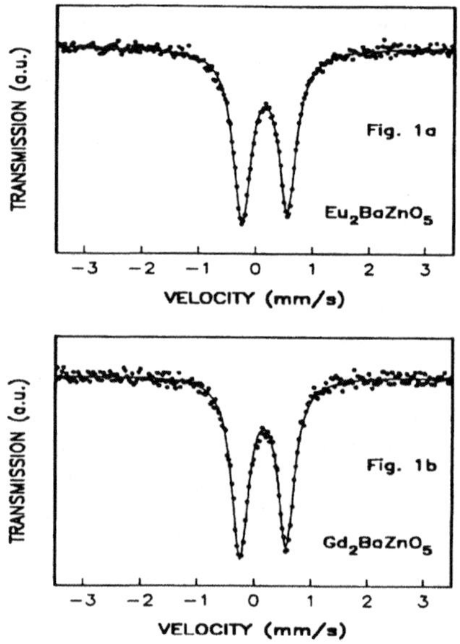

Fig. 1. Mössbauer spectra for Fe-doped R_2BaZnO_5 compounds (R = Eu and Gd).

Discussion

Prior to our work[5] it was suggested that the presence of a magnetic transition metal was necessary to induce the long range ordering in the rare-earth sublattice[8], because no ordering had been noticed in all the members of the R_2BaZnO_5 investigated to temperatures down to 1.5K. However, we were able to observe a 3D magnetic order in Gd_2BaZnO_5 at 2.3 ± 0.1K through susceptibility measurements. Since the Gd_2BaCuO_5 compound has a Néel temperature of 12K, to understand the type of R-M interaction it is most important to follow the increase of the ordering temperature from the pure Zn oxide to a compound with a magnetic localized moment at the metal site.

The Mössbauer signals observed could arise from Fe atoms placed at different substitutional sites of the structure or even to interstitial positions. However, according to the spectra of the present studied series, these were reduced to only two quadrupole signals.

Studying in detail the spectrum of $Gd_2BaZn_{0.99}Fe_{0.01}O_5$ it was found that its quadrupole

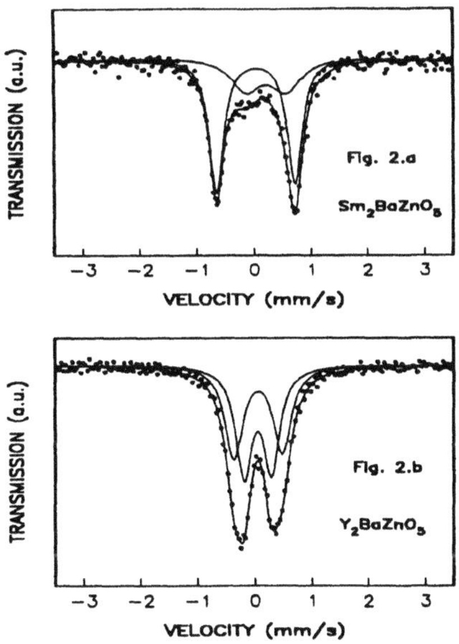

Fig. 2. Mössbauer spectra for Fe-doped R_2BaZnO_5 compounds (R – Sm and Y).

splitting is $\Delta=0.88\pm0.01$mm/s. Since this is the only signal that the spectrum displays, that in addition has a relatively narrow linewidth of 0.34 ± 0.02mm/s, it is reasonable to assume that Fe is located at only one site in this compound. To assign the location of Fe, we performed EPR measurements on $Gd_2BaZn_{0.99}Fe_{0.01}O_5$ finding that g=6 is a remarkable feature observed [9]. This type of unusual value of g has been noticed for other series of compounds with an atypical S=3/2 electronic spin state for Fe^{3+}, caused by a negative value of the crystalline field constant D, that makes the $|\pm3/2>$ spin state to lie below the $|\pm1/2>$ one, causing $g_{\parallel}$ to adopt the value of 6, while $g_{\perp}=0$[10]. This means that even for polycrystalline specimens the only component contributing to g is the parallel one and its average value is also 6. This same state of the Fe spin was found by us in the substitution of Fe for Cu in Fe-doped Y_2BaCuO_5[3]. This spin state, thus seems to be the one preferred by Fe when located at the base of a square pyramidal coordination polyhedron. The EPR measurements have enabled us to assign the Mössbauer quadrupole splitting $\Delta\approx0.8$ mm/s to the location of iron at the Zn site.

In the case of Sm with one signal of $\Delta=1.37\pm0.02$mm/s, it is likely that this is a signature of iron replacing the rare earth ion, since this a much more distorted structural site. Besides, considering its $\delta=0.13\pm0.01$mm/s, the hyperfine parameters are consistent with a high spin Fe^{III} state, that would reinforce the idea of a distorted environment for this site.

Currently susceptibility measurements are being carried out and Mössbauer measurements at low temperatures are being planned to establish the possible orderings of these samples at low temperatures.

The correlation between the macroscopic and hyperfine measurements will allow to advance in the knowledge of R-M magnetic interaction in this family of compounds.

Conclusions

The quadrupole signal of $\Delta \approx 0.8$mm/s found for most of the members of this series of oxides, has been unambiguously assigned to Fe at the Zn site. The signal of $\Delta = 1.37$mm/s is likely arising from Fe atoms located at Sm sites.

Table I : Mössbauer hyperfine parameters. Δ is the quadrupole splitting; δ is the isomer shift referred to α-Fe and Γ is the linewidth. Errors are quoted between parenthesis.

Fe-doped compound	Δ (mm/s)	δ (mm/s)	Γ (mm/s)	%
Y_2BaZnO_5	0.85(3)	0.16(1)	0.34(2)	51(3)
	0.49(2)	0.15(1)	0.37(2)	48(3)
Gd_2BaZnO_5	0.81(2)	0.26(2)	0.34(2)	100
Dy_2BaZnO_5	0.77(2)	0.21(1)	0.43(2)	100
Eu_2BaZnO_5	0.79(2)	0.26(2)	0.36(2)	100
Ho_2BaZnO_5	0.83(2)	0.26(2)	0.50(3)	100
Sm_2BaZnO_5	1.37(2)	0.13(1)	0.30(2)	65(3)
	0.71(5)	0.32(2)	0.60(3)	34(2)

References

[1] R. C. Mercader, G. Punte and E. Baran, Solid State Comm. 74, (1990) 943.

[2] S. J. Stewart, G. F. Goya, R. C. Mercader, A. E. Goeta, G. Punte and F. Sapinya Navarro, Hyperfine Int. 83, (1994) 421 .

[3] G. F. Goya, S. J. Stewart, A. E. Goeta, A. G. Grunfeld, G. Punte and R. C. Mercader, J. Mag. Mag. Mat. 138, (1994) 147.

[4] G. F. Goya, S. J. Stewart and R. C. Mercader, Solid State Comm. 96, (1995) 485.

[5] G. F. Goya, R. C. Mercader, L. B. Steren, R. D. Sánchez, M. T. Causa, and M. Tovar, J. Phys. Cond. Matt. 8, (1996) 4529.

[6] A. Sakthivel and R. A. Young, DBWS-9006PC, "Program for Rietveld analysis of X-ray powder diffraction patterns", Georgia Institute of Technology, Atlanta, (1991).

[7] R. E. Vandenberghe, E. DeGrave and P.M.A. de Bakker, Hyp. Int. 83, (1994) 29.

[8] J. Hernández-Velasco, A. Salinas-Sánchez, F. Fernández and R. Sáez-Puche, J. Alloys and Compounds 203, (1993) 15.
R. Sáez-Puche and J. Hernández-Velasco, Adv. Materials Research 1, (1994) 65.

[9] G. F. Goya, M. Tovar, M. T. Causa and R. C. Mercader, to be published.

[10] P. Ganguli, V. R. Marathe, and S. Mitra, Inorg. Chem. 14, (1975) 970.

A MÖSSBAUER STUDY OF Sm_2Fe_{17} NITRIDES

C.E. Rodríguez Torres, M.B. Fernández van Raap, F.H.Sánchez and L.A. Mendoza Zélis.
Departamento de Física, Universidad Nacional de La Plata,
c.c. 67 (1900) La Plata, Argentina.

Y.D. Zhang, J.I. Budnick and N.X. Shen.
Department of Physics, University of Connecticut, Connecticut 06269, Storrs, USA.

Mössbauer spectra of $Sm_2Fe_{17}N_x$ ($0 \le x \le 2.61$) have been measured at 85 K and analyzed using a non homogeneous nitrogenation model which assumes the simultaneous existence of nitrided and unnitrided regions for intermediate N content. This model is in good agreement with previously reported NMR and electron microprobe results.

Introduction

Recently, it has been found that the compound Sm_2Fe_{17} may be nitrided to form the intermetallic phase $Sm_2Fe_{17}N_x$ improving the magnetic properties like anisotropy field, saturation magnetization, and Curie temperature[1]. Such excellent intrinsic magnetic properties make this phase a good candidate for preparing high performance permanent magnets.

It is now widely accepted that nitrogen atoms enter the Sm_2Fe_{17} 2:17 lattice through a diffusion process, mostly occupying the octahedral interstitial sites. Very little is known concerning the N diffusion mechanism at an atomic level into the 2:17 system. Initially the nitrogenation was treated as a free diffusion process of N atoms in the 2:17 lattice, resulting in a continuous solid solution distribution (CSSD) profile of N atoms in the sample particles[2,3]. Later subsequent NMR studies resulted inconsistent with the free N diffusion model[4]. N diffusion did not lead to the CSSD pattern, instead, it was interpreted a nitrided/unnitrided configuration; i.e., the N atoms absorbed in a R_2Fe_{17} particle are concentrated in a nitrided region, leaving the remaining part of the particle unnitrided. In this model the nitrogenation process is the growth of the nitrided region while the N distribution in the nitrided region does not change. In order to check the validity of this analysis for Sm_2Fe_{17}, ^{57}Fe Mössbauer effect measurements have been performed at 85 K on $Sm_2Fe_{17}N_x$ ($0 \le x \le 2.61$).

Experiment

The Sm_2Fe_{17} ingots, made by arc melting, have the rombohedral Th_2Zn_{17} structure. Nitrogenation was performed by heating powders with different particle sizes under high purity nitrogen gas flow at 460°C for up to 17 h, thereby producing $Sm_2Fe_{17}N_x$ samples with x=0.28, 0.66, 1.31, 1.79, 2.26, and 2.61.

^{57}Fe Mössbauer spectra were collected using a conventional constant acceleration spectrometer with a 20 mCi source of ^{57}Co in rhodium. For the low temperature measurements a liquid nitrogen cryostat, was employed. The velocity scale was calibrated using an α-Fe absorber at room temperature. Absorbers of ≅20 mg/cm^2 thickness were prepared from powdered samples.

Results and Discussion

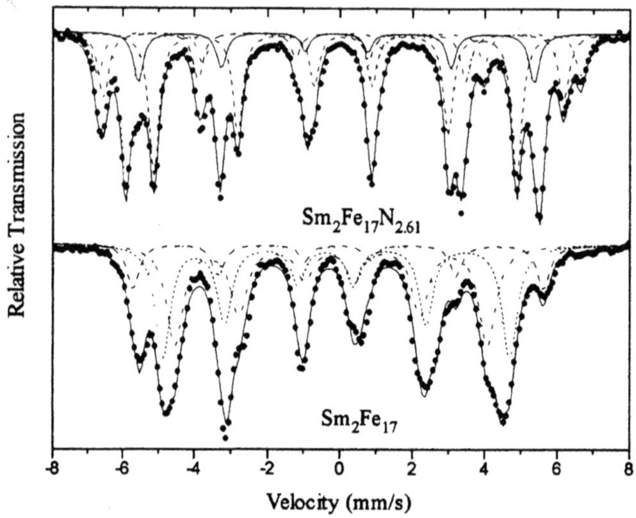

Fig.1- 57Fe Mössbauer spectra of Sm2Fe17 and Sm2Fe17N2.61 at 85K with the fit shown as a full curve. The broken lines refer to subspectra proposed for the four inequivalent Fe sites.

Table 1
Hyperfine parameters obtained at 85K and Wigner-Seitz cell volumes for Sm_2Fe_{17} and $Sm_2Fe_{17}N_{2.61}$. Typical hyperfine parameters errors are 1 kOe and 0.01 mm/s for H_{hf} and δ_{IS} respectively. Isomer shifts are refereed to α-Fe at room temperature.

Parameter	Compound	6c	9d	18f	18h
H_{hf} (kOe)	Sm_2Fe_{17}	352	309	298	268
	$Sm_2Fe_{17}N_{2.61}$	416	396	355	311
δ_{IS} (mm/s)	Sm_2Fe_{17}	0.230	-0.080	0.046	0.028
	$Sm_2Fe_{17}N_{2.61}$	0.260	0.021	0.128	0.206
V_{ws} 295°K	Sm_2Fe_{17}	12.34	11.2	11.76	12.05
($Å^3$)[a]	$Sm_2Fe_{17}N_{2.61}$	12.71	11.8	11.69	12.37

[a]Data from ref.[7].

The Sm_2Fe_{17} and $Sm_2Fe_{17}N_{2.61}$ spectra and the corresponding fits, are shown in Fig.1. In general the Mössbauer spectra of the R_2Fe_{17} compounds could be resolved into subspectra associated with the four non-equivalent iron sites. Whenever the magnetic anisotropy is along the c-axis, a single hyperfine pattern can be assigned to each iron site. However, when the magnetization is in the basal plane the hyperfine pattern associated with the f and h-iron nuclei site splits up and the Mössbauer spectra should be fitted with six different subspectra. According to Gubbens et al[5] the efg at 9d iron site lies along the c-axis (as it happens with the 6c sites) leading to non splitting of the associated hyperfine pattern. However other researchers do propose the splitting of this pattern[6,7]. It must be mentioned that other physical effects may induce additional splitting of the hyperfine interaction[8]. Furthermore, it is known that in Sm_2Fe_{17} the easy magnetization axis lies on the basal plane of the unit cell, but there is some controversy concerning its precise direction ([100][5] or [120][6]).

Given this rather confusing situation, and the extra complexity of the spectra obtained from partially nitrogenated samples, we have chosen to describe the non nitrogenated phase with only four sets of hyperfine interactions, corresponding to the four iron inequivalent sites. The hyperfine parameters are listed in Table I.

There is no doubt that the 6c dumbbell site should have the largest hyperfine field since it has the most iron (13), and the fewest rare-earth (1) near neighbors(nn). Similarly, the 18f sites (10 nn_{Fe}, 1 nn_{Re}) has a larger hyperfine field than the 18h (9 nn_{Fe}, 3 nn_{Re}). The 9d site (10 nn_{Fe}, 2nn_{Re}) can be distinguished by its intensity from the 18f and 18h sites. From these considerations combined with the fit results we find that hyperfine fields decrease in the order 6c>9d>18f>18h. This order is consistent with neutron diffraction results on Nd_2Fe_{17}[9] and band calculations on Y_2Fe_{17}[10].

In Table I Wigner-Seitz cell volumes (V_{ws})[7] associated to the different crystallographic sites are listed, useful information for the discussion of the isomer shifts. As expected, the best fit results are obtained with the larger and smallest isomer shifts corresponding to 6c and 9d sites respectively, which have the largest and smallest V_{ws} respectively.

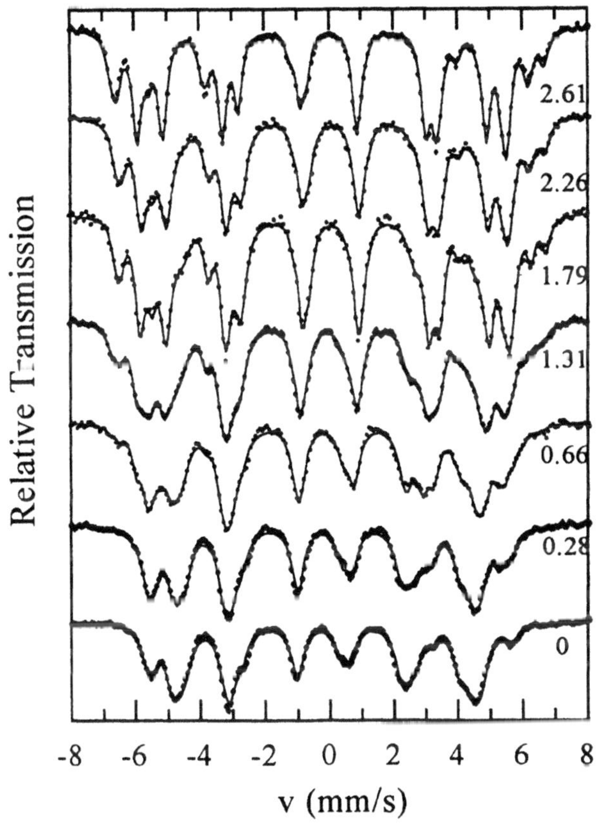

Fig.2- ^{57}Fe Mössbauer spectra of $Sm_2Fe_{17}N_x$ with the fits shown as a full curve.

$Sm_2Fe_{17}N_{2.61}$ has an easy magnetization c-axis and the spectrum can be fitted properly with five sets of interactions, four corresponding to the crystallographic inequivalent sites in the 2:17 structure and one additional to α-Fe. In general grounds this description is consistent with considerations of lattice expansion under nitrogen intake and nitrogen instertice occupation, and in good agreement with other authors description[7].

The spectra with intermediate N concentration (0<x<2.61), were fitted with two groups of interactions. One that accounts for an unnitrided region (with hf parameters similar to those of x=0 sample) and the other corresponding to a fully nitrogenated region (hf parameters similar to x=2.61 sample). By this means a satisfactory set of fits was obtained (Fig.2). Within each group of interactions the hyperfine parameters remain approximately constant and so do the relative intensities. This results supports the initial description of the partially nitrogenated system in terms of two regions.

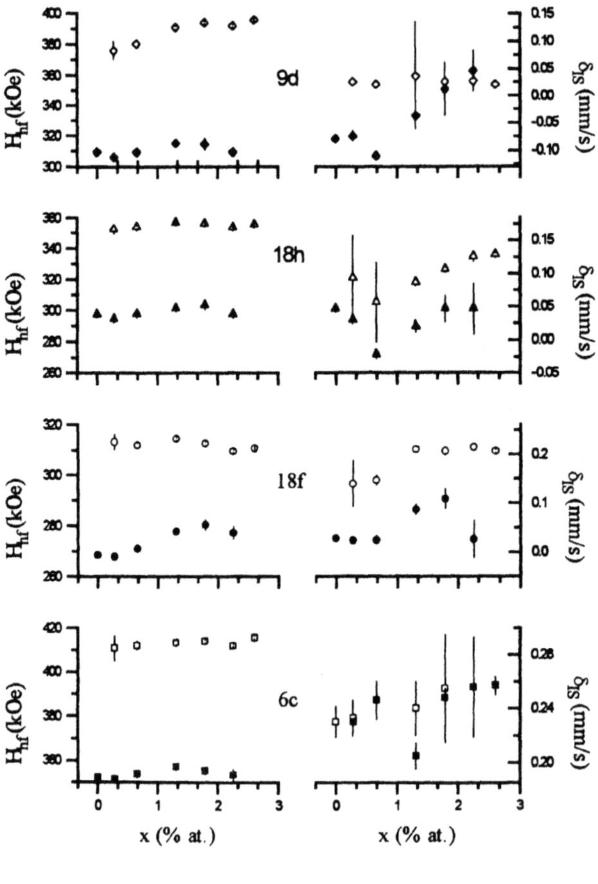

Fig.3- Hyperfine fields and isomer shifts (relative to metallic iron at room temperature) at the four inequivalent Fe sites of $Sm_2Fe_{17}N_x$ ($0 \le x \le 2.61$) Solid and open symbols correspond to unnitrided and nitrided regions respectively.

Therefore we may describe the process by the solid state reaction

(eq.1) $Sm_2Fe_{17} + xN \Rightarrow a'Sm_2Fe_{17}N_y + aSm_2Fe_{17} + bSmN + cFe$,

where the disproportionation observed during the initial nitrogenation stage has been considered.
The coefficients a, a' and c are obtained from the fits and x is known. Thus, both the fraction of nitrogenated sample a'/(a+a') and y can be calculated from the experimental results (Fig.4). Initially y increases rapidly to a value of about 2.5 and then remains almost constant being the main effect the growth of the nitrogenated fraction. This is in good agreement with the results reported in references[4] and[11].

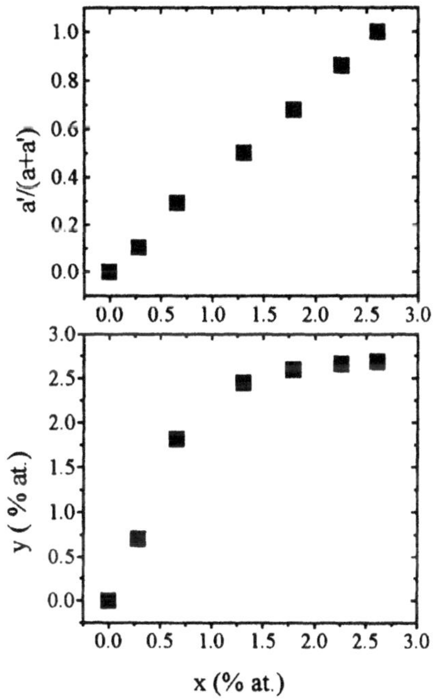

Fig.4- Top: fraction of nitrided phase ($Sm_2Fe_{17}N_y$) calculated from experimental data and eq.1 versus total N concentration x.. Bottom: N concentration y of the nitride phase calculated from experimental data and eq.1 versus total N concentration x.

Summary and Conclusions

The ^{57}Fe Mössbauer spectra for intermediate nitrogen concentration can be simulated by superposition of two patterns, one for the pure phase and the other for the nitrogen saturated phase.
This results would not support the CSSD model for nitrogen diffusion. A more detailed discussion will appear in a forthcoming paper.

Acknowledgments

Authors are grateful to the European Commission and the Consejo Nacional de Investigaciones Científicas y Técnicas de Argentina (through Programa TENAES) for partial economics support.

References

[1] J.M.D. Coey and H. Sun, J. Magn. Magn. Mater., **87** (1990) L251.
[2] R. Skomski and J.M.D. Coey, J. Appl. Phys., **73** (1993) 7602.
[3] J.M.D. Coey, R. Skomski and S. Wirth, IEEE Trans. Mag., **28** (1992) 2332.
[4] Y.D. Zhang, J.I. Budnick, D.P. Yang, G.W. Fernando, W.A. Hines, T.D. Xiao and T. Manzur, Phys. Rev. B **51** (1995) 12 091.
[5] P.C.M. Gubbens, J.J. Van Loef and K.H.J. Buschow, Journal de Physique **12** (1974) C6 617.
[6] Bo-Ping Hu, Hong-Shuo Li, Hong Sun and J.M.D. Coey, J. Phys., **3** (1991) 3983.
[7] G.J. Long, S. Mishra, O.A. Pringle, F. Grandjean, and K.H.J. Buschow, J. Appl. Phys. **75** (1994) 5994.
[8] W. Steiner and R. Haferl, Phys. Status Solidi **a 42** (1977) 739.
[9] J.F. Herbst, J.J.Croat, R.W. Lee and W.B. Yelon, J. Appl. Phys. **53** (1982) 250.
[10] B. Szpunar and J.A. Szpunar, J. Appl. Phys. **57** (1985) 4130.
[11] C.C. Colucci, S. Gama, C.A. Ribeiro, and L.P. Cardoso, J. Appl. Phys. **57** (1985) 6003.

SPINODAL DECOMPOSITION IN A Fe-30.8Cr-12.2Co ALLOY INVESTIGATED BY MOSSBAUER SPECTROMETRY

S. Alleg[1], B. Bouzabata[1], and J.M. Greneche[2]

1 Unité de Recherche: Matériaux et Métallurgie, Institut de Physique BP12 Annaba 23000 Algérie

2 Laboratoire de Physique de l'Etat Condensé, URA CNRS 807 Université du Maine F72017 Le Mans Cedex France

Abstract:

We report first results obtained by Mössbauer spectrometry on an Fe-30.8Cr-12.2Co (weight %) alloy annealed at two different temperatures for several times in order to characterize the disorder-order transformation. The evolution of the average values of the magnetic hyperfine field and of the distribution of hyperfine fields are consistent with a spinodal decomposition within Fe-30.8Cr-12.2Co alloy.

Introduction

Much effort, both experimental and theoretical, has been devoted to studies of the interesting ferromagnetic properties as well as the phase decomposition in binary Fe-Co and Fe-Cr alloys [1-4] and in ternary Fe-Cr-Ni and Fe-Cr-Co alloys [5-9], essentially the kinetics of the disorder → order transformation. ^{57}Fe Mössbauer spectrometry and ^{59}Co NMR provide complementary informations on magnetic properties of FeCo-containing alloys and are able to distinguish between spinodal decomposition and nucleation growth, thanks to their respective local probe behaviour. Indeed, the chemical ordering and its kinetics can be followed thanks to the evolutions of both isomer shift (total s density) and the hyperfine field (mainly due to the Fermi contact interaction, sensitive to the local magnetic moment) at iron sites. Previous studies indicated that the spinodal decomposition was not fully developed in Fe-26Cr-14.2Co alloys [6]. From literature, one can conclude that slight changes in the composition, the conditions of melting and quenching and those of aging can induce different mechanisms of transformation. The present study aims to characterize the spinodal decomposition by Mössbauer spectrometry in Fe-30.8Cr-12.2Co (weight %) alloy as a function of aging time.

Experimental section

The Fe-30.8Cr-12.2Co (weight %) ternary alloy results from levitation melting under vacuum Fe, Cr, and Co powders (99,9%) with appropriate concentrations and quenched in a water cooled copper ingot-mould. Cylinders 1-3 μm thick were taken from the ingot and then polished by a mechanical procedure to get disks 40-50 μm thick. The disks were annealed at 520°C and 560°C in a furnace under vacuum to prevent oxidation, for several aging times: 1, 2, 5, 15, 30, 60, 300, 720, 1440, and 2880 min. The bcc structure was checked by X-ray diffraction.

Mössbauer spectra were registered at 300K and 77K in transmission geometry with a ^{57}Co source diffused into a Rh matrix. The normal of the disk was oriented parallel to the γ-beam. The hyperfine parameters were refined with the least squares MOSFIT program [10].

Results and discussion

Some typical Mössbauer spectra obtained either at 300K or at 77K on an as-prepared disk, and disks annealed at 520°C and 560°C for 5 min and 30 min are illustrated in figs. 1a, 2a, and 3a, respectively.

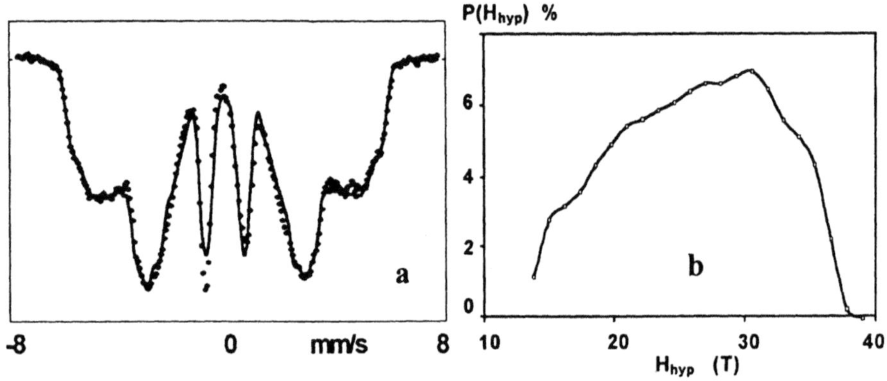

Fig. 1: 300K-Mössbauer spectrum on as-prepared Fe-30.8Cr-12.2Co alloy (a) and corresponding P(Hhyp) (b)

Mössbauer spectra exhibit magnetic sextets with broad lines and the presence of weak sharp lines which increase with aging time. The presence of broad lines is consistent with random atomic arrangements and the changes of the hyperfine structure with local atomic rearrangement. In the present approach, the fitting procedure of these Mössbauer spectra consists in considering a superposition of sextets with lorentzian lines, i.e. a discrete distribution of hyperfine fields $P(H_{hyp})$; both the isomer shift and quadrupolar shift were fitted in the same way for all magnetic sextets. Figs. 1b, 2b, and 3b show $P(H_{hyp})$ corresponding to spectra of figs. 1a, 2a, and 3a, respectively.

One observes a slight misfit of Mössbauer spectra: improvements would require a more complex fitting procedure based on correlations between isomer shift and hyperfine field. In addition, we report on fig. 4 the time dependence (in semi-logarithmic scale) of the mean hyperfine field ($<H_{hyp}>$) value for the two considered annealing temperatures.

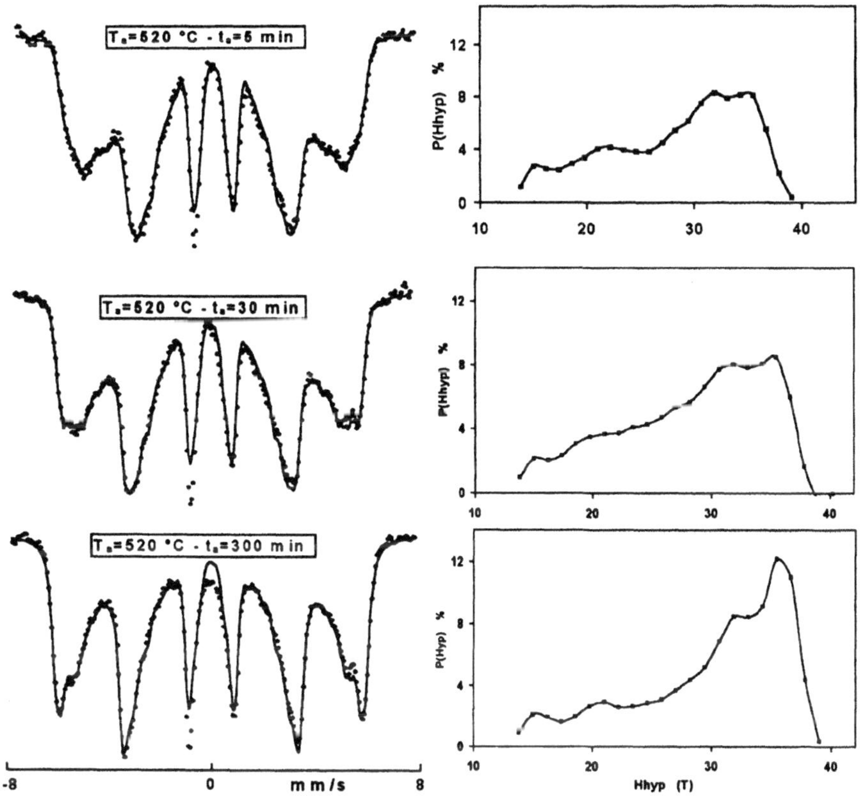

Fig.2: (a) 300K Mössbauer spectra (left) and (b) corresponding P(Hhyp) (right) for samples annealed at 520 °C for different aging times.

The experimental features reveal different phenomena. (i) For short aging times (t < 5min at Ta = 560°C; t < 15 min at Ta = 520°C), a slight broadening of $P(H_{hyp})$ and an increase of $<H_{hyp}>$ are evidenced and can be attributed to structural relaxation phenomena within the crystalline matrix. (ii) For higher aging times, one observes a shift of $P(H_{hyp})$ towards high fields and an increase of $<H_{hyp}>$. Indeed, due to their electronic structure, Cr is well known to reduce the hyperfine field at iron nucleus in metallic alloys drastically; the effect of Co is weak but opposite. So, the lower values hyperfine values (ranging from 13 to 30T) may be assigned to the presence of Cr in the immediate iron nuclei surrounding (as first and second neighbour) while the high field values to a relative enrichment of Co. As quoted in [11], H_{hyp} values of 37T (see P(Hyp) presented fig. 2b and 3b) suggest a concentration of Co in the range 20-30 at.%, for both the disordered and ordered atomic phases. (iii) The evolutions of both $P(H_{hyp})$ and $\langle H_{hyp} \rangle$ are more pronounced for 560°C, indicating that the aging temperature influences the kinetics of phase transformation. For large aging times (t > 24 h), $P(H_{hyp})$ exhibit similar shapes and $\langle H_{hyp} \rangle$ tend to the same value.

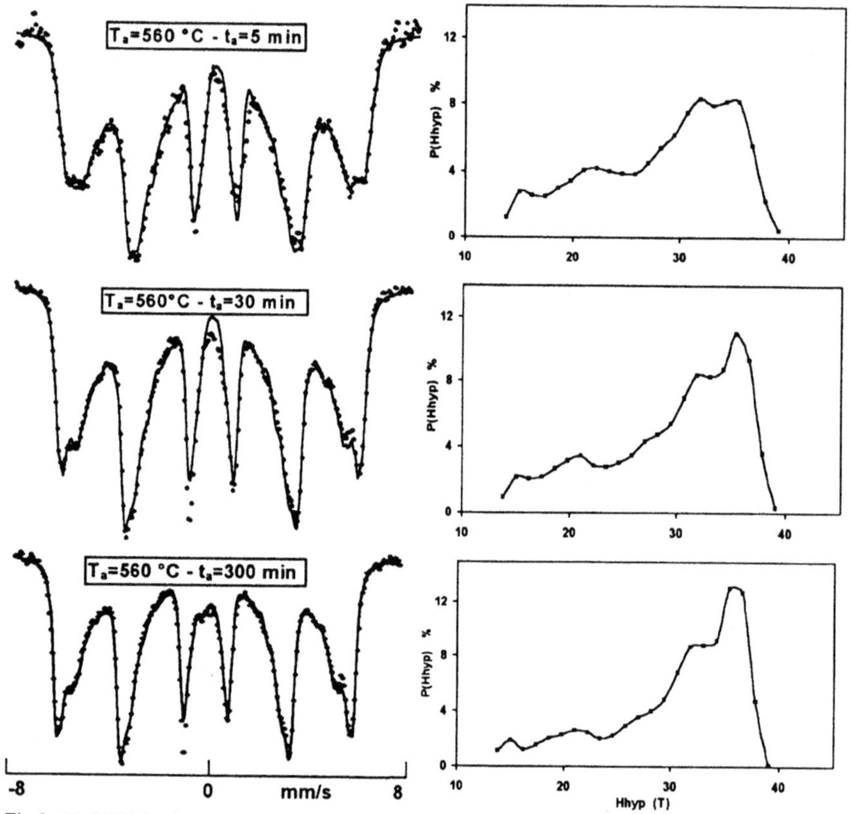

Fig.3: (a) 77K Mössbauer spectra (left) and (b) corresponding P(Hhyp) (right) for samples annealed at 520 °C for different aging times.

Conclusions

These preliminary results report on the evolution of the magnetic hyperfine field distributions at two aging temperatures and give evidence for the role of Cr and Co atoms. One can conclude that the disorder-order transformation in Fe-30 8Cr-12.2Co (weight %) alloy can be described by a spinodal decomposition mechanism. New perspectives are offered to the present study: (i) a better fitting procedure of Mössbauer spectra involving distributions of both hyperfine fields and isomer shift, consistently with electronic structure and (ii) a comparison of Mössbauer hyperfine data at iron nuclei with NMR data at cobalt nuclei. Such developments are in progress.

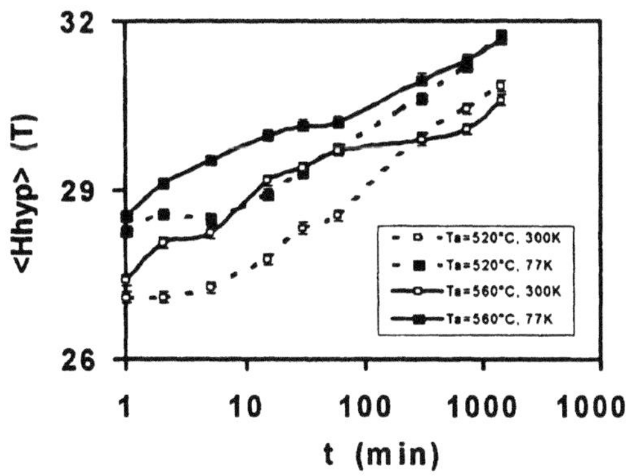

Fig. 4: Time dependence of <Hhyp> obtained at 300K and 77K for two annealing temperatures: 520°C and 560°C.

References:

[1] H. Kuwano, Trans. Jap Inst. Met. 26 (1985) 473; 26 (1985) 482; 26(1985) 721

[2] T. de Nys and P.M. Gielen, Metall. Trans. 2 (1971) 1423

[3] H. Pollak, U. Karfunkel, J.A. Lodya, and N. Mala, Hyp. Int. 94 (1994) 2355

[4] H. Kuwano and K. Ono, J. Phys. Soc. Jpan 42 (1977) 72

[5] H. Kuwano, Y. Nakamura, K. Ito, and T. Yamada, Il Nuevo Cimento 18 (1996) 259

[6] M. Eibschütz, G.Y. Chin, S. Jin, and D. Brasen, Appl. Phys. Lett. 33 (1978) 362

[7] M.E. Houghton, P.L. Rossiter, and P.E. Clark, J. Mater. Sci. 13 (1978) 155

[8] H.H. Hamdeh, J. Okamoto, and B. Fultz, Phys. Rev. B 42 (1990) 6694

[9] B. Fultz, Phys. Rev. B 44 (1991) 9805 and references therein

[10] J. Teillet, and F. Varret, unpublished Mosfit program

[11] H.H. Hamdeh, B. Fultz, and D.H. Pearson, Phys Rev. B 39 (1989) 11233

VI. Physical Metallurgy and Materials Sciences

MÖSSBAUER STUDY OF Fe-Mn-Al-C STEELS

V. F. Rodríguez, A. J. Sánchez, A. Bohórquez, and G. A. Pérez Alcázar
Departamento de Física, Universidad del Valle, A. A. 25360, Calı, Colombia

ABSTRACT

Experimental studies by x-ray diffraction and Mössbauer spectroscopy, at room temperature on disordered Fe-Mn-Al-C alloys, are reported. The Fe concentration is between 50 to 60 at%, the Al concentration is between 0 and 10 at%, the C concentration is between 0 and 1 at%, and the Mn concentration is the balance. All the alloys are in the fcc structural phase with the lattice parameter near 3.6 Å, which increases linearly with the Al content. This alloys are antiferromagnetic and this behaviour decreases with the Fe or Al concentrations increase. For fixed Al content the antiferromagnetic character saturates for high Mn concentration to a hyperfine field value that depends on the Al content. In general the mean saturation hyperfine field is near 28 kOe for Mn contents near or higher 40 at%.

1. INTRODUCTION

Simultaneously with development of stainless steels there exists big interest in obtaining corrosion resistant alloys with good mechanical properties but without using Cr and Ni strategic elements, which are substituted for Al and Mn, respectively. Fe-Mn-Al alloys in the fcc disordered phase have been little studied. Chakrabarti [1] obtained the phase diagram for the ternary system, for alloys quenched from 1000 °C, and this diagram shows that the fcc phase is stable for low Al concentration (<15 at%) and for less that 60 at% Mn. Another work [2] on the same system on their magnetic properties showed that the system is antiferromagnetic and this behaviour is diluted by Fe or Al atoms. Depending on the composition range for each alloying element these alloys present different types of antiferromagnetism depending on the spin orientation (collinear or non collinear) [3].

In this work we describe the experimental results obtained by Mössbauer spectroscopy and x ray diffraction of the Fe-Mn-Al-C system in the fcc disordered phase. We analyze the effect of the different elements in the antiferromagnetic character.

2. EXPERIMENTAL PROCEDURE

The alloys were melted in an arc furnace under argon atmosphere using Fe, Mn, Al and C with more than 99.9% purity. After wards, they were sealed in a quartz tube and homogenized by annealing at 1000 °C during a week. Then, the samples were quenched in ice water. Powdered samples for Mössbauer spectroscopy and x ray diffraction measurements were prepared by mean a diamond file. The obtained spectra were fitted with a hyperfine field distribution using the NORMOS program [5].

3. RESULTS AND DISCUSSION

Table 1 displays the nominal concentration, the hyperfine parameters obtained by Mössbauer spectroscopy , and all parameter from x ray diffraction data for all the studied samples. Figure 1 shows the Mössbauer spectra obtained at room temperature and the corresponding hyperfine field distribution (HFD) for the alloys of the A series of Table 1. The obtained spectra for the other series are similar to that showed in figure 1. In this figure the effect of substituting Mn atoms for Al atoms

Sample series #	Fe (at%)	Mn (at%)	Al (at%)	C (at%)	IS (mm/s)	MHF (T)	Cell dim. (Å)
A1	50	49.5	0	0.5	-0.22(1)	2.94(2)	3.621(1)
A2	50	47	2.5	0.5	-0.22(1)	2.69(2)	3.627(1)
A3	50	44.5	5	0.5	-0.20(1)	2.31(2)	3.641(1)
A4	50	39.5	10	0.5	-0.18(1)	2.56(2)	3.654(1)
B1	50	49	0	1	-0.21(1)	2.38(2)	3.626(1)
B2	50	46.5	2.5	1	-0.22(1)	2.55(2)	3.636(1)
B3	50	44	5	1	-0.19(1)	2.29(2)	3.648(1)
B4	50	39	10	1	-0.17(1)	2.29(2)	3.661(1)
C1	55	44.5	0	0.5	-0.23(1)	2.97(2)	3.621(1)
C2	55	42	2.5	0.5	-0.21(1)	2.86(2)	3.635(1)
C3	55	39.5	5	0.5	-0.20(1)	2.24(2)	3.641(1)
C4	55	34.5	10	0.5	-0.18(1)	1.72(2)	3.657(1)
D1	55	44	0	1	-0.22(1)	2.93(2)	3.637(1)
D2	55	41.5	2.5	1	-0.22(1)	2.91(2)	3.643(1)
D3	55	39	5	1	-0.20(1)	2.17(2)	3.635(1)
D4	55	34	10	1	-0.19(1)	1.96(2)	3.645(1)
E1	60	39.5	0	0.5	-0.22(1)	2.78(2)	3.612(1)
E2	60	37	2.5	0.5	-0.21(1)	2.37(2)	3.624(1)
E3	60	34.5	5	0.5	-0.19(1)	1.86(2)	3.636(1)
E4	60	29.5	10	0.5	-0.17(1)	1.49(2)	3.648(1)
F1	60	39	0	1	-0.23(1)	2.86(2)	3.619(1)
F2	60	36.5	2.5	1	-0.21(1)	2.46(2)	3.622(1)
F3	60	34	5	1	-0.19(1)	1.47(2)	3.623(1)
F4	60	29	10	1	-0.17(1)	1.45(2)	3.645(1)

Table 1. Atomic composition, isomer shift relative to ^{57}Fe at 300 K, mean hyperfine field and lattice parameter of the Fe-Mn-Al-C alloys.

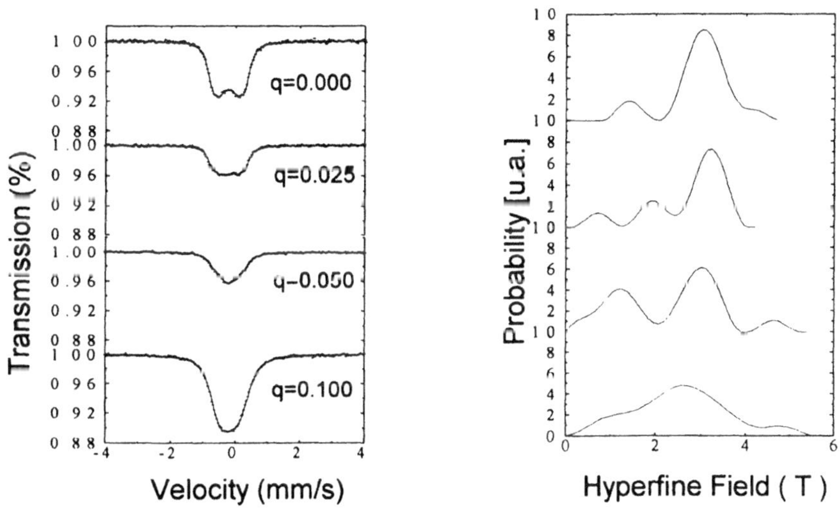

Figure 1 Mossbauer spectra and their corresponding hyperfine field distribution for the alloys series $Fe_{0.500}Mn_{0.495}Al_qC_{0.005}$

is showed. It can be noted that the HFDs present three well defined peaks and that the intensity of the peaks for low fields increase while that of the peaks for high fields decrease when the Al contents decrease. This result is similar for A and B series. Analogous results are obtained for the other alloys series but in this case the HFDs present only two peaks. The presence of this well-defined peaks confirm the existence of some more probable atom configurations around the Fe atoms and then an evidence of the localized character of the magnetic moment in this alloys. Up to now it was not possible to elucidate which atoms configurations are those appear in the present HFDs. This result is in accordance with previous reports [2,4].

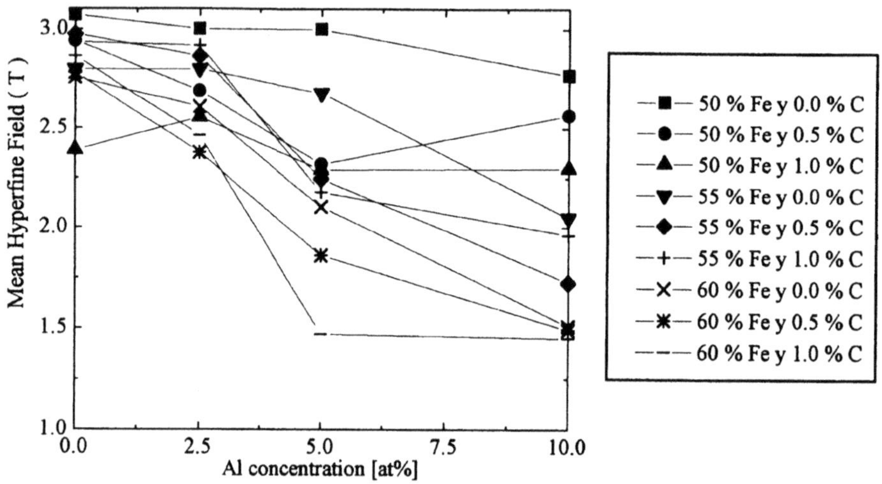

Figure 2. Mean hyperfien field vs. Al concentration for 50, 55 and 60 at% Fe and different C concentrations. The error bars of the MHF are smaller than the size of the symbols.

In figure 2 is shown the variation of the mean hyperfine field (MHF) with the Al concentration (q) for the six studied alloys series. It can be noted the effect of substituting Mn for Al atoms. In this case the MHF decreases with the Al increase and furthermore, this decreases is bigger for those alloys with more Fe and C contents. The decrease of the MHF can be explained as due the Al effect which dilute the antiferromagnetic interactions (Fe-Mn and Mn-Mn) and also expand the lattice as can be noted in Table 1. This also explain the tendency to obtain paramagnetic line with the increase of Al contents.

Figure 3 shows the variation of MHF with Fe concentration (p). It can be noted that for low Al contents (0, 2.5 and 5 at%) and low C contents (0 and 0.5 at%) the MHF is nearly constant for 50 and 55 at% of Fe and that it decreases when the Fe contents increases to 60 at%. There is an exception for this behaviour and this is presented for the alloys with low Al contents (0 and 2.5 at%) and 1 at% C for which MHF initially increases and then decreases. Otherwise for higher Al content (10 at%) MHF continuously decreases when the Fe contents increases.

The fact that MHF is nearly constant is in accord with Endoh's [3] results, in which the Mn atoms produce antiferromagnetic interactions with the Fe atoms and this interaction is saturated with the Mn increase in such way that from certain Mn concentration the MHF remains nearly constant. The necessary Mn contents to saturate the field in each of the alloy series depends on the other elements' contents. For example, for those alloys whose MHF continuously decrease the saturation is not achieved while for alloys with 0 at% Al and 0.5 at% C the saturation is obtained for 35 at% Mn in accordance with Figure 3.

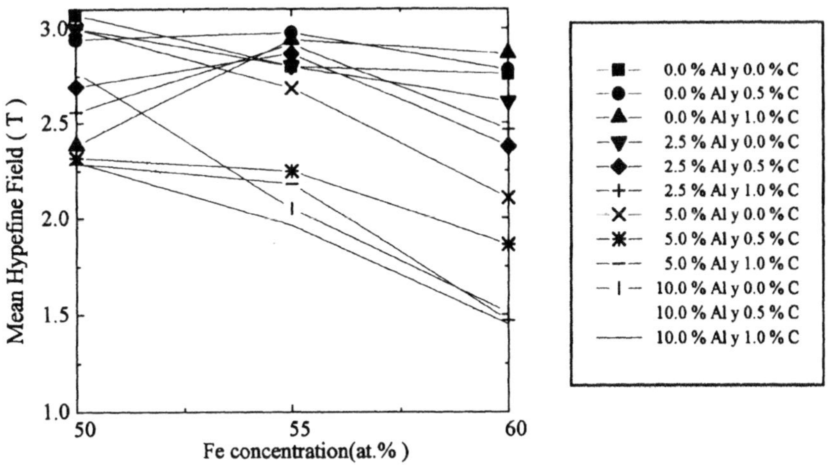

Figure 3. Mean hyperfine field vs. Fe concentration for 0, 0.5 and 1 at% C and different Al concentrations. The error bars of the MHF are smaller than the size of the symbols.

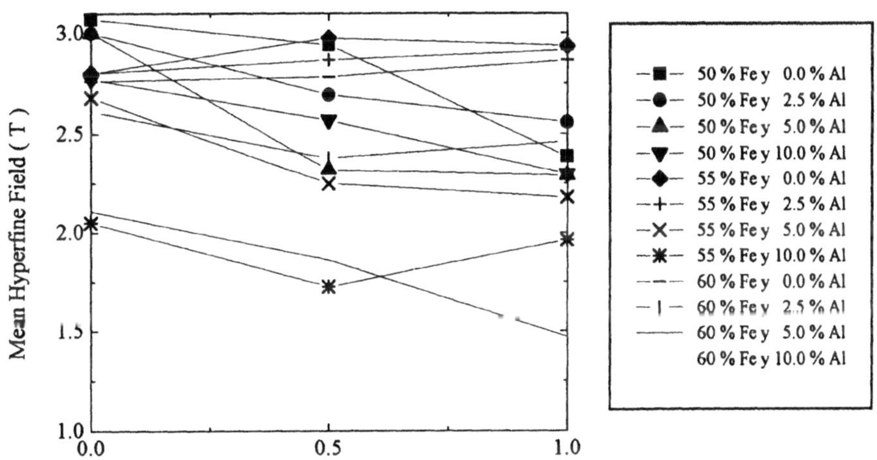

Figure 4. Mean hyperfine field vs. C concentration for 50, 55 and 60 at% Fe and different Al concentrations. The error bars of the MHF are smaller than the size of the symbols.

Figure 4 shows the variation of the MHF with the C contents. It can be noted that for high Fe contents (55 and 66 at%) and 0, 2.5 and 10 at% Al the MHF is nearly independent of the C contents. It has a value of 2.85 T for alloys with 55 at% Fe and 0 at% Al and a value of 2.5 T for 60 at% Fe and 2.5 at% Al.

ACKNOWLEDGMENTS

The authors would like to thank COLCIENCIAS, the UNIVERSIDAD DEL VALLE and the ALAMED program of the European Community for their financial support.

REFERENCES

[1] D. J. Chakrabarti, Met. Trans. B **8** (1977), 121

[2] G. A. Pérez Alcázar, E. Galvão da Silva and C. Paduani, Hyp. Inter. **66** (1991), 221

[3] Y. Endoh and Y. Ishikawa, J. Phys. Soc. Jap. **30** (1968) 1614

[4] Y. Ishikawa and Y. Endoh, J. Phys. Soc. Jap. **23**, (1967) 205

[5] R. A. Brand, Nucl. Instrum. Methods, Phys. Res. B **28** (1987), 398 and Nucl. Instrum. Methods. Phys. Res. B **28** (1987), 417

THE INFLUENCE OF THE SAMPLE PROPERTIES ON THE ELECTRICAL HYPERFINE PARAMETERS IN SYNTHETIC ALUMINOGOETHITES[1]

C.A. Barrero, R.E. Vandenberghe*, E. De Grave*,2, and A.L. Morales***

Department of Subatomic and Radiation Physics, University of Ghent, Proeftuinstraat 86, B-9000 Ghent, Belgium.
***Departamento de Física, Universidad de Antioquia, Medellín, Colombia, A.A. 1226.*

Mössbauer spectra (MS) of 26 synthetic aluminogoethites have been collected at 80 K and 400 K with the aim to search for possible multiple linear correlations between some sample properties on the one hand, and their electrical hyperfine parameters on the other hand. The electrical hyperfine parameters at 80 K are more regularly influenced by the aluminum content than by the other sample properties. Adequate linear relationships between the magnitude of the maximum-probability and average of the quadupole shift and the aluminum substitution and other structural properties could be derived.

INTRODUCTION

It has long been known that the Mössbauer hyperfine parameters of aluminogoethites depend to some extent to the properties of the sample. In that respect, however, a systematic investigation has been carried out only for the magnetic hyperfine measured at low temperatures, in relation to the specific surface area, *S*, the average structural-domain size and the aluminum content, Al_s, [1,2]. But until now, according to the best our knowledge, a systematic work searching for the possible linear correlations between the sample properties and the electrical hyperfine parameters has not been reported yet. The aim of the present work is to investigate these features. In addition to the three aforementioned sample properties, also the excess hydroxyl and the presence of structural defects have been considered. The latter one is quantified by Δa, the difference between the observed unit-cell parameter *a* and the value predicted by Vegard's law [3].

EXPERIMENTAL

Four differently prepared series of synthetic aluminogoethites have been investigated. Series 31, 12 and 3+4 are described in references [3] and [4]. The sample properties for these series were taken from the same references. The sample series GA were synthetized from ferrihydrite following the method described by Schwertmann and Cornell [5]. The average structural-domain sizes along the [111] and [100] directions for the series GA were determined from X-ray diffraction measurements. All MS at 80 K were fitted with hyperfine field distributions [6] using linear correlations between the field value and the quadrupole shift, and all MS at 400 K were fitted with quadupole splitting distributions.

---

[1] Contribution No SSF96-09-03 from the Department of Subatomic and Radiation Physics.

[2] Research Director, National Fund for Scientific Research, Belgium.

RESULTS AND DISCUSSION

TABLE 1. Sample properties of series GA.

Sample	*S* (m^2/g)	$\Delta OH^{\$}$	x*	y*	z**	MCD(111) nm	MCD_a nm
GA/0	33	0.34	0.000	-0.033	-0.011	35.4	19
GA/1	37	1.46	0.016	0.042	0.030	33.0	18
GA/2	41	1.98	0.025	0.087	0.053	26.8	15
GA/3	36	1.43	0.030	0.054	0.047	36.7	19
GA/4	37	1.34	0.046	0.043	0.060	36.7	20
GA/5	29	1.18	0.048	0.043	0.062	45.1	28
GA/8	28	1.16	0.077	0.044	0.091	62.0	30
GA/12	25	0.88	0.113&	0.034&	0.123	70.9	39

$: ΔOH, the excess hydroxyl, is the amount of additional hydroxyl and surface water content and was calculated as the difference between the measured (wt.% OH + wt.% H_2O) and the stoichiometric amount of water estimated from the formula $Fe_{1-x}Al_xOOH$ [3].

*: these values were calculated using the goethite's new formula α-$(Fe_{1-x}Al_x)_{1-y/3}O_{1-y}(OH)_{1+y}$ and the measured values for wt.% OH and for wt.% Al. Taking into account the amount of chemisorbed water (wt.% of H_2O) did not appreciably influence the results.

**: z =x + [(1-x)y/3]. This variable, that was directly derived from the goethite's new formula, takes into account the replacement of the iron ions by the aluminum and the hydroxyl ions at once.

&: contains ca. 2 % hematite. The measured x and y were corrected accordingly.

Some of the sample properties for series GA are summarized in Table 1. A slight increase in the mean crystallite dimension in the [111] and [100] directions, MCD(111) and MCD_a respectively, and a slight decrease in the surface area and in the excess hydroxyl with increasing Al substitution, except for the first two samples, are observed.

TABLE 2. Electrical hyperfine parameters for some selected samples at 80 K and at 400 K.

	80 K			400 K		
Sample	$\lvert(2\varepsilon)_m\rvert$	$\lvert(2\varepsilon)_{av}\rvert$	δ	$(\Delta E_Q)_m$	$(\Delta E_Q)_{av}$	δ
31/0	0.242	0.245	0.364	0.459	0.482	0.196
31/8	0.265	0.279	0.356	0.492	0.539	0.183
GA/0	0.244	0.244	0.362	0.459	0.488	0.199
GA/12	0.267	0.286	0.356	0.458	0.521	0.185
12/0	0.243	0.245	0.362	0.458	0.523	0.196
12/30	0.265	0.295	0.356	0.493	0.563	0.176
3/0	0.237	0.252	0.362	0.536	0.658	0.186
4/20	0.237	0.257	0.360	0.520	0.630	0.191

$\lvert(2\varepsilon)_m\rvert$: the magnitude of the quadrupole shift corresponding with the field of maximum probability in mm/s; $\lvert(2\varepsilon)_{av}\rvert$: the magnitude of the average quadrupole shift in mm/s; δ: centre shift in mm/s (relative to Rh); $(\Delta E_Q)_m$: quadrupole splitting of maximum probability in mm/s; $(\Delta E_Q)_{av}$: average quadrupole splitting in mm/s.

The MS for all the studied samples consist of only a sextet at 80 K and only a doublet at 400 K. The obtained electrical hyperfine parameters for some selected samples are listed in Table 2. It is observed that the magnitude of the average quadrupole shift is close to one-half of the average quadrupole splitting above T_N. This indicates that the principal axis of the EFG remains perpendicular to the direction of the effective magnetic hyperfine field when the aluminum content and/or the crystallinity changes.

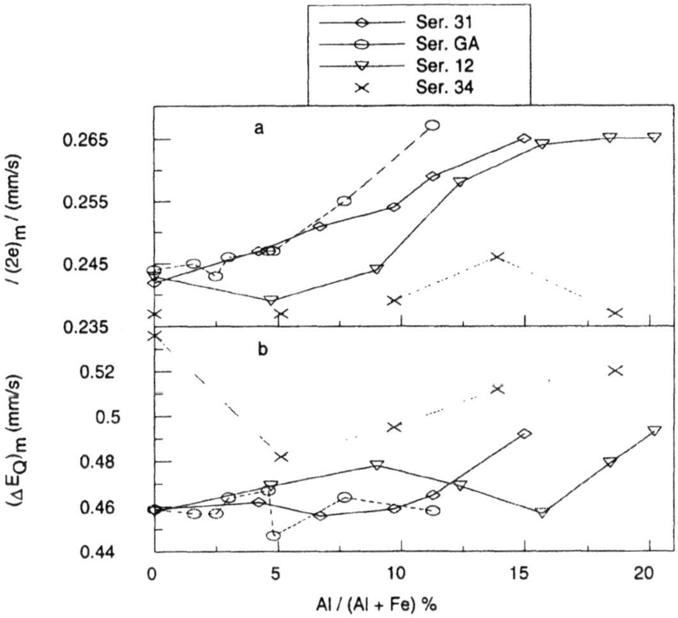

Figure 1. Effect of the Al content upon (a) the magnitude of the maximum-probability quadrupole shift at 80 K and (b) the maximum probability quadrupole splitting at 400 K, for all series.

Figure 1 shows the effect of the aluminum content upon the quadrupole interaction both at 80 K and at 400 K for all the studied samples. Except for the series 3+4, the goethites exhibit a slight increment in the magnitude of the maximum-probability quadrupole shift at 80 K with increasing Al content. This is most probably due to the reduction of the unit cell dimensions and the deformation of the lattice with the presence of the Al ions. Nevertheless, no regular behaviour is observed at 400 K.

Adequate linear relationships (correlation coefficients above 0.9) between maximum-probability and average quadrupole shift and the aluminum substitution and structural-domain size or surface area or structural defects can be derived (see Table 3). The obtained linear equations for the maximum-probability quadrupole shift as a function of two variables, with the largest correlation coefficients are:

$$|(2\varepsilon)_m(80\ \mathrm{K})| = 0.250(2) + 0.12(2)\ \mathrm{Al_s} - 0.28(4)/\mathrm{MCD}(111) \qquad (R = 0.915) \qquad (1)$$

$$|(2\varepsilon)_m(80\ K)| = 0.247(2) + 0.14(2)\ Al_s - 0.13(2)\times10^{-3}\ S \qquad (R = 0.919) \qquad (2)$$
$$|(2\varepsilon)_m(80\ K)| = 0.258(4) + 0.17(2)\ Al_s - 7(1)\ \Delta a \qquad (R = 0.942) \qquad (3)$$

the number in parenthesis is the uncertainty in the last digit. From these equations, it is possible to deduce that the decrease of the structural-domain size or the increment of the surface area or the structural defects leads to a reduction upon the magnitude of the maximum-probability quadupole shift. However, the increment in the Al content leads to an increment upon this parameter.

TABLE 3. Correlation coefficients and number of data pairs corresponding to the multiple linear regression analysis among the electrical hyperfine parameters to some of the sample properties.

	80 K			400 K		
	$(2\varepsilon)_m$	$(2\varepsilon)_{av}$	δ	$(\Delta E_Q)_m$	$(\Delta E_Q)_{av}$	δ
S	-0.420	-0.063	-0.047	0.762	0.879	-0.033
	(25)	(25)	(25)	(25)	(25)	(25)
Al_s	0.628	0.819	-0.771	0.348	0.377	-0.672
	(26)	(26)	(26)	(26)	(26)	(26)
Δa	-0.314	-0.028	-0.072	0.662	0.781	0.121
	(18)	(18)	(18)	(18)	(18)	(18)
S, Al_s	0.919	0.909	0.784	0.763	0.879	0.662
	(25)	(25)	(25)	(25)	(25)	(25)
S, Δa	0.557	0.456	0.158	0.716	0.861	0.283
	(17)	(17)	(17)	(17)	(17)	(17)
Al_s, Δa	0.942	0.905	0.853	0.670	0.805	0.720
	(18)	(18)	(18)	(18)	(18)	(18)

In conclusion, among the sample properties, the electrical hyperfine parameters at 80 K are more regularly influenced by the aluminum content. In contrast to the results at 400 K, adequate linear relationships between quadrupole interaction measured at 80 K and the aluminum substitution and structural-domain size or surface area or structural defects can be derived.

ACKNOWLEDGEMENTS. We wish to thank Prof. U. Schwertmann for kindly providing the sample series 31, 12 and 3+4, Prof. R. Vochten for performing the TGA and the wt. % aluminum determination on the series GA, and Prof. L.H. Bowen for performing the surface area measurements on the same series. This work was partially supported by COLCIENCIAS (Colombia) and the Fund for Joint Basic Research (Belgium).

REFERENCES.

[1] E. Murad and U. Schwertmann, Clay Min. 18 (1983) 301.
[2] D.C. Golden, L.H. Bowen, S.B. Weed, and J.M. Bigham, Soil Science Soc. Am. J. 43 (1979) 801.
[3] D.G. Schulze and U. Schwertmann, Clay Min. 19 (1984) 521.
[4] D.G. Schulze. Clays Clay Miner. 32 (1984) 36.
[5] U. Schwertmann and R.M. Cornell, in: Iron Oxides in the Laboratory. Preparation and Characterization (VCH Verlagsgesellschaft mbH, Weinheim, 1991) p. 61.
[6] R.E. Vandenberghe, E. De Grave, and P.M.A. de Bakker, Hyp. Int. 83 (1994) 29.

VII. Solid State Reactions in Mechanical Milling and Ionic Implantation

SHORT RANGE ORDER EVOLUTION IN ORDER-DISORDER TRANSITIONS IN THE AlFe SYSTEM

M. MEYER, L. MENDOZA-ZELIS and F.H. SANCHEZ
Departamento de Física, Universidad Nacional de La Plata, C.C. 67,
1900 La Plata, Argentina.

Results on the order-disorder transition promoted by ball milling in the intermetallic compound AlFe, at stoichiometry and two neighbouring compositons, are presented. The fully ordered compounds were ground in a horizontal oscillatory mill and their evolution under mechanical work has been examined by transmission Mössbauer spectroscopy and X-ray diffraction. Short and long range disorder was observed: the different local configurations produced by milling were charactherized by the evolution of the Mössbauer spectra and correlated with the long range order information coming from the X-ray diffractograms. Results are compared with preceding information on related systems and with predictions based on a simple model for randomly distributed antisite disorder.

Introduction

Disorder induced by mechanical work in *bcc* intermetallic compounds with the B2 structure, like AlFe, is a well known phenomenum. Nearly thirty years ago, Huffman et al [1] studied by Mössbauer effect (ME) spectroscopy the modifications induced by cold working in a series of AlFe alloys with different Al content. The stability of these compounds under ball-milling has been examined in the last decade and recently Pochet et al [2] presented a detailed study of the order-disorder transition induced by milling in B2-AlFe. It is rather well stablished that this compound, as well as B2-AlRu [3], does not amorphize under milling but undergoes a transition to a partially disordered steady state with a high concentration of antisite defects. This transformation is accompanied by an increase in lattice parameter and a reduction in particle and domain sizes.

Mössbauer spectroscopy, may prove useful in the study of the evolution of local configurations during these mechanically induced transformations due to their ability of sensing the surroundings of Fe atoms. We present here some results on the mechanical grinding of Al_{1-x} Fe_x compounds, with x = 0.45, 0.50 and 0.55, obtained with ME spectroscopy and X ray diffraction (XRD) with the aim of correlating the evolution of short- and long range order during the mechanically induced transformation to a metastable disordered state.

Experimental

The starting compounds were prepared by heating in vacuum a mixture of pure Al (0.995) and Fe (0.995) powders in appropriate proportion. The resulting ingots were ground and annealed in vacuum for a few days at 550 °C in order to remove the remaining stress and to improve the order. Samples of 250 mg were sealed into cylindrical vials under Ar atmosphere after several Ar whashings and mechanically ground with a Retsch MM2 horizontal oscillatory mill operating at 25 Hz. We used a stainless steel 5cm^3 vial and one ball (ϕ = 9 mm) of the same material. The estimated milling intensity [2] under these conditions was 380 m/s^2. After each milling process of a given duration t_m, the samples were taken off the vial and characterized by XRD and ME spectroscopy.

XRD patterns were obtained in reflection geometry using a Philips PW1710 diffractometer with Cu Kα radiation. For the analysis of these diffractograms we use a full pattern fitting procedure DBWS-9411 [4] which allows the simultaneous refinement of the lattice parameter **a**, the average size of the coherent regions for diffraction L, the mean square deviation of the unit volume deformation (microstrain) ε and the atomic fraction of Al and Fe in both sublattices of the B2 structure.

The absorbers for the ME measurements were prepared by spreading about 10 mg of material over an area of 1.5 cm^2, yielding an effective thickness of 1.1 x 10^{18} $^{57}Fe/cm^2$. The experiments were carried out with a conventional spectrometer under transmission geometry, employing a ^{57}CoRh radioactive source. Quoted isomer shifts are relative to α-Fe.

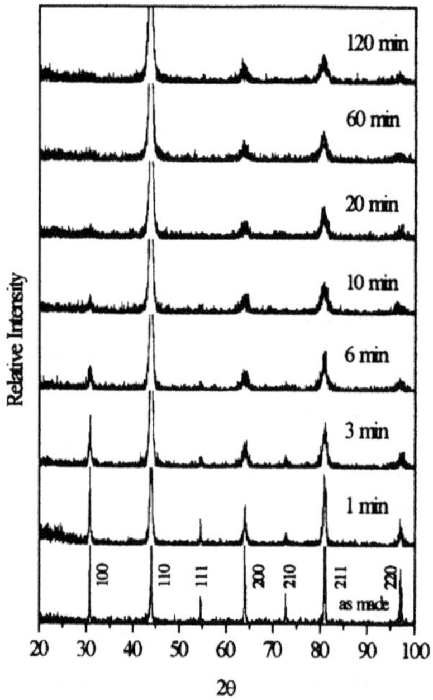

Fig 1: X-ray diffractograms of $Al_{50}Fe_{50}$ after different milling times.

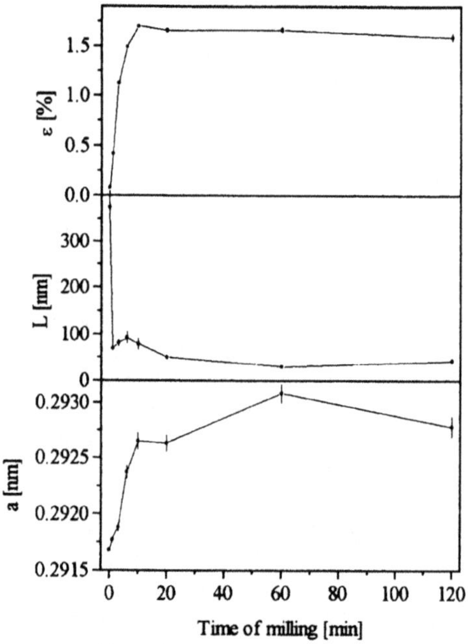

Fig 2: Evolution of a, L and ε with milling time, for $Al_{50}Fe_{50}$.

Results

As shown in Figure 1 for $Al_{50}Fe_{50}$ intermetallic compound, the XRD patterns of the starting compounds display the reflections characteristic of the *bcc* phase with B2 ordering. As milling proceeds, the intensity of the superstructure reflections decreases as a consequence of disorder. Additionally, all the peaks broaden, reflecting the grain size (L) reduction as well as the increase in stored microstrain (ε). The fitted parameters **a**, L and ε evolve rapidly with t_m, as may be seen in Table 1 and Figure 2 for $Al_{50}Fe_{50}$ intermetallic compound. The increase in the lattice parameter (a) is a clear indication of the dominant role of antisite defects in the observed disorder [5,6]. Then, from the fitted atomic fractions the long range order parameter (S), for each composition has been evaluated (see Figure 3). It may be seen how S reduces monotonically with t_m and reaches a steady state value after about 1h of milling.

It is worth noticing that the observed reduction in grain size is not as severe as reported in previous work on similar systems [2,3,5,6] in which a final size of less than 15 nm is attained. This is probably due to the characteristics of our milling device. On the other hand, the time required to get a steady state of disorder is considerably shorter in our case.

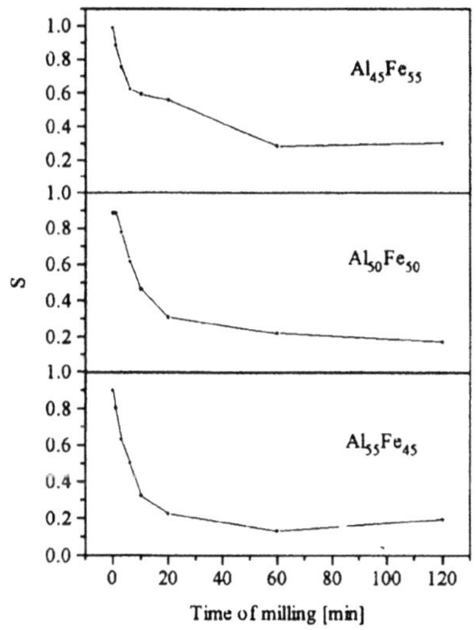

Fig 3: Evolution of the LRO Parameter with milling time for $Al_{1-x}Fe_x$ (x=0.45, 0.5, 0.55).

The ME spectra obtained after various grinding times, displayed in Figure 4 for the sample $Al_{50}Fe_{50}$, show a broad absorption peak at low velocities, characteristic of paramagnetic Al-Fe phases.

In order to analyse these unresolved doublets, several kind of fits were attempted using standard least squares routines. Good fits (shown in Fig. 4) were obtained with Voigt-shaped doublets which give account of a Gaussian distribution of the quadrupole splitting defined by its mean value $\overline{QS}$ and standard width σ. The proposed fitting function also allows for a linear correlation between the isomer shift (IS) and QS: $IS = IS_0 + \alpha\, QS$. The resulting parameters for $Al_{50}Fe_{50}$ intermetallic compound are quoted in Table I.

The $\overline{QS}$ and $\overline{IS}$ values resulting from these fits for $Al_{55}Fe_{45}$ and $Al_{50}Fe_{50}$ are shown in Figure 5 as a function of t_m. In general both $\overline{QS}$ and σ increase rapidly with t_m reflecting the departure from cubic symmetry of the local charge distribution around Fe sites and the wide variety of local configurations produced by milling. Also α changes very rapidly approaching 0. from negative values, probably as a consequence of the increasing chemical homogeneity of the Fe environments. In turns, $\overline{IS}$ decreases from values around that expected for the ordered B2 phase (0.28 mm/s) [7] towards a steady state value which depends on composition. The evolution of IS is also connected with the increase in local disorder but mainly in the chemical one, its value being closely related to the number of Al neighbours.

Table I: Fitted XRD and ME parameters for $Al_{50}Fe_{50}$ after different milling times.

t_m [min]	$\overline{QS}$ [mm/s]	$\overline{IS}$ [mm/s]	σ [mm/s]	α	S	a [nm]	L [nm]	ε [%]
0	0.042_2	0.274_3	0.078_3	-0.693_{37}	0.885_1	0.29168_1	374_{36}	0.08_1
1	0.054_2	0.269_1	0.132_2	-0.335_{11}	0.885_2	0.29177_3	69_5	0.42_2
3	0.089_2	0.256_1	0.185_1	-0.264_3	0.779_2	0.29188_4	81_8	1.12_2
6	0.135_1	0.241_1	0.216_1	-0.198_2	0.615_2	0.29237_6	92_{14}	1.19_2
10	0.187_1	0.231_1	0.218_2	-0.153_2	0.462_2	0.29265_8	79_{13}	1.70_2
20	0.217_1	0.226_1	0.220_2	-0.127_2	0.308_3	0.29263_8	50_6	1.66_3
60	0.250_1	0.217_1	0.237_2	-0.100_2	0.221_4	0.29308_9	30_3	1.66_4
120	0.250_2	0.217_1	0.247_3	-0.091_2	0.173_4	0.29278_9	42_6	1.59_4

The behaviour of $\overline{QS}$ and $\overline{IS}$ for $Al_{45}Fe_{55}$ compound, departs from the one described above, probably due to the presence of magnetic interaction, and needs further confirmation.

Discussion

As previously proposed [8], after a few minutes of milling in which most of the grain fragmentation occurs, the observed LR disorder should be mainly associated with lattice (intragrain) disorder, as indicated by the important increase in ε that accompanies the reduction in S. ME results are consistent with this interpretation. First there is some similarity in the evolution of $\overline{QS}$ and ε with t_m, both being related with SR disorder. On the other hand, the reduction in IS, indicating the variation in the number of Al nearest neighbours of Fe atoms., supports the idea of an increase in lattice disorder.

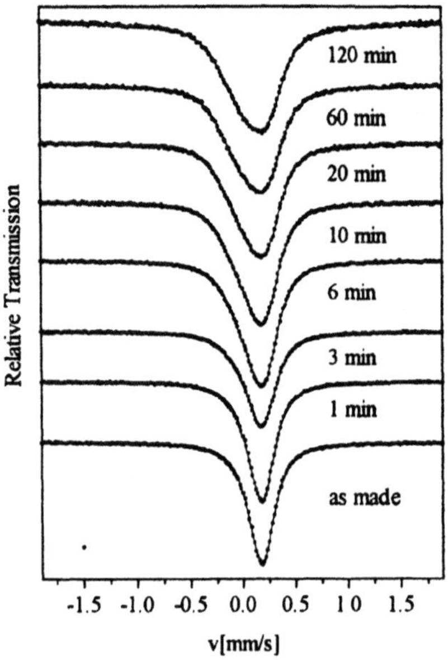

Fig 4: Mössbauer effect spectra for $Al_{50}Fe_{50}$ after different milling times.

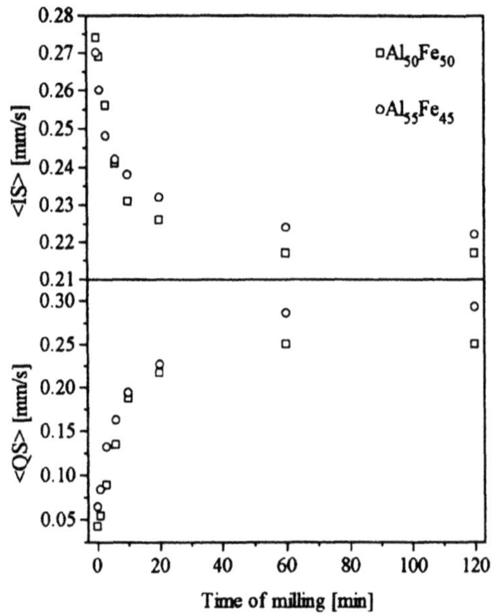

Fig 5: Evolution of QS and IS with milling time for $Al_{55}Fe_{45}$ and $Al_{50}Fe_{50}$ alloys.

In a very simple model, IS for a given Fe site may be expressed as: $m\delta_2 + n\delta_2$, where m(n) is the number of Al *nn* (*nnn*) and δ_1 (δ_2) the corresponding contribution to IS of each Al atom. The value of δ_1 and δ_2 may be adjusted to fit the measured value of IS in B2-AlFe (m=8, n=0) and in a set of *bcc* $Al_{1-x}Fe_x$ random solid solutions ($\overline{m}=8x$, $\overline{n}=6x$) [9]. If we assume that in our milled compounds disorder is solely due to antisite defects statistically distributed, n and m may be calculated from the observed value of S and a predicted value of IS results. It may be easily shown that in this case:

$$IS = (1 - x)(8\delta_1 + 6\delta_2) + xS^2(8\delta_1 - 6\delta_2)$$

The experimental values follow the predicted trend with S but are generally higher (see Figure 6). This difference may be attributed to the lattice expansion for which a positive contribution to IS is expected [10] and/or to a non uniform distribution of disorder as suggested by calorimetric studies on similar samples [8].

Calculations to improve this simple model by taking into acount i) the volume effects on IS and ii) the expected value of QS arising from the possible local configurations for the Fe neighbourhood in a point charge scheme, are presently in course.

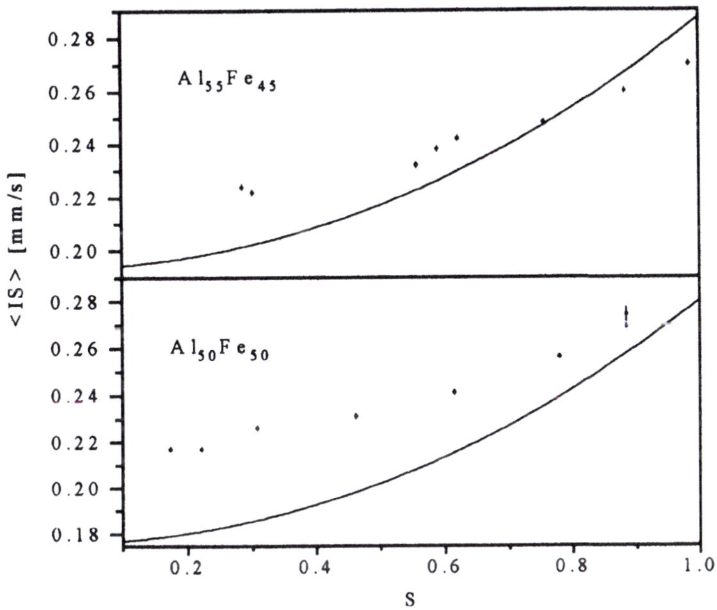

Fig 6: Comparison of the observed IS with the model described in text.

Conclusions

Room temperature milling of $B2\text{-}Al_{1-x}Fe_x$ compounds near stoichiometry leads to a steady partially disordered state. After a first stage of grain fragmentation disorder is dominated by antisite defects, probably originated in the high dislocation density introduced by the mechanical work. The short range order evolution also indicates highly disordered local environments and is consistent with a non statistical order distribution throughout the sample.

References

[1] G.P. Huffman and R.M. Fisher, *J. Appl. Phys.* **38** (1967) 735.
[2] P. Pochet, E. Tominez, L. Chaffron and G. Martin, *Phys. Rev.* **52** (1995) 4006.
[3] E. Hellstern, H.J. Fecht, Z. Fu and W.L. Johnson, *J. Appl. Phys* **65** (1988) 305.
[4] R.A. Young, A. Sakthivel, T.S. Moss, C.O. Paiva-Santos, *J. Appl. Phys.* **28** (1995) 36.
[5] I. Boerner and J. Eckert, *Mat. Sci. Forum*, **225-227**, (1996) 377.
[6] S. Suriñach, S. Gialanella, X. Amils, L Luterotti and M.D. Baró *Mat. Sci. Forum*, **225-227**, (1996) 395.
[7] S. Nasu, U. Gonser and R.S. Preston, *J. Physique* **41** (1980) C1-385.
[8] M.T. Clavaguera-Mora, J. Zhu, M. Meyer, L. Mendoza-Zélis, F.H. Sánchez and N. Clavaguera in: Proceedings of ISMANAM 96, Rome 1996, (Mat. Sci. Forum, Trans. Tech. Publications , 1996).
[9] J.H. Hsu and C.L. Chien, *Hyp.Int.* **69** (1991) 451.
[10] R. Ingalls, *Solid State Comm.* **14** (1974) 11.

STABILITY OF PHASES AND GRAIN GROWTH IN Fe-DOPED CUPRIC OXIDE.

S. J. Stewart, G. F. Goya, G. Punte, L. Mendoza Zélis
and R. C. Mercader.
Departamento de Física, Universidad Nacional de La Plata
C.C. 67, 1900 La Plata, Argentina.

Abstract

Prolonged ball-milling of α-Fe_2O_3 in a matrix of CuO lead to a broad distribution of very small particle sizes of superparamagnetic hematite and tenorite. The CuO particles grow after each annealing while α-Fe_2O_3 particles increase their size when the higher annealings temperatures are achieved. At 1000 K the spinel phase is formed. Within the limits of sensitivity of Mössbauer spectroscopy, no intermetallic compound or inclusion of Fe into the CuO matrix can be observed.

1. Introduction

In a previous work[1] we showed that by solid-state reaction iron oxide embedded in a cupric oxide matrix forms $CuFeO_2$ (delafossite) and cubic and tetragonal $CuFe_2O_4$ (spinel), depending on the annealing atmosphere. When the precursor oxides are ball-milled and annealed at high temperatures, the specimens display Mössbauer spectra with a new paramagnetic signal. Although the ball-milling process can lead to metastable phases even of elements without chemical affinity like Cu and Fe[2], there is no clear evidence about the origin of this new signal and therefore a more detailed study of this process is necessary.

To complement our work we report a Mössbauer and XRD study of the phases formed during a prolonged ball-milling performed on Fe-doped CuO, their evolution and the grain growth under different annealing temperatures.

2. Experimental

Stoichiometric amounts of CuO and α-Fe_2O_3 (^{57}Fe enriched to $\approx$ 95%) to obtain nominal $Cu_{0.995}Fe_{0.005}O$ were mechanically ground during 48 h in a cylindrical steel vial with one steel ball using a vibratory mixer mill (Retsch MM2).

The *as-milled* sample (S_0) was subsequently annealed at 600, 750, 800, 850 and 1000 K (yielding samples S_1, S_2, S_3, S_4 and S_5, respectively). Each treatment was performed in air during 40 h with heating and cooling rates of 8 K/min and 1 K/min, respectively.

The samples were characterized with XRD performed in a Philips PW-1710 Diffractometer using Cu-Kα radiation. The x-ray data were refined by Rietveld profile analysis using DBWS-9006PCS [3].

Mössbauer measurements were taken in transmission geometry with a 25mCi ^{57}Co source in a Rh matrix, with a multiscaler of 512 channels in constant acceleration mode. The amount of sample used to take each spectrum was $\approx$20mg/cm^2. A non-linear least-squares program was used to fit the spectra to Lorentzian line shapes. In some cases it was necessary to use a shape-independent distribution fitting program[4]. The isomer shifts are referred to α-Fe at 300 K.

3. Results and discussion

The XRD pattern for the *as-milled* sample (S_0) shows broadening of the most intense CuO lines indicating that the prolonged milling has reduced the particle size (Fig.1) The mean particle diameter estimated from x-ray data is ≈ 3.5 nm. This rather small particle size is not surprising, since both substances, CuO and α-Fe_2O_3, have a considerable degree of abrasiveness.

The room temperature Mössbauer spectrum for S_0 (Fig.2) displays a doublet and a small percentage of a magnetic signal(M1) whose hyperfine parameters correspond to α-Fe_2O_3 bulk. Such a doublet consists of broadened lines indicating the presence of a distribution of Fe^{3+} quadrupole splittings with an average value of about 1.2 mm/s. Taking into account the small particle size that resulted after the ball-milling, the quadrupole split doublet has been also fitted assuming two sites belonging to bulk and surface iron atoms, giving $\Delta_B = 0.97(3)$ and $\Delta_S = 1.55(2)$ mm/s respectively with isomer shifts characteristic of Fe^{3+}. The results are similar to those of Kündig et al.[5] assigned to superparamagnetic hematite with an average particle diameter of about 3 nm. In our case, the coexistence of a magnetic signal belonging to bulk atoms in the same spectrum is indicative of a broad distribution of particle sizes.

The XRD patterns of the annealed samples show a decrease in the width of the lines belonging to CuO (Fig.1), indicating the increase in the particle size as a result of the annealing. The mean particle diameter estimated for S_2 and S_4 are ≈ 2 and 7 times larger than S_0, respectively. The small quantity of iron in the samples does not allow the detection of iron phases by XRD.

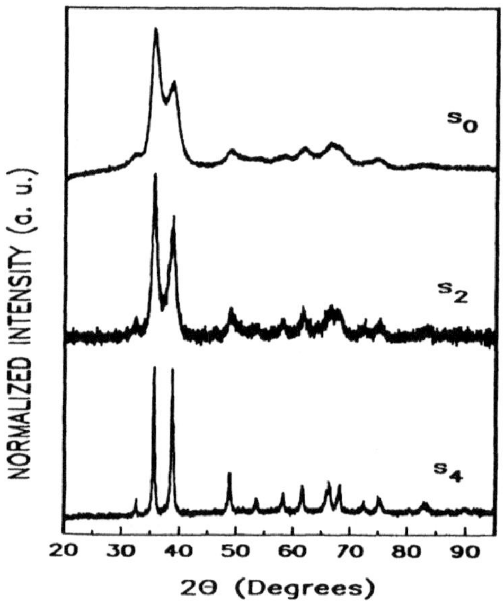

Fig.1. X-ray diffraction data for samples S_0 (as-milled), S_2 (final annealing at 750 K) and S_4 (final annealing at 850 K).

The Mössbauer spectra show that the superparamagnetic hematite doublet is present as a majority Fe-phase for the S_1, S_2, and S_3 specimens (Figs. 2 and 3), while the percentage of the magnetic signal remains almost constant. The fitting procedure with distributions yields a quadrupole splitting $\Delta \approx 1.1$ mm/s and $\Delta_B = 0.86(2)$ and $\Delta_S = 1.45(2)$ mm/s when the spectra are analyzed with two quadrupole sites. Only a decrease in the Mössbauer signal linewidths can be observed, and a slight decrease in the hyperfine parameters of the doublet compared with those of the *as-milled* sample, likely due to a structural relaxation produced in the annealings.

In S_4 the magnetic signal M1 has increased up to 15 per cent of the Fe-phases, while a new magnetic signal M2 is appearing (Figs. 2 and 3). This new signal can be fitted with very broad linewidths indicating a hyperfine field distribution, centered at $\approx$ 46.4T, and becomes the majority Fe-phase for S_5. The temperature at which it appears, 850K, is approximately the same as the formation temperature of the spinel $CuFe_2O_4$ phase in the Cu-Fe-O system [6]. This fact, jointly with the fitted hyperfine parameters, makes the assignment of this signal to $CuFe_2O_4$ a direct matter. Besides, the superparamagnetic hematite doublet is still present for this sample in a very small percentage and another Fe^{3+} doublet with $\Delta \approx 0.6$ mm/s can be distinguished.

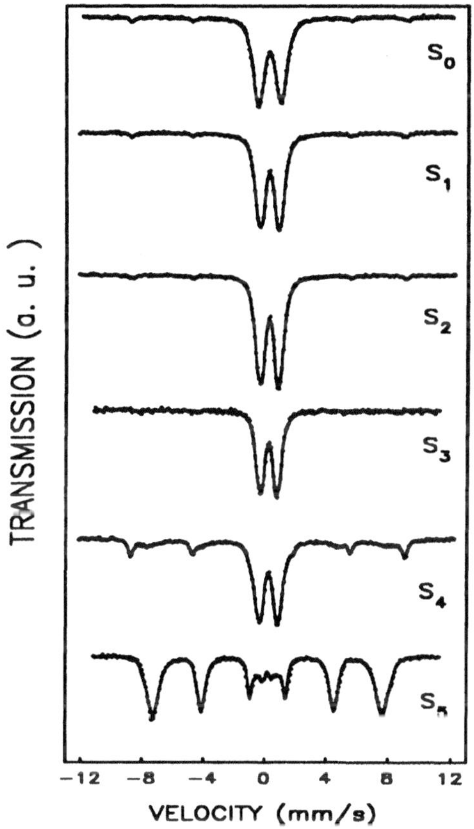

Fig.2. Mössbauer spectra for S_0, S_1, S_2, S_3, S_4 and S_5 samples. Treatments are described in the experimental section.

These results might be interpreted as follows: the prolonged ball-milling has caused a decrease of CuO particle sizes, revealed by the broadening of the x-ray lines, as well as those of

α-Fe_2O_3 evidenced by the presence of superparamagnetic signals of this compound in the Mössbauer spectra. No formation of phases between precursor oxides can be appreciated. Previous works on a similar system, namely, ball milling of a mixture of CuO and Fe, report that a displacement reaction is initiated to form Fe_3O_4 and Cu when the milling is performed under Ar atmosphere [7-8]. Although Mössbauer Spectroscopy which could allow to identify small quantities of Fe-phases was not used, also formation of phases between the reactants were not observed.

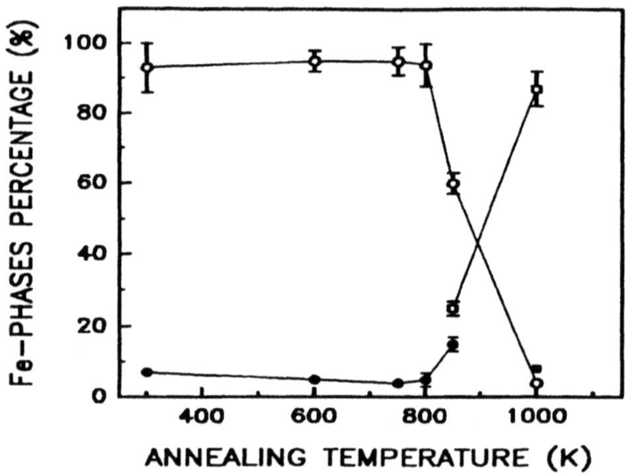

Fig.3. Fe-phases percentage. ○ Superparamagnetic hematite; ● Hematite bulk(M1); □ Spinel(M2) and ■ Fe^{3+} $\Delta \approx 0.6$ mm/s signal.

Particle sizes of CuO grow after each annealing, a fact confirmed by the linewidths decrease of the x-ray lines. However, the similarity of the Mössbauer spectra for S_1, S_2 and S_3 permit us to conclude that α-Fe_2O_3 particles have not grown as much as CuO. Probably the dispersion of hematite particles in the cupric matrix produced by the milling process and the low thermal energy supplied in the treatments up to 750 K is not enough to promote the diffusion and further sintering of hematite particles. Neither any Cu-Fe-O compound can be formed at these annealing temperatures.

The increase of signal M1 (bulk hematite) and the appearance of the M2 signal (assigned to cubic spinel $CuFe_2O_4$) in S_4, are indicative that hematite diffusion in the CuO matrix as well as the spinel formation seem to arise.

At the highest annealing temperature reached (1000 K, sample S_5) the hematite diffusion has practically been completed as evidenced by the great increase of the spinel signal. The fact that the spinel has a hyperfine field corresponding to a cubic (and not a tetragonal) structure would indicate the presence of oxygen vacancies. These vacancies would mean that $CuFeO_2$ delafossite may have formed in S_5 where a remainder Fe^{3+} doublet with $\Delta \approx 0.6$ mm/s can be observed. However, these results do not allow us to discard that this signal might be attributed to superparamagnetic spinel.

This work was supported by CONICET and U.N.L.P., Argentina.

References

[1] S. J. Stewart, G. F. Goya, G. Punte and R. C. Mercader, J. Phys. Chem. Solids, 58, 73 (1997).
[2] J. Eckert, J. C. Holzer and L. Johnson, J. Appl. Phys. 73, (1993) 131.
[3] A. Sakthivel and R. A. Young, DBWS-9006PC, Program for Rietveld analysis of x-ray powder diffraction patterns, Georgia Institute of Technology, Atlanta (1991).
[4] R. E. Vandenberghe, E. DeGrave and P.M.A. de Bakker, Hyp. Int. 83, (1994) 29.
[5] W. Kündig, K. J. Ando, R. H. Lindquist, and G. Constabaris, Czech. J. Phys.B 17, (1967) 467.
[6] D. S. Buist, A. M. M. Gadalla, and J. White, Mineral Mag. 103, (1966) 731 .
[7] J. S. Forrester and G. B. Shaffer, Metall. Trans. A, 26, (1995) 25.
[8] M. Pardavi-Horvarth and L. Takacs, J. Appl. Phys. 73, (1993) 6958.

THE CONCOMITANT EFFECT OF NOBLE GAS BOMBARDMENT AND TEMPERATURE ON THE PHASE TRANSFORMATIONS IN THE Fe-N SYSTEM

E.C. Moreira, G. Sánchez, L. Amaral and M. Behar

Instituto de Física, UFRGS, Caixa Postal 15051, 91501-970 Porto Alegre, Brasil

Abstract.

In the present work we show for the Fe-N system that the combined effect of ion irradiation and thermal annealing can lower the temperature of a given phase transformation. By Ar bombarding at 250°C and 300°C during 1 h, N implanted, Fe samples we were able to induce the $\varepsilon\text{-}Fe_2N \Rightarrow \varepsilon\text{-}Fe_{2+x}N + \varepsilon\text{-}Fe_{3.2}N$ and $\varepsilon\text{-}Fe_2N \Rightarrow \varepsilon\text{-}Fe_3N + \gamma'\text{-}Fe_4N$ phase transformations respectively. These temperatures are 50°C lower than the ones needed to produce the same transformations by single thermal annealing. In addition we have studied the phase transformation dependence on the irradiation time. The results are interpreted according to previously reported theoretical models.

Introduction.

Nitrogen implantation is a well known technique commonly used in industrial processes in order to improve tribological, mechanical and chemical properties of Fe and different kinds of steels. Several experiments related with the nitrogen implantation in Fe were reported in the literature [1-4] being the most relevant features summarized in the review article by Marest [5]. As a result of this systematic study several general features have been established. Between them, the phase formation and phase transformations resulting from annealing procedures. It was shown [3,5,6] that the Fe_2N phase is formed whenever an Fe sample is N implanted with a low energy (E $\cong$ 50 keV) and relatively high dose ($\phi \geq 3\times10^{17}$ at./cm^2). Further thermal annealing (for typically 1 h) performed on the as implanted Fe sample produce the following phase transformations.

$$\varepsilon - Fe_2N \begin{cases} \underset{250^\circ C}{\Rightarrow} \varepsilon - Fe_2N + \varepsilon - Fe_{2+x}N \\ \underset{300^\circ C}{\Rightarrow} \varepsilon - Fe_{2+x}N + \varepsilon - Fe_{3.2}N \\ \underset{350^\circ C}{\Rightarrow} \varepsilon - Fe_{3.2}N + \gamma' - Fe_4N \\ \underset{400^\circ C}{\Rightarrow} \gamma' - Fe_4N \end{cases}$$

This means that below 300°C the thermal annealing of the $\varepsilon\text{-}Fe_2N$ phase basically induces the $\varepsilon\text{-}Fe_2N \Rightarrow \varepsilon\text{-}Fe_{2+x}N$ transformation (with $0<x<1$). The $\varepsilon\text{-}Fe_{3.2}N$ phase only appears after 300°C annealing, and the $\gamma'\text{-}Fe_4N$ only after 1 h, 350°C annealing. Finally, thermal treatments performed at 400-420°C result in the complete dissolution of the Fe-N precipitates.

On the other hand, the influence of ion bombardment on nitride and carbonitrides produced by ion implantation in Fe and steels has been much less studied. Nevertheless, a systematic study performed in recent years [6-8] have shown that noble-gas (He, Ar, Kr and Xe) bombardment of the precipitates induces nitride or carbonitride dissolution and reprecipitation process.

In the present work we have extended the above results by studying the Fe-N phase transformations produced by the combined use of temperature and ion bombardment. We will show that the concomitant use of ion irradiation and thermal annealing can lower by at least 50 °C the temperature of a given phase transformation.

Experimental Procedure.

Samples of 1.5 cm diameter were cut from 99.999% pure Fe samples and mechanically polished to yield mirror clean surfaces. The N ions were implanted at 50 keV with a fluence of $\phi = 4\times10^{17}$ at./cm^2, which led to a N peak concentration of $c_p \cong 50$ at.%. The irradiations were performed with a 300 keV Ar beam. In this way the Ar ions are left beyond the region where the N ions have been implanted [$R_p(N) = 55$ nm; $R_p(Ar) = 140$ nm]. The Ar irradiation time was fixed at 1 h with a current of i = 1.5 μA.

The N implanted Fe samples were Ar bombarded at temperatures which were varied between 200 and 400°C. For comparison we have performed two additional experiments. A first one in which the as implanted samples were only submitted to 1 h thermal annealing in a 200-400°C temperature range. A second experiment, in which the as implanted samples were Ar bombarded at room temperature. In this way it was possible to characterize the ballistic effects, the pure thermal ones and both effects acting simultaneously. In addition we have studied the time dependence of the phase transformations by performing Ar irradiations at 250 °C during different time intervals varying them between 10 and 120 minutes.

At each stage of the experiment the nitride phase characterization was done by means of the conversion electron Mössbauer spectroscopy (CEMS) technique at room temperature with a ^{57}Co-Rh source in a standard set up.

Results.

In the CEMS spectra presented below the following notation was used: H_1 is the magnetic field of the α-Fe phase, Q_1 is the quadrupole splitting corresponding to the ε-Fe_2N phase , H_2 and Q_2 are the magnetic field and the quadrupole splitting, respectively, of the ε-$Fe_{2+x}N$ phase, finally H_3 and H_4 are the magnetic fields of the ε-$Fe_{3\,2}N$ and γ'-Fe_4N phases, respectively. The Mössbauer hyperfine parameters used in the present work are those reported in ref. [5].

The CEMS spectrum of the as N implanted sample (not shown here) displays the characteristic sextet of the martensitic phase plus the doublet typical of the ξ or ε-Fe_2N phase [5]. The subsequent 250°C 1 h anneal produces a partial ε-$Fe_2N \Rightarrow$ ε-$Fe_{2+x}N$ dissolution as shown in Figure 1(a). A 300°C, 1 h anneal on the as implanted sample induces the ε-$Fe_2N \Rightarrow$ (ε-$Fe_{2+x}N$ + ε-$Fe_{3\,2}N$) phase transformation – see Figure 1(c). At 350°C this transformation is completed and the γ'-Fe_4N phase appears. Finally the 400°C anneal dissolve almost all the precipitates leaving a small quantity of γ'-Fe_4N nitrides in the CEMS analyzed region. In Figure 2(a) are shown the normalized CEMS spectral areas for the matrix and the nitrides as a function of the annealing temperature.

On the other hand the Ar irradiation on the as implanted sample performed at room temperature only induces a partial ε-$Fe_2N \Rightarrow$ ε-$Fe_{2+x}N$ transformation feature which is in agreement with previous results [7].

In what follows, we are going to describe the results of the combined action of ion bombardment and temperature on the N implanted samples. In a first step we have bombarded the nitrides at a fixed time (1 hour) changing the temperature of bombardment between 200 and 400 °C

The results for the Ar irradiated samples at 200 and 225°C are very similar between them. Starting from the as-implanted N sample (with 55% of ε-Fe_2N) the almost total ε-$Fe_2N \Rightarrow$ ε-$Fe_{2+x}N$ phase transformation (15% of ε-Fe_2N and 40% of ε-$Fe_{2+x}N$) is produced. On the other side, the Ar bombardment performed on the as implanted sample at 250°C induces a drastic change in the CEMS spectrum as illustrated in Figure 1(b). An analysis of the spectrum shows three interesting features: a) complete dissolution of the ε-Fe_2N precipitates, b) a drastic reduction in the amount of ε-$Fe_{2+x}N$ nitrides (22% of the total CEMS spectrum) and c) a significative amount of ε-$Fe_{3\,2}N$ precipitates (35% of the total CEMS spectrum).

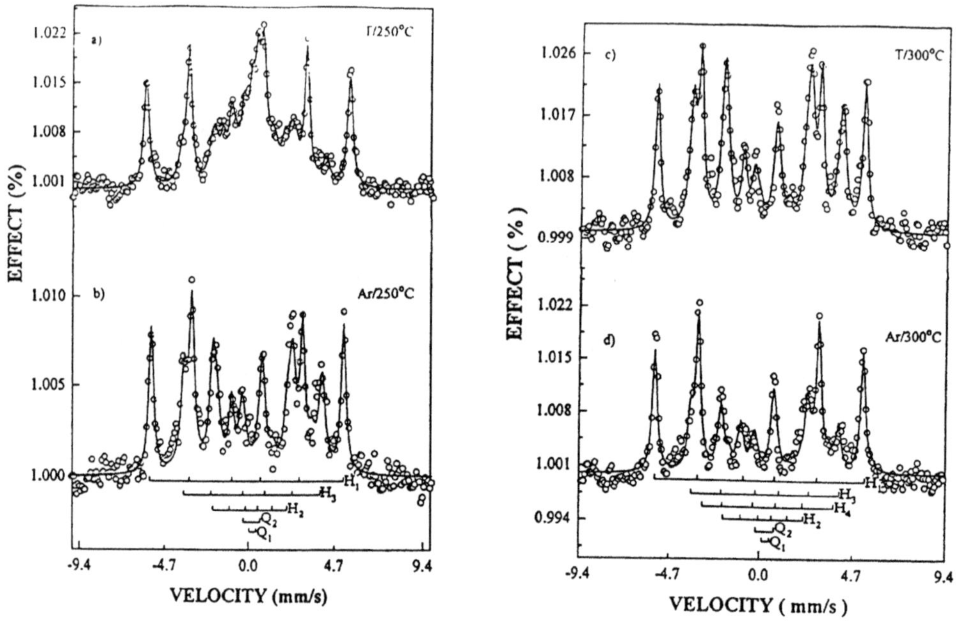

Figure 1: CEMS spectra of sample (a) thermal annealed at 250 °C, (b) Ar-irradiated at 250 °C, (c) thermal annealed at 300 °C and (d) Ar-irradiated at 300 °C.

The Ar bombardment on the N implanted sample performed at 300°C (see Figure 1 (d)), basically induces the ε-Fe_2N ⇒ ε-$Fe_{3.2}N$ + γ'-Fe_4N phase transformation. It should be stressed that the amount of ε-$Fe_{2+x}N$ is strongly reduced (less than 5%) while for the first time appears the γ'-Fe_4N phase (≅14%).

The Ar bombardments performed at 350°C and 400°C do not bring significative new results. At 350 °C there is again an ε-Fe_2N ⇒ ε-$Fe_{3.2}N$ + γ'-Fe_4N phase transformation. However, in this case the γ'-Fe_4N phase is the one that predominates.

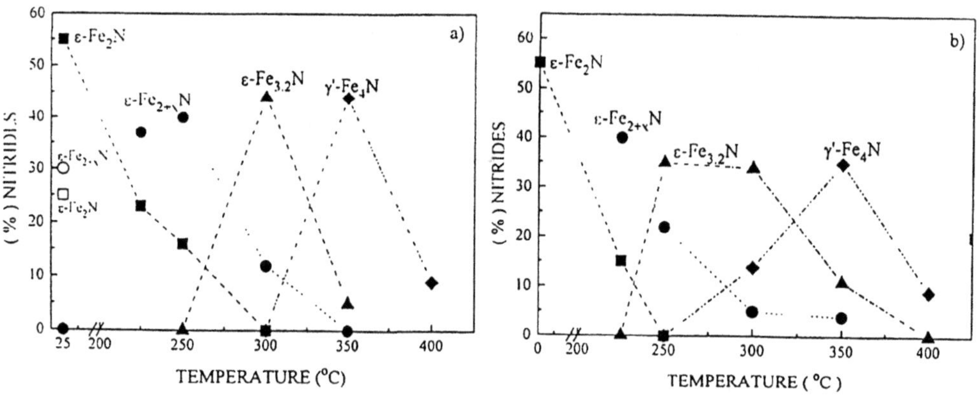

Figure 2: Nitride retention as a function of the temperature for (a) thermal annealed and (b) Ar-bombarded samples. The open symbols represent the nitride retention after Ar-bombardement at RT.

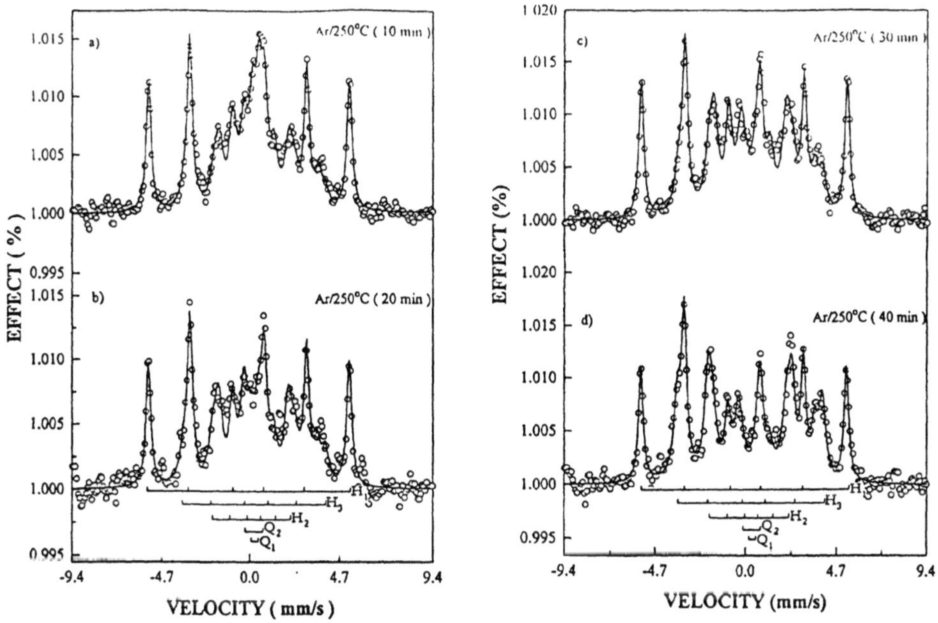

Figure 3: CEMS spectra of samples Ar-irradiated at 250 °C during (a) 10 min, (b) 20 min, (c) 30 min and (d) 40 min.

Finally at 400°C, almost all the nitrides are dissolved remaining a little amount of γ'-Fe_4N precipitates.

In Figure 2(b) we summarize the above mentioned data. There are shown the relative amount of each type of nitride as a function of the temperature at which the samples were bombarded.

In a second step we have studied the nitride evolution as a function of the irradiation time. With this purpose we have changed the time of bombardment between 10 min and 120 min maintaining the samples at a fixed temperature (250 °C).

After 10 minutes of irradiation, the ε-Fe_2N phase is partially dissolved into ε-$Fe_{2+x}N$ and ε-$Fe_{3\ 2}N$ phases, as can be seen in Figure 3 (a) where the characteristic doublet Q_1 of the phase ε-Fe_2N is clearly distinguished.

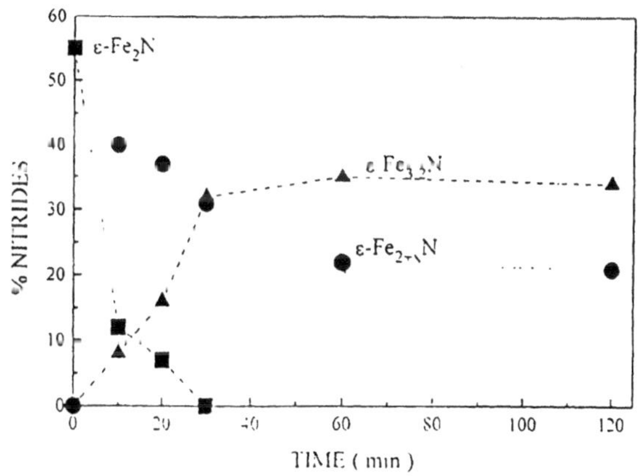

Figure 4: Dependence of the nitride retention on the time of irradiation.

With increasing the irradiation time there is a decreasing in the amount of ε-Fe_2N precipitate and simultaneously, an increasing of the ε-$Fe_{3.2}N$ phase. After 30 minutes of irradiation the ε-Fe_2N phase is totally dissolved remaining just the ε-$Fe_{2+x}N$ and ε-$Fe_{3.2}N$ phases (see Figure 3 (c)). Finally, after 40 minutes of irradiation, the ε-$Fe_{3.2}N$ phase predominates -see Figure 3 (d), at the expense of the ε-$Fe_{2+x}N$ phase . We have not observed any additional changes when the irradiation time was further increased, feature that indicate that a steady state was reached after 40 minutes of irradiation.

The above mentioned results are summarized in Figure 4, where it can be observed that a saturation process of the phase ε-$Fe_{3.2}N$ is reached after 60 minutes of irradiation.

Discussion.

A comparison between the above described results and the ones obtained by single annealing reveals that the combined effects of Ar bombardment and temperature are able to reduce the temperature for a given phase transformation in at least 50°C. In fact an inspection of Figure 2 shows that the Ar bombardment performed at 250°C produces similar phase transformation that the single thermal annealing done at 300°C. A similar feature is observed when the Ar bombardment is performed at 300°C, as compared with the result obtained by the thermal treatment performed at 350°C: a reduction in the amount of the most rich nitrides (ε-$Fe_{2+x}N$ and ε-$Fe_{3.2}N$) and the appearance of less rich N precipitates (γ'-Fe_4N).

The present results can be explained on the basis of the model proposed by Nelson et al. [11] which basically states that when metals or alloys containing precipitates are irradiated, the thermodynamics can be changed by two processes: First, dissolution of the precipitates through cascade dissolution and/or disordering effects and second, radiation enhanced diffusion of the solute within the host matrix.

In the present case the single Ar bombardment on the ε-Fe_2N precipitates may cause simultaneously a partial dissolution of the nitrides and a N radiation enhanced diffusion toward the precipitates leading to the ε-$Fe_2N \Rightarrow$ ε-Fe_2N + ε-$Fe_{2+x}N$ reprecipitation process. Therefore, according to Nelson et al. [11], the fact that the Ar bombardment at high temperatures leads to the formation of less rich ε-$Fe_{3.2}N$ and γ'-Fe_4N nitrides is an indication that the concomitant action of the temperature and Ar irradiation inhibit the diffusion toward the precipitates, enhancing instead the N diffusion out of the implanted region. The same conclusion can be reached if one compares the results of the single thermal annealing performed at 250°C with the one obtained with the Ar bombardment performed at the same temperature – see Figure 2. As can be observed in the second case we have obtained less rich nitrides because: a) we have dissolved completely the ε-Fe_2N precipitates, b) we obtained a smaller amount of ε-$Fe_{2+x}N$ nitrides and c) we induced, for the first time, the formation of ε-$Fe_{3.2}N$ precipitates.

As can be observed by inspection of Figure 4, exists a minimum irradiation time (approximately 25 min) in order to induce the complete dissolution of the ε-Fe_2N phase, the most stable one under Ar-irradiation. We can interpret this result as follows: with increasing the irradiation time the nitrogen out diffusion of the precipitates mechanism predominates on the N enhanced diffusion toward the nitrides. As a consequence the precipitates become (very likely) smaller and less N rich. However for a given precipitate's size an equilibrium is reached being the N out diffusion compensated by the N in diffusion. Of course, this is a qualitative picture and calculation should be done in order to confirm this description.

Conclusions.

We have demonstrated that the combined use of ion irradiation and temperature for the Fe-N system can lower the temperature for a given phase transformation. We have shown that by Ar bombarding at 250 and 300°C N implanted Fe samples, we were able to reduce the phase transformation temperature in at least 50°C as compared with the one induced only by simple thermal annealing. The simultaneous action of the particle bombardment and the temperature seems to induce the N diffusion out of the implanted region, favoring the formation of less rich nitrides.

Also, we have observed that the phase transformations are produced progressively during the ion bombarding until a steady state, is reached after 60 minutes of irradiation.

References.

1. G. Principi, P. Matteazi, E. Ramous, and G. Longworth, *J. Mat. Sci.* **15**, 2665 (1980).
2. D. Firrao, M. Rosso, G. Principi, and R. Grattini, *J. Mat. Sci.* **17**, 1773 (1982).
3. C.A. dos Santos, M. Behar, and I.J.R. Baumvol, *J. Phys. D* **17**, 969 (1984).
4. N. De Cristoforo and R. Kaplow, *Mat. Trans.* A **8A**, 425 (1977).
5. G. Marest, in *Ion Implantation* 1988, F.H. Wöhlbier (Ed.) (Trans-Tech. Publication, 1988) and references therein.
6. M. Kopcewicz, J. Japielbki, A. Turos and D.L. Williamson, *J. Appl. Phys.* **71**, 4217 (1992).
7. B. Rauschenback, A. Kolitsch and K. Hohmuth, *Phys. Stat. Solid A* **80**, 471 (1983).
8. M. Behar, P.J. Viccaro, M.T.X. Silva, A. Vasquez, C.A. dos Santos, and F.C. Zawislak, *Nucl. Instr. and Meth.* **B19//20**, 132 (1987).
9. S.M.M. Ramos, L. Amaral, M. Behar, G. Marest, A. Vasquez and F.C. Zawislak, *Radiation Effects and Defects in Solids* **110**, 355 (1989).
10. C.E. Foerster, L. Amaral, N. Moncoffre and M. Behar, *Nucl. Instr. and Meth. B* **80/81**, 313 (1993) and references therein.
11. R.S. Nelson, J.A. Huddson, and D.Y. Mazey, *J. Nucl. Mat.* **44**, 318 (1972).

SOLID STATE REACTION IN FE_{75} GE_{25} BY MECHANICAL WORK

A.F. Cabrera, F.H. Sánchez and L. Mendoza Zélis
Departamento de Física , F.C.E, UNLP, La Plata, Argentina

Abstract

Mixtures of elemental constituents with nominal composition $Fe_{75}Ge_{25}$ were mechanically alloyed in a horizontal oscillatory ball mill under Ar atmosphere.

The alloying reaction was studied by ^{57}Fe Mössbauer effect spectroscopy and Cu Kα X-ray diffractometry. The evolution of the system can be described in terms of two processes, for t < 4h, the diffusion of Ge into bcc-Fe leads to a dilute solid solution $Fe_{1-x}Ge_x$ and to the formation of a paramagnetic phase. For $t \geq 4$ h a random, supersaturated bcc-$Fe_{75}Ge_{25}$ solid solution forms. Between 4h and 20h the Mössbauer spectra remain in a pseudostationary state and the grain size at about 7 nm. Finally, for t = 30h the system amorphizes.

Introduction

Mechanical alloying is a suitable tool for producing metaestable intermetallic phases. In particular extended range solid solutions[1] and out of equilibrium superstructures[2] can be prepared using this technique combined with appropriate thermal treatments.

In this paper the formation of bcc-$Fe_{75}Ge_{25}$ supersaturated solid solution is reported. We also report that after sufficiently long times the system appears to undergo an amorphization transition

Experimental procedure

$Fe_{75}Ge_{25}$ alloys were prepared by mechanical alloying (MA) from elemental constituents Fe (foil, 1mm thick, 99,999%) and Ge (ingot , 99,998%). Ball milling, for different times , was performed at about 30 Hz with a Retsch MM2 horizontal vibratory mill using a 5 cm^3 stainless steel vial with a 9 mm ball of the same material. Samples of 300 mg were sealed into the milling vial under Ar (0.9998) atmosphere in a globe bag connected to a vacuum pump and to the Ar supply.

The alloys were characterized by means of Mössbauer Effect (ME) experiments and X-ray diffractometry (XRD).

The Mössbauer spectra were recorded in transmission geometry using a constant acceleration spectrometer with a 20 mCi ^{57}Co*Rh* source. The spectra were fitted using a large number of theoretical subspectra separated by regular hyperfine magnetic field (HMF) intervals (hyperfine field distributions), a linear relationship between the isomer shift (δ) and HMF (H) has been assumed . All isomer shifts quoted are relative to α-Fe. XRD patterns were obtained in reflection geometry using a Philips PW1710 diffractometer with Cu-Kα radiation. All measurements were performed at room temperature.

Results and Discussion

We analyzed the $Fe_{75}Ge_{25}$ system, as a function of milling time (t) in order to monitor the changes during the MA.

The evolution of the system can be described in three distinctive steps. The first one (t < 4h) is shown in the fig.1a. The spectra were processed with two distributed sites, one of them a dipolar magnetic interaction and the other an electric quadrupolar interaction. This can be interpreted as the reaction advance by means of two simultaneous processes: the diffusion of Fe into Ge (paramagnetic phase) and the diffusion of Ge into bcc-Fe producing a dilute solid solution (magnetic phase).

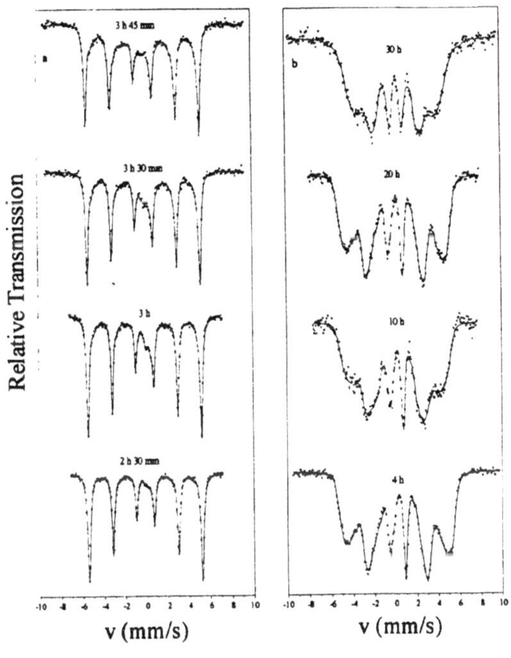

Fig1.ME spectra obtained for different time of mechanical work

The broad quadrupolar doublet with $\Delta Q \approx 0.52$ mm/s and $\delta \approx 0.36$ mm/s is similar to that reported by Hamdeh et al.[3] for amorphous Fe_xGe_{1-x} ($0.18 \leq x \leq 0.27$) and hence it may be correspond to a disordered Ge-rich phase.

With increasing milling time the fraction of probes in paramagnetic environments (quadrupolar doublet) increases while the HMF distribution broadens and shifts to lower H ranges. The behaviour observed for the evolution of the mean values of H and δ agrees with that reported by Vincze et al.[4] for the dilute alloys.

The reaction accelerates drastically in the interval 3 h 45 min -4 h, as can be observed from the sudden change of the ME spectra and the bcc lattice parameter (see Fig.3).

In the second stage ($4h \leq t \leq 20h$, fig. 1b) supersaturated bcc-$Fe_{75}Ge_{25}$ random solid solution was formed. The measurements can be fitted with one distributed site (H from 10T to 40T), the associated isomer shifts are in accord with those reported by Fultz et al.[2] for the different environments of ^{57}Fe atoms in a partially ordered solid solution . Both the ME and XRD patterns present only subtle changes suggesting the achievement of a pseudostationary state. As it will be discussed below in this stage the solid state reaction has been completed.

Finally, for t = 30 h the Mössbauer spectra becomes symmetric. At the same time the XRD patterns shows only two very broad reflections, suggesting that the system has been amorphized.

Representative X-ray diffraction patterns are presented in Fig.2.

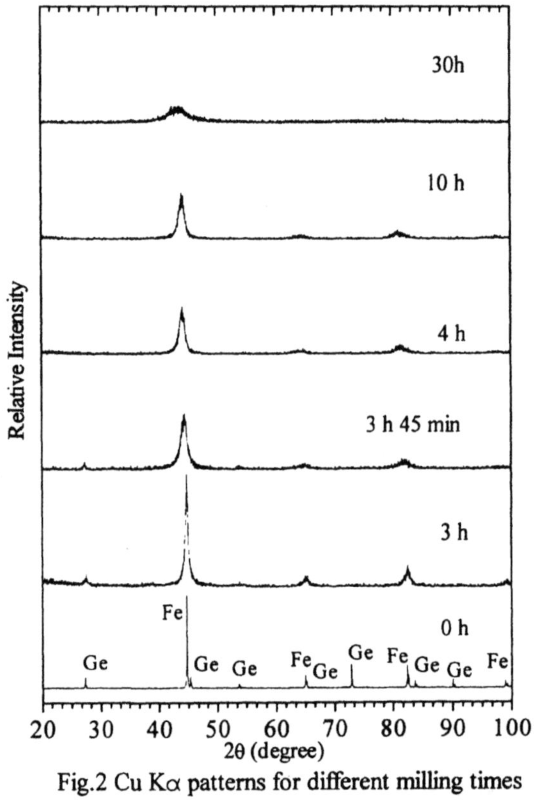

Fig.2 Cu Kα patterns for different milling times

During the first stage these data show the diffraction lines corresponding to bcc-Fe and to fcc-Ge. The diffraction peaks of bcc lattice become broader while the peaks of Ge disappear for milling times greater than 3 h 45 min, indicating that Ge has diffused completely into bcc-Fe. Between t= 4 h and 20 h the diffractograms correspond to a random supersaturated bcc-$Fe_{75}Ge_{25}$ solid solution. These patterns can be distinguished from the pure bcc-Fe one by a systematic shift of the lines to lower angles. The lattice parameter a, (Fig. 3), were obtained by the Nelson-Riley extrapolation method[5] with the (110), (200) and (211) diffraction. For t = 20h we estimated a = 2.90 Å.

The linewidths were used to estimate the means squared strain and grain size by the integral width method[6] . The grain size decrease from about 210 nm (2 h 30 min) to 7 nm (10 h), this value was in accord with the reported by Bansal et al.[7] .

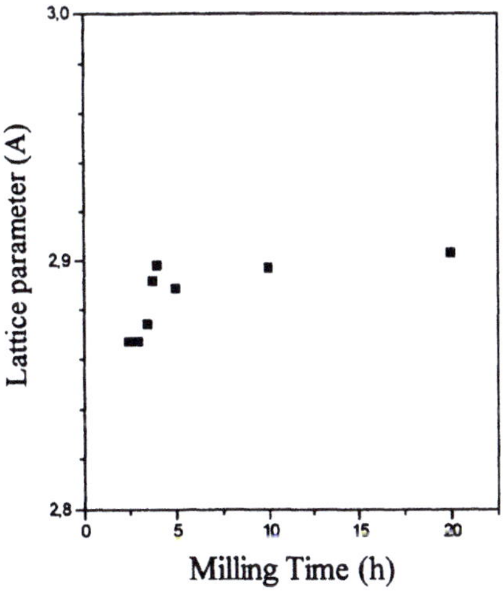

Fig3. Dependence of the bcc lattice parameter on milling time

Summary and Conclusions

It is evident that, under ball milling, the system goes through three distinct stages. The first one (t< 4 h) is dominated by the interdiffussion of the constituent elements yielding one paramagnetic and one magnetic phases. These two phases correspond to diffusion of :

1. Fe into Ge, characterized by an electric quadrupolar interaction
2. Ge into Fe, characterized by a hyperfine magnetic field distribution.

The grain size decreases drastically in this stage from about 210 nm (2 h 30 min) to about 10 nm (3 h 45 min).

In the next stage (4 h $\leq$ t $\leq$ 20h) when the interdiffussion has completed, the alloying process reaches a pseudostationary state. The Mössbauer spectra can be reproduced by a broad hyperfine field distribution and the XRD patterns present the three peaks corresponding to a random bcc-$Fe_{75}Ge_{25}$ supersaturated solid solution. The lattice parameter of the bcc structure has been measured to be a $\approx$ 2.90 Å (20 h).

During the last stage (t $\approx$ 30 h) the system appears to undergo an amorphization transition, characterized by the broadening of the two main bcc reflections.

References

[1] A.F. Cabrera, F.H. Sánchez and L.Mendoza Zélis, Phys. Rev. B 53 (1996) 8378
[2] B. Fultz, Z.Q. Gao, H.H. Hamdeh and S.A. Oliver, Phys. Rev. B 49 (1994) 6312.

[3] H.H. Hamdeh, M.R. Al-Hilali, N.S. Dixon and L.S. Fritz, Phys. Rev. B 45 (1992) 2201.
[4] I. Vincze and T. Aldred, Phys. Rev. B 9 (1974) 3845.
[5] H.P. Klug and L.E. Alexander, in X-Ray Diffraction Procedures (Wiley, New York, 1974) 594.
[6] H.P. Klug and L.E. Alexander, in X-Ray Diffraction Procedures (Wiley, New york,1974) Chap 9.
[7] C. Bansal, Z.Q. Gao, L.B. Hong and B. Fultz, J. Appl. Phys 76 (1994) 5961.

MÖSSBAUER SPECTROSCOPY AND X-RAY DIFFRACTION STUDY OF Fe_3ZnN NITRIDE OBTAINED BY MECHANOSYNTHESIS

R. S. de Figueiredo[1,2], J. Foct[2] and M. A. B. de Araújo[1]

[1]Grupo de Magnetismo e de Materiais Magnéticos
Dep. Física - UFCe 60.451-970 Fortaleza CE Brazil
[2]Laboratoire de Métallurgie Physique, URA CNRS 234
Bât. C6 2ème étage - USTL 59655 Villeneuve d'Ascq France

ABSTRACT :

The Fe_3ZnN nitride is obtained by grinding iron nitride and metallic zinc powders. X-ray diffraction reveals that this compound is two phased. Mössbauer spectroscopy reveals that one of the phases is hexagonal disordered iron based nitride and that the magnetic hyperfine field is reduced by the presence of zinc. After thermal ageing at 820 K it is observed that this compound becomes Perovskite-like (γ'-Fe_3ZnN) and ordering mechanisms take place reducing the magnetic moment of iron atoms. Lattice parameter increases with Zn substitutions and changes from 0.3784 nm (γ'-Fe_4N) to 0.3799 nm (γ'-$Fe_{3.4}Zn_{0.6}N$).

INTRODUCTION :

The possible applications of iron nitrides are very important mainly because of well-established reasons: a) high nitrogen steels present good mechanical and corrosion resistance; b) transition metal nitrides may also be used as hard metals in order to make wear resistant materials; c) some iron nitrides have high saturation magnetization and low coercivity, which point them out as good candidates for magnetic recording heads, as well as other magnetic applications. Here we are interested in combined effects of nitriding and galvanising treatments of iron [1]. Up to now the ternary phase diagram (Fe-Zn-N) is not known. Due to the fact that the stability of iron nitrides is rather low [2] in comparison with that of Ti, V, Cr, etc., the synthesis of Fe-N alloys is to some extent difficult. This paper shows that mechanical alloying, or mechanosynthesis, constitutes an unique technique for making these alloys. In fact it has been demonstrated that many binary and ternary iron nitrides can be obtained by mechanosynthesis [3,4]. Its has also been shown that under grinding the Perovskite γ'-Fe_4N nitride becomes hexagonal (ε) [3] and other Fe_XN alloys, $X \geq 5$ become cubic [5]. In this paper we will discuss the formation of γ'-Fe_3ZnN phase by mechanosynthesis, the ageing effects, its crystallographic structure, Mössbauer parameters and ordering mechanisms.

EXPERIMENTAL METHOD :

Commercial iron powder (50 μm and 99.9% of purity) was gas nitrided in order to obtain Fe_2N powder, with 15 μm particle size and well known ζ structure verified by X-ray diffraction and Mössbauer spectroscopy. The ζ-Fe_2N nitride was ground in a SPEX 8000 mill under inert atmosphere with iron (99.9%) and zinc (99.5%) commercial powders at Fe_3ZnN stoichiometry. The ball to powder mass ratio was maintained near 10. In order to homogenise the Fe_3ZnN compound it is necessary a milling time over 64 h which is much greater than the time necessary to homogenise the $2Fe+Fe_2N$ binary mixture (16 hours). The milled powders were analysed by X-ray diffraction in a Bragg-Bretano geometry using a 0.02 degree as 2θ step with a Co anticathode, and by Mössbauer spectroscopy in transmission geometry using a ^{57}Co source in Rh matrix. All isomer shifts quoted are relative to α-Fe.

RESULTS AND DISCUSSION:

Figure 1 presents the X-ray diffraction pattern of Fe_3ZnN milled after 64 h of milling (A) and then after a thermal treatment under inert atmosphere at 830 K for 1 minute (B); 8 min. (C); 20 min. (D) and 30 min. (E). It is clear that a crystallographic transformation occurs and seems to be over for times greater than 8 min. Virgin milled powders present two phases, a typically hexagonal close packed phase, and probably the Γ1 phase, however due to the very broad and not well defined X-ray diffraction peaks an unnambiguous identification was not possible. Moreover, the poor quality of the diffractograms indicates that this compound is nanocrystalline with a grain size about 20 nm. After thermal ageing the minor Γ1 phase volume decreases rapidly for $t < 8$ min. and then asymptotically for ageing times greater than 8 min. The major phase is indexed as a Perovskite-like with a lattice parameter a = 0.3799 nm. The fitting of these diffractograms, using Rietveld method, reveals no changes in lattice parameter for ageing time greater than 8 min. and also that around 60% of (0,0,0) position are occupied by Zn atoms if no Zn is supposed to be at center of the face, (½, ½, 0) position. Figure 2 presents the Mössbauer spectra of virgin Fe_3ZnN milled for 64 h (A) and after a thermal treatment under inert atmosphere at 830 K for 1 min. (B); 8 min. (C); 20 min. (D) and 30 min. (E). Fitting of these spectra was based on a hyperfine magnetic field (H_{HF}) distribution with fixed quadrupole splitting and linear correlation between isomer shift and H_{HF} by histogram method. Data obtained by these fitting are summarised in Tab. I. As it can be easily observed, the Mössbauer spectroscopy reveals a crystallographic transformation. The most important changes appeared between 2C) and

Fig.	Sub-Spectrum	H_{HF} (T)	QS (mm/s)	IS (mm/s)	S (%)
2 A)	***	<21.1>	-0.01	<0.140>	100
2 B)	***	<21.0>	-0.04	<0.200>	100
2 C)	I	~31.5	-0.01	0.133	20
	II	~20.5	-0.01	0.205	68
	III	~8	-0.01	0.300	12
2 D)	Doublet	***	0.840	0.232	37
	O	~31.0	0.03	0.103	15
	II	~18.0	0.03	0.209	48
2 E)	Doublet	***	0.802	0.236	45
	O	~30.0	-0.06	0.164	9
	II	~17.0	-0.06	0.210	46
as E)	O	33.0	-0.057	0.230	10
at 15 K	II	16.1	-0.079	0.312	90

Tab. I - Mössbauer data of Fig. 2. H_{HF} is magnetic hyperfine field. QS is the quadrupole splitting. IS is the isomer shift, quoted relative to α-Fe and S is the abundance.

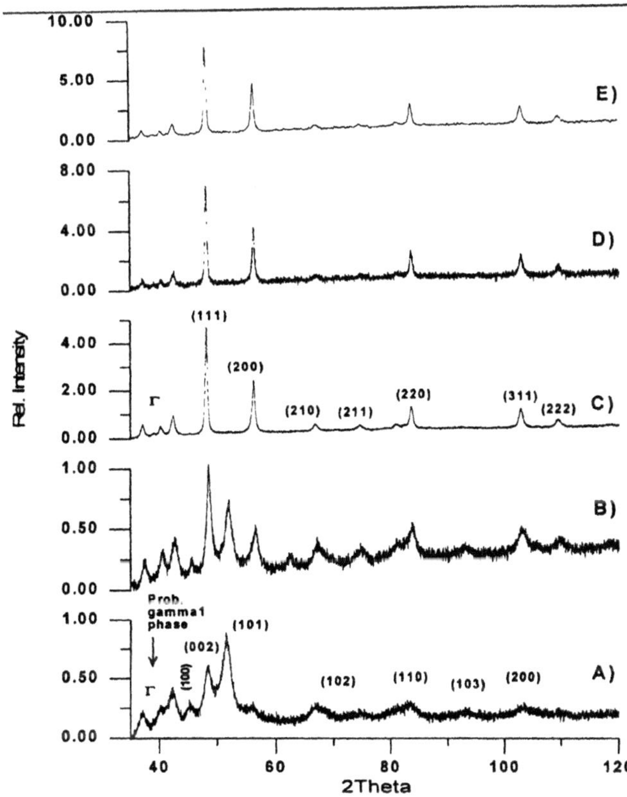

Fig. 1 - XRD pattern of milled and aged Fe_3ZnN powders. A) Virgin 64 hours milled, showing a hcp major phase and a Γ1 minor phase; B) aged for 1 min at 830 K, showing a partial crystallographic transformation; C) aged for 8 min; D) aged for 20min; E) aged for 30 min, showing a complete crystallographic transformation hcp → Perovskite-like.

2D) (ageing time between 8 and 30 min.) and correspond to a "magnetic→non-magnetic" transformation. The quadrupole splitting observed in a paramagnetic doublet 2D) (~0.8 mm/s) is larger than the one observed in the magnetic spectrum 2C) (~ 0) suggesting that this compound presents a spin rotation, despite the greater error in quadrupole splitting value induced by distribution method, which has been observed in Zn-Mn-N nitride [6]. When the ageing time is equal or longer than 8 min. it is possible to separate the magnetic hyperfine field distribution in three different blocks: O, II and III representing 0, 2 and 3 nitrogen nearest neighbour respectivelly.

It is clear that mechanosynthesis produces a two phased compound (Fe_3ZnN), made up of a small amount of Zn rich phase (probably Γ1 phase) and of a hcp phase iron nitride rich phase. The existence of an ε-$Fe_{4-X}Zn_XN$ phase can be explained by the same reason as for ε-Fe_4N. In fact, impacts generated during milling induce many defects as stacking faults in ABCA package of fcc structure which becomes ABAB corresponding to a local hcp structure. This local hcp structure can be relatively stable.

It seems that nitrogen reduces the stacking fault energy in austenite [7,8], but substitutional atoms also play an important role. The substitutional atoms can act in two different ways. First by changes in N distributions. In fact it has been observed that some substitutions such as Cu, favour an interstitial ordering reducing the stability of II sites [9]. Second, by increasing the free energy $\Delta G_{\gamma \rightarrow \varepsilon}$ that occurs probably during Sn and In substitutions [4]. Furthermore, Zn substitution does not change the nitrogen distribution in virgin milled iron nitrides powders, as shown in Fig. 2 A), since ground ε-Fe_4N and ε-Fe_3ZnN presents basically the same magnetic hyperfine field distributions, centered in Fe II environment (~21T) [3]. So it is expected that the same mechanisms occur during mechanosynthesis of binary iron nitrides and mechanosynthesis of ternary iron-zinc nitrides (Zn in minority). The most important difference between virgin milled Fe_4N and Fe_3ZnN is given by X-ray diffraction pattern which reveals that Fe_3ZnN is two phased in contrast with of the single phased Fe_4N nitride. By thermal ageing the ε-$Fe_{4-X}Zn_XN$ + $\Gamma 1$ compound evolves, Figs. 1 and 2, as observed for Fe_4N case. The volume reduction of $\Gamma 1$ phase during first steps of ageing is caused by an increasing Zn solubility in the Perovskite phase. The enhancement of solubility in milled powder has been observed in other compounds [4,10] and has given reasons to explain why it is possible to obtain alloys with positive enthalpy of mixture by milling [11,12]. For ageing time longer than 20 min., the volume of $\Gamma 1$ phase reduces very slightly, which can be explained by zinc vapour pression of $\Gamma 1$, higher than that of γ' at 830K, since the lattice parameter of Perovskite phase did not change. The second and the most important phase detected by X-ray diffraction also changes during thermal ageing, a hcp→Perovskite-like transformation is clearly observable in Fig. 1. This transformation is fast as indicated by the appearance of γ'-200 peak, Fig. 1 B). Although the transformation caused by thermal ageing seems to be over after 8 minutes, as shown in Fig. 1, the Mössbauer spectra shown in Fig. 2 exhibit a continuous change due to ordering phenomena. As observed in several iron-nitrogen compounds, hyperfine field is mainly determined by the N nearest neighbours, therefore it is possible to evaluate the number of nitrogen nearest neighbour and the nitrogen content [3,13,14]. The first modification observed in magnetic hyperfine field distribution is an important decreasing of site I abundance, in Fig. 2. This change in the interstitial ordering corresponds to $\varepsilon \rightarrow \gamma'$ transformation. After the disappearance of site I the most important changes are the vanishement of site III and the gradual decreasing of O site, Fig. 2 and Tab. I. The iron nitrogen bond which is stronger than zinc-nitrogen bond, leads to an ordered occupation of 0,0,0 position (O environment) by Zn atoms. The driving force for ordering arises from local free enthalpy and involves diffusion of substitutional atoms which explain why the kinetics is slower. Since Zn atoms move to crystallographic sites for whose distance to N atoms is maximum, the changes in the Mössbauer spectra show a drop in the value of median magnetic hyperfine field. It is reasonable to suppose that for 30 min. ageing time the totality of Zn atoms are positioned at O site, so X=0.64 since 9% of iron atoms still occupy this position (Tab. I). This value of X obtained from Mössbauer spectroscopy agrees rather well with XRD results (X = 0.60). In order to understand the variation of quadrupole splitting between magnetic sextet and paramagnetic doublet, Fig. 2 E) and Tab. I, Mössbauer measurement at 15 K was carried out, which shows that at low temperature this nitride is magnetic as it is the γ'-Fe_4N nitride. Moreover Tab. I shows that quadrupole splitting of this nitride measured at 15 K, remains low, in other words, the magnetic transition corresponds also to a variation of quadrupole splitting which seems to vanish when the temperature is lowered. In fact the quadrupole splitting (QS) measured by Mössbauer spectroscopy depends of angle (β) between magnetic easy axis and electric field gradient. The combined effects of magnetic and electric quadrupole splitting ($H_{ef} \gg QS$) can be well described as in [15], where the expression is proportional to $3\cos^2\beta - 1$ indicating that the QS value becomes zero if β approaches to 0.955 rad, i. e. the magnetic easy axis point along the 111 direction. We can then conclude that Zn substitutions lead to a rotation of magnetic easy axis.

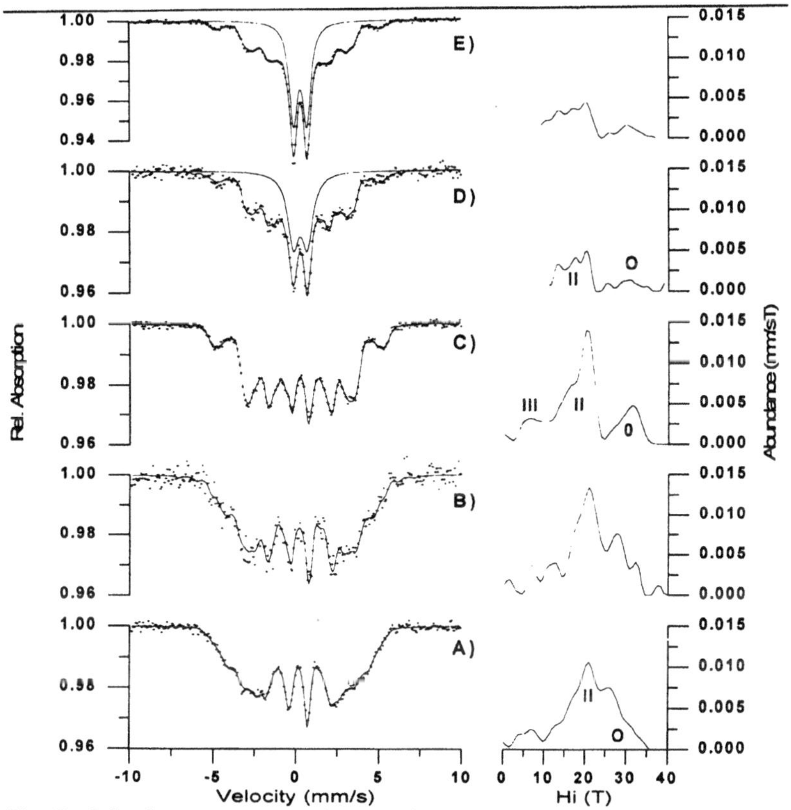

Fig. 2 - Mössbauer spectra and magnetic hyperfine fiel distribution at room temperature of : A) Virgin 64 hours milled; B) aged for 1 min at 830 K; C) aged for 8 min; D) aged for 20min; E) aged for 30 min. The different iron sites are identified in magnetic hyperfine field distribution (O to III).

CONCLUSIONS :

Here the γ'-$Fe_{3.4}Zn_{0.6}N$ nitride is obtained after milling and then submitted to thermal ageing. X-ray diffraction and Mössbauer analysis show that milling of Fe_3ZnN compound is very similar to milled Fe_4N compounds. The existence of an ε phase (hcp) is in accordance with mechanisms proposed for the congruent transformation $\gamma' \rightarrow \varepsilon \rightarrow \gamma'$ observed during milling of γ'-Fe_4N and 2α-Fe + ζ-Fe_2N. After a thermal treatment the hcp phase becomes Perovskite-like and ordering mechanisms take place. The first one is an ordering of interstitial nitrogen atoms, which is very fast and the second one is driven by local free enthalpy, which is slow and replaces the Zn atoms at the corner of the cube. We found that the limit of Zn solubility in γ'-$Fe_{4-X}Zn_XN$ phase is X=0.6, so mechanosynthesis of Fe_3ZnN compound leads to a two phased powder. The most abundant phase is a disordered ε-$Fe_{4-X}Zn_XN$ (hcp) as demonstrated by X-ray diffraction and Mössbauer spectroscopy. Mössbauer results also suggest that magnetic easy axis in γ'-$Fe_{3.4}Zn_{0.6}N$ points along of 111 direction. The existence of relatively stable iron-zinc-nitrogen phases opens a research field on properties of superficial layers as corrosion protection based on two combined superficial treatments (nitriding and galvanising).

REFERENCES :

1 J. Foct , French Patent N° 9005428 (1990).

2 R. Juza, *Advances in Inorganic Chemistry and Radiochemistry*, H. J. Emléus, A. G. Sharpe, Academic Press, **9** (1966) 81.

3 J. Foct and R. S. de Figueiredo, *NanoStructured Materials*, **4**(6) (1994) 685.

4 R. S. de Figueiredo and J. Foct, *Proceedings of ICAME-95*, **50**, (1996) 509 Rimini - Italy.

5 R. S. de Figueiredo and J. Foct, *ISMANAM-94 in Materials Sci. Forum*, **179-181** (1995) 219.

6 P. l'Heritier, Dr. Thesis, INP Grenoble - France (1980).

7 R. E. Schramm and R. P. Reed, *Met. Transactions A*, **6A** (1974) 1975.

8 R. Taillard and J. Foct, HNS-88 (1988) Lille France.

9 R. S. de Figueiredo, A. V. dos Santos and C. A. Kuhnen, *10th International Conference on Hyperfine Interaction in Hyperfine Interaction*, (1995) Leuven - Belgium.

10 R. S. de Figueiredo and J. Foct, *Fourth Seeheim Worshop on Mössbauer Spectroscopy* (1994) Seeheim - Germany.

11 A. R. Yavari, P. J. Desré and T. Benameur, *Phys. Rev. Let.*, **68**(14) (1992) 2235 .

12 A. R. Yavari, *Mat. Science and Eng.*, **A179/A180** (1994) 20.

13 J. Foct, J. P. Senateur, J. M. Dubois and G. Le Caer, *J. de Physique Coll. C2*, **40**(3) (1979) 647.

14 J. Foct and P. Rochegude, *Hyperfine Interactions*, **28** (1986) 1075.

15 C. A. Kuhnen, R. S. de Figueiredo, V. Drago and E. Z. da Silva, *J. of Mag. and Mag. Materials*, **111** (1992) 95.

MÖSSBAUER STUDY OF MECHANICALLY ALLOYED INVAR Fe-Ni ALLOYS

S. K. Xia, R. B. Scorzelli, I. Souza Azevedo and E. Baggio-Saitovitch

Centro Brasileiro de Pesquisas Fisicas, Rua Dr. Xavier Sigaud 150, 22290-180, Rio de Janeiro, RJ, Brazil

The mixture of Fe-Ni elemental powders with Ni concentration 40 at.%, which is in the invar composition region, was mechanically alloyed. The structural evolution was followed by X-ray diffraction and Mössbauer spectroscopy. The milled powder was then annealed at 350° C for different periods of time. The results showed that, after 10 h milling, a disordered invar alloy was formed. The annealing at 350° C of the milled samples leads to segregation of fcc phases with different Ni concentration.

I. INTRODUCTION

A disordered Fe-Ni alloy in invar composition range (30-50 at.% Ni) is in a metastable state below about 700K [1]. These alloys usually segregate into a non-magnetic Fe-rich phase (Ni ≤30%) and the Ni-rich phase which, in the equilibrium state, is an ordered and stoichiometric $Fe_{50}Ni_{50}$ phase with the CuAu type superstructure[2]. However, it is difficult to observe such phase separation experimentally because of the slow diffusion rate at the transition temperature (T_c= 320° C).

The equilibrium Fe-Ni phases in the invar composition range have been observed in some iron-nickel bearing meteorites [3]. In this case, the formation of these equilibrium phases results from the extremely slow cooling rate (~1° C/10^6 years). In order to attain a state closer to equilibrium at reasonable experimental times, it is necessary to use special techniques to enhance diffusion. For example, electron and neutron irradiation and ion bombardment proved to be powerful techniques for promoting phase segregation [4,5].

It is known that mechanical alloying (MA) is a high energy ball milling process performed at low temperatures. It causes grain refinement and produces a great amount of lattice defects [6]. Therefore, one can expect that MA can greatly enhance the diffusivity of the powders. As an alternative method, in this work, we obtained a mechanically alloyed $Fe_{60}Ni_{40}$ powder and studied the phase separation process during a subsequent heat treatment, by X-ray diffraction (XRD), and Mössbauer spectroscopy.

II. EXPERIMENTAL DETAILS

A mixture of Fe (99.9%) and Ni (99.9%) with a composition of 40 at.% Ni was ball milled in Ar gas atmosphere in a vial (70 cm^3 in volume) with 4 balls (two with 1 cm diameter and two with 0.5 cm diameter). Both the vial and balls were made of hardened stainless steel. The weight ratio between balls and powder was 7:1. The milling was performed on a SPEX-8000 mixer. Samples were taken from the vial in some intermediate periods of time with the vial opened inside a glove-box filled with purified Ar gas. The longest milling time was 90 hours.

For the subsequent heat treatment, the powder milled for 90 h was sealed in quartz tubes with a vacuum of 8x10^6 torr. These samples were heated at 350° C for different annealing times. XRD experiments were carried out using Cu-Kα radiation performed on a Philips X-ray diffractometer equipped with a LiF monochromator. Mössbauer spectra were obtained at room temperature (RT) using a ^{57}Co source in Rh matrix. The velocity scale for the Mössbauer spectra was calibrated related to α-Fe at RT. The composition of the powder after milling was checked by EDS, which indicates that there is no contamination from the tools.

III. RESULTS AND DISCUSSION

XRD patterns indicate that, after 10 h milling the mixture of Fe-Ni elemental powders transform into a fcc disordered alloy. In order to obtain a well homogeneous sample, milling was performed until 90 hours. The effective crystallite size (L) was estimated to be 12 nm from the XRD.

The Mössbauer spectra of the starting powder and milled for different hours are shown in Fig. 1. The initial stage of the milling show the alloying process. The spectrum of 5 h milled powder is composed of three subspectra which result from a mixture of several phases with different compositions. This is due to the mutual interdiffusion, i. e. Fe diffuses to Ni matrix and Ni diffuses to Fe matrix, leading to the coexistence of Fe-rich and Ni-rich alloys. After 10 h milling, the fcc disordered alloy is formed and further milling does not change the alloy structure. The disordered alloy has an average hyperfine field of 31T.

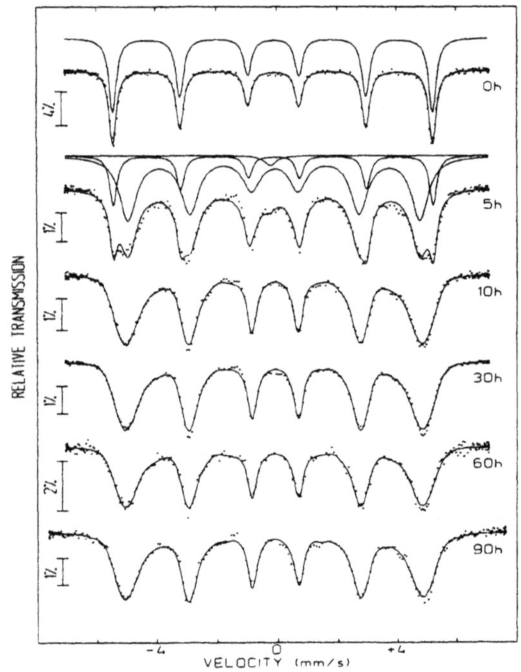

Fig. 1 - RT Mössbauer spectra of starting and mechanically alloyed Fe-Ni powders with Ni content 40 at.% at the indicated milling times

Figure 2 shows the Mössbauer spectra of the powders milled for 90 h and submitted to an annealing at 350 °C for different periods of time. After 20 h annealing, the segregation starts to occur as indicated by the appearance of a non-magnetic singlet. This singlet is attributed to a fcc non-magnetic phase with Ni concentration ≤30 at.%. After 100 h annealing, segregation becomes more clear. Beside the non-magnetic singlet, a component with a lower field of about 20 T can be seen, attributed to an fcc phase with a composition around 33 at.% Ni [7]. It is interesting to note that this component does not change its hyperfine field with further annealing. This component has also been observed by Toselli et al. [8] after irradiating $Fe_{65}Ni_{35}$ multilayers with Kr ions. Thus, this phase may be more stable compared with other disordered invar alloys. For the powders annealed for 250 h and 500 h, the best fit of the Mössbauer spectra indicated that another component may exist, with an hyperfine field of 29 T and quadrupole splitting QS ~ 0.2 mm/s, typical of the ordered $Fe_{50}Ni_{50}$ phase. This seems to be a sign of formation of this phase in the segregation process [9].

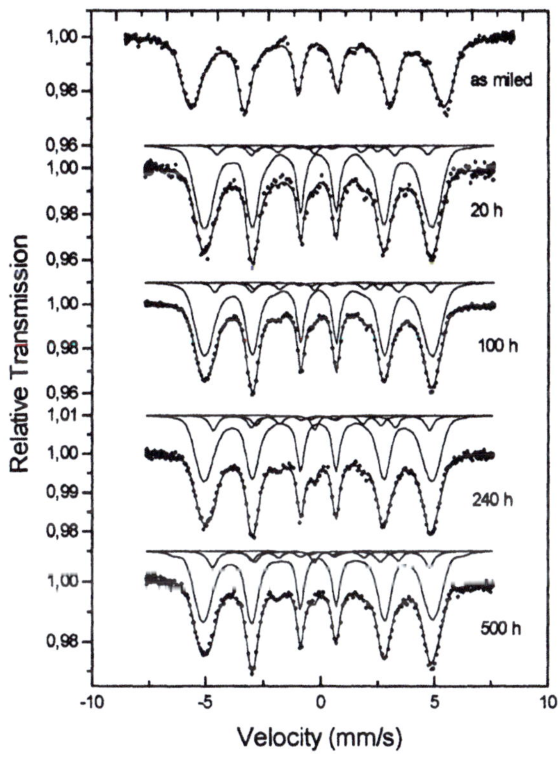

Fig. 2 - RT Mössbauer spectra of 90 h milled $Fe_{60}Ni_{40}$ alloy submitted to annealings at 350 °C for the indicated times.

In conclusion, as an alternative method for obtaining the Fe-Ni equilibrium phase, MA has been performed on a mixture of Fe and Ni elemental powders with Ni content of 40 at.%. A defective nano-crystalline Fe-Ni disordered alloy was obtained which is believed to enhance the diffusion rate in these alloys. Under an annealing at 350 °C, a thermodynamic phase segregation was observed forming fcc phases with different Ni concentration.

REFERENCES

[1] Y. Tanji, Y. Nakagawa, Y. Saito, K. Nishimura and K. Nakatsura, Phys. Status Solidi A 56 (1979) 513.
[2] A. Chamberod, J. Laugier and J. M. Penisson, J. Magn. Magn. Mater. 10 (1979) 139.
[3] J. Danon, R. Scorzelli, I. Souza Azevedo, W. Curvello, J. F. Albertsen and J. M. Knudsen, Nature 277 (1979) 283.
[4] K. C. Russel and F. A. Garner, Metall. Trans. A 21 (1990) 1073.
[5] M. R. Gallas, L. Amaral and J. A. H. da Jornada, J. Appl. Phys. 70 (1991) 131.
[6]. C. C. Koch, Mater. Sci. Forum 88-90 (1992) 243.
[7] J. B. Muller and J. Hesse, Z. Phys. B54 (1983) 43.
[8] C. Toselli, F. Ferrari, R. Brand, W. Keune, G. Marest, M.A. El Khakani, J. Parellada, G. Principi, S. Lo Russo, V. Rigato and S. Enzo, Nucl. Instr. Meth. Phys. Res. B80/82 (1993) 459.
[9] Y. Gros and J. Pauleve, J. de Phys. 31 (1970) 459.

VIII. Superconducting Materials, Thin Films and Artificially Structured Materials

NEW IRON SPECIES OBSERVED BY MÖSSBAUER SPECTROSCOPY IN Y-Sr-Cu-O COMPOUNDS CONTAINING OXYANIONS

Angel Bustamante D.[1,2], *R.B. Scorzelli*[1] *and E. Baggio Saitovitch*[1]
1 Centro Brasileiro de Pesquisas Fisicas, Rua Dr. Xavier Sigaud 150, 22290-180 Rio de Janeiro, Brazil
2 Facultad de Ciencias Físicas, Universidad Nacional Mayor de San Marcos, Apartado Postal 14-0149, Lima-14, Perú.

Incorporation of sulphate (SO_4^{2-}), phosphate (PO_4^{3-}) and borate (BO_3^{3-}) ions stabilizes $YSr_2Cu_3O_7$ under ambient conditions. X-ray structural analysis and ^{57}Fe Mössbauer spectroscopy measurements were performed on samples doped with 0.3, 1.0 and 2.0 at.% ^{57}Fe showing species similar to Y-Ba-Cu-O compounds and other species related to the presence of the oxyanions.

1. INTRODUCTION

Undoped $YSr_2Cu_3O_7$ has been synthesized only under high pressure (7 GPa) conditions [1]. Cation substitution in the chain Cu(1) sites however stabilizes $YSr_2Cu_3O_7$ in the triperovskite structure under ambient conditions [2]. Particularly interesting is the partial substitution of oxyanions such as CO_3^{2-}, SO_4^{2-}, PO_4^{3-} and BO_3^{3-} ions [3]. Such oxyanion derivatives of $YSr_2Cu_3O_7$ have been rendered superconducting (Tc~35-40K) by the partial susbtitution of Y by Ca, Sr by Ba or by the substitution of the carbonate group by the nitrate group [4].

We report here X-ray structural analysis and Mössbauer spectroscopy results obtained on ^{57}Fe: Y-Sr-Cu-O related compounds containing oxyanions sulphate (SO_4^{2-}), phosphate (PO_4^{3-}) and borate (BO_3^{3-}).

2. EXPERIMENTAL

The samples have been prepared by the conventional solid solution method with 0.3, 1.0 and 2.0 at.% ^{57}Fe substituting Cu ions. Appropriate amounts of high purity Y_2O_3, $SrCO_3$, $CaCO_3$, $BaCO_3$, $Y_2(SO_4)_3.8H_2O$, CuO, $NH_4H_2PO_4$, H_3BO_3 and $^{57}Fe_2O_3$ compounds were well mixed in stoichiometric amounts and pressed into pellets. Intimately mixed powders were heated in flowing oxygen at 1065°C for 14h, for (SO_4^{2-}) and (PO_4^{3-}) groups and 1000°C for (BO_3^{3-}) group; given a similar heat treatment, and finally slowly cooled (50°C/h).

The analysed representative samples are shown below:

S1FE= $YSr_2Cu_{2.78}Fe_{0.01}(SO_4)_{0.21}O_7$
S2FEN= $[Y_{0.84}Sr_{0.16}]Sr_2\{Cu_{2.77}Fe_{0.01}(SO_4)_{0.22}\}O_{6.12}$
P1FE= $YSr_2Cu_{2.78}Fe_{0.01}(PO_4)_{0.21}O_7$
P2FEN= $[Y_{0.70}Ca_{0.30}]Sr_2Cu_{2.78}Fe_{0.01}(PO_4)_{0.21}O_{6.16}$
B1FEN=$Y(SrBa)\{Cu_{2.49}Fe_{0.01}(BO_3)_{0.50}\}O_{5.50}$
B2FEN=$[Y_{0.85}Ca_{0.15}](SrBa)\{Cu_{2.49}Fe_{0.01}(BO_3)_{0.50}\}O_{5.50}$

The samples were characterized for phase identification, homogeneity and lattice parameters by X-ray diffraction patterns obtained using HZG-4 powder difractometer with Cu-K_α radiation (λ=1.5418Å) and Co-K_α radiation (λ=1.7902Å).

The Mössbauer spectra (MS) were recorded using a ^{57}Co source in Rh matrix at room temperature (RT); the fits were performed assuming superposition of quadrupole doublets of equal intensities and linewidths for all the samples.

3. RESULTS AND DISCUSSION

The X-ray structural analysis showed that these compounds are isostructural to the superconductor $YBa_2Cu_3O_{7-\delta}$. The samples containing sulphate and phosphate group have orthorhombic (Pmmm) structure with lattice parameters in agreement with the reported values [5] showing a reduction of the c/a and c/b values as compared to the Y-Ba-Cu-O system (a=3.820Å, b=3.88Å, c=11.68Å e V=173.116Å [6]). The powder XRD pattern of the borate samples confirm that the majority of the diffraction lines can be indexed with a tetragonal unit cell with space group P4/mmm. The lattice parameters of the compounds are in accordance with the ones already reported [7].

Room-temperature (RT) ^{57}Fe Mössbauer spectroscopy measurements of the same samples are shown in figs.1and 2. No impurity phases, such as Fe_2O_3, $CaFeO_3$ or $SrFeO_3$, were detected in any of the MS. In Table I we summarize the Mössbauer parameters obtained from the fits.

Table I

Mössbauer parameters at RT for the different species in the $YSr_2Cu_3O_{7-\delta}$ system containing oxyanions. IS=Isomer Shift relative to metallic Fe at RT (±0.002mm/s), QS=Quadrupole Splitting (±0.01mm/s), Γ=full width half maximun linewidth (±0.01mm/s) and A=relative area (±3%).

Samples	Species	IS (mm/s)	QS (mm/s)	Γ (mm/s)	A (%)
S2FEN	A	0.085	1.97	0.37	23
	B	0.058	1.04	0.41	43
	C	0.156	0.44	0.37	21
	D'	-0.097	1.33	0.40	13
S1FE	A	0.069	1.97	0.37	17
	B	0.141	1.04	0.37	31
	C	0.160	0.45	0.37	25
	D'	-0.052	1.20	0.40	27
P2FEN	A	0.111	2.05	0.40	22
	B	0.104	1.02	0.46	28
	F	0.231	0.54	0.40	30
	H	-0.166	0.97	0.54	20
P1FE	A	0.089	2.01	0.44	26
	B	0.173	1.02	0.40	37
	F	0.228	0.49	0.35	24
	H	-0.249	1.02	0.48	13
B2FEN	A	0.142	2.10	0.44	22
	D	-0.126	1.60	0.48	14
	F	0.277	0.67	0.32	58
	I	0.181	1.54	0.45	06
B1FEN	A	0.096	2.02	0.47	28
	D	-0.120	1.70	0.39	13
	F	0.289	0.66	0.38	46
	I	0 260	1.55	0 50	13

We have assigned subspectra A, B and D to iron substitution at the Cu(1) site in fourfold planar, fivefold pyramidal, and fourfold tetrahedral (twin-plane associated) oxygen coordination, respectively [8], and subspectra C and F to iron substitution at the Cu(2) sites in quasi-octahedral and fivefold pyramidal oxygen coordination, respectively. The subspectra E represents Fe clustering.

The MS of the samples S1FE and S2FEN containing (SO_4^{2-}), were analysed with species A, B, C and D'. Species A and B are also found in the Y-Ba-Cu-O system. Species B has an a-axis oxygen bridging chains of the $Cu(1)O_x$ plane. These a-axis oxygens may be common to two slightly different B species, forming an Fe-O-Fe bridge, or they may form an Fe-O-Cu bridge within the $Cu(1)O_x$ plane.

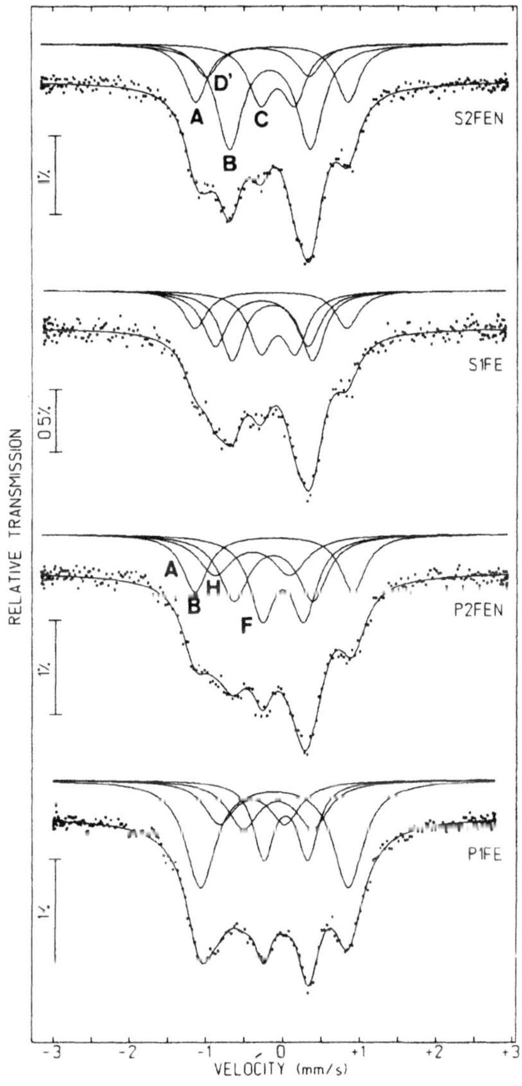

Fig. 1 Room temperature ^{57}Fe Mössbauer spectra of selected samples containing sulphate (SO_4^{2-}) and phosphate (PO_4^{3-}) ions.

For these samples the more interesting feature is a high population of species C (~20%), which can be assigned to iron in the $Cu(2)O_2$ plane; generally this species is present in the Y-Ba-Cu-O system containing above 10 at.% of ^{57}Fe [9]. Species D' can be related to the presence of sulphur (S) as the first Fe near neighbour. This species was also detected in samples containing iron sulphide (FeS) like $YBa_2[(CuO)_{1-x}(FeS)_x]_3O_{4-\delta}$ with x=0.10 and 0.15 [10]. These compounds are superconductors with T_c~35K.

The MS of samples P1FE and P2FEN, containing (PO_4^{3-}), exhibited the presence of species A, B, F and H. Species F corresponds to Fe at the Cu(2) sites in fivefold pyramidal oxygen coordination and species H can be associated to the presence of phosphorous (P) as the first Fe near neighbour. Species F atributed to Fe in Cu(2) sites, where it is five-fold coordinated, was also observed in compounds like $YSr_2Cu_{3-x}Fe_xO_y$ [11], $Y_{1-x}Ca_xSr_2Cu_{3-y}Fe_yO_{6+z}$ [12], $Y_{1-x}Ca_xSr_2Cu_{2.5}Fe_{0.50}O_{6+z}$ and $Y_{1-x}Ca_xBa_2(Cu_{1-x}Fe_x)O_{6+z}$ [13-14].

The MS of samples B1FEN and B2FEN containing (BO_3^{3-}), displayed the species A, D and F already described above and a new species, associated to the presence of boron (B) as Fe first near neighbour. The high population of species F, 56% (B1FEN) and 48 % (B2FEN), indicates that besides the partial substitution of Y^{3+} for Ca^{2+} the presence of the (BO_3^{3-}) group also favours the formation of this species.

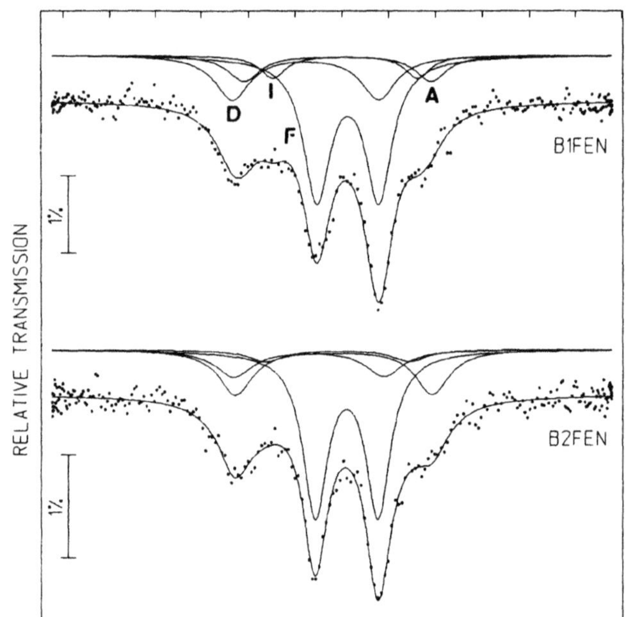

Fig. 2 Room temperature ^{57}Fe Mössbauer spectra of selected samples containing borate (BO_3^{3-}) ions. The sample B2FEN contains 15 at.% of Ca substituting Y

CONCLUSIONS

In the Y-Sr-Cu-O system containing oxyanions, the presence of species A, B, C, D, E and F have been observed, similarly to Y-Ba-Cu-O system and related compounds.

New species denominated D', I and H have also been detected and have been attributed to the signature of a perturbation introduced by the doping with (SO_4^{2-}), (PO_4^{3-}) and (BO_3^{3-}).

The incorporation of these oxyanions seems to favour the presence of species C and F, which represent Fe in Cu(2) sites in quasi-octahedral and fivefold pyramidal oxygen coordination respectively, in high proportion.

REFERENCES

[1] B. Okai, Jpn. J. Appl. Phys, 29 (1990) L2180.

[2] P.R. Slater and C. Greaves, Physica C 180 (1991) 299.

[3] C. Greaves, M. Al-Mamouri, P. R. Slater and P. P. Edwards, Physica 235-240 (1994) 158.

[4] A. Maignan, M. Hervieu, C. Michel and B. Raveau, Physica C, 208 (1993) 116.

[5] P.R. Slater, C. Greaves, M. Slaski and C. Muirhead, Physica C 208 (1993) 193.

[6] V.P.S. Awana and A.V. Narlikar; Phys. Rev. B 49 (1994) 6353.

[7] P. Kurmaev, S.N. Shamin, P.R. Slater and C. Greaves, Physica C 227 (1994) 309.

[8] E. Baggio-Saitovitch, I. Souza Azevedo, R.B. Scorzelli, H. Saitovitch, S.F. da Cunha, A.P. Guimarães, P.R.Silva and A. Takeuchi, Phys. Rev. B 37 (1988) 7967.

[9] E. Baggio Saitovitch, Hyp. Int., 66 (1991) 231.

[10] Angel Bustamante D, R. B. Scorzelli, E. Baggio Saitovitch, J. A. Aguiar and J. M. Ferreira in: Proc. Int. Conf. on Appl. Mössbauer Effect, Rimini 1995, ed. I. Ortalli (Italian Physical Society) p. 559.

[11] M. Pissas, G. Kallias, E. Moraitakis, D. Niarchos e A. Simopoulos; Physica C 234 (1994)127.

[12] V. Terziev, R. Suryanarayanan, Mamidanna S. R. Rao, L. Ouhammou, O.Gorochov J.L. Dormann; Physical Review B 48(1993)13037.

[13] M.G. Smith, J.B. Goodenough, R.D. Taylor, J.J. Neumeier, J. O. Willis; Physica C 222 (1994) 157.

[14] M.G. Smith, R.D. Taylor, H. Oesterreicher, Phys. Rev. B 42 (1990) 4202.

HYPERFINE PROPERTIES OF 2Fe/Pd AND 2Pd/Fe MULTILAYERS

A. V. dos Santos
Instituto de Física Universidade Federal do Rio Grande do Sul. CP 15051, 91501-970 e-mail vando@if.ufrgs.br

ABSTRACT

In this work we analyze the magnetic moment and hyperfine properties of 2Fe/Pd and 2Pd/Fe multilayers and the influence of pressure using the Linear Muffin-Tin Orbital (LMTO) method. At theoretical equilibrium volume we found a high magnetic moment the Fe site for 2Pd/Fe and 2Fe/Pd multilayers. The hyperfine field at iron nuclei is near -39.0 T for 2Pd/Fe multilayers and has a lower value of -21.0T for the 2Fe/Pd system. The results for the magnetic moments and hyperfine parameters at different lattice spacing show a strong dependence of the magnetic properties of the 2Fe/Pd system on pressure.

1. INTRODUCTION

In the last years there is a growing interest in the study of electronic and magnetic properties of thin film and multilayers, that are composed of periodic structures with different materials alternating in the layers. Due to the great technological interest awaked for this new class of materials, this kind of structure has been studied in theoretical [1-6] and experimental works [7-11]. Through theoretical studies applying local spin density, we can get a good description of the hyperfine structure and magnetic properties of the thin films and multilayers.

In the present work we investigate magnetic properties and hyperfine structure of 2Fe/Pd and 2Pd/Fe multilayers employing self-consistent calculations of band structure, using the Linear Muffin-Tin Orbital (LMTO) method of Andersen [12] in the Atomic Sphere Approximation (ASA). In the study of magnetic properties and hyperfine structure of 2Fe/Pd and 2Pd/Fe multilayers, we used spin polarization and the von Barth and Hedin [13] parametrization for the exchange-correlation energy of the electron gas. We don't consider spin-orbit interactions but include the combined corrections terms. For the paramagnetic calculations we used the Hedin-Lundquist [14] parametrization. Our solutions of the Schrödinger equation used s,p,d and f of the basis functions for the LMTO method. Self-consistent potentials converged for 776k-points and 1500 points in the energy window using the same radius for the Wigner-Seitz spheres for both Fe and Ni atoms. The density of states was calculated as the δ function convoluted with a Gaussian function for a fixed number of points in each energy.

The calculation were made for a structural mode where these multilayers grow along the (0,0,1) direction according to an ABA plane piling sequence. For the 2Fe/Pd multilayers the Fe atom positions are (0,0,0) and (0,1/2, -1/2) while that of Pd atom is (0, 1/2, 1/2). For the 2Pd/Fe sistem the positions are (0,0,0) and (0,1/2,-1/2) for the Pd atoms, and (0, 1/2, 1/2) position for Fe atom. respectively. The assumed structure has tetragonal cell with c/a=2.23. For this structure we note that the neighborhood of the Fe atom in the 2Fe/Pd multilayers is same which implies that the sites are equivalent. This holds also for the 2Pd/Fe multilayers where now we have two equivalent Pd sites. In the next sections we present a discussion of the main results in this work.

2.RESULTS AND DISCUSSION

2.1 Magnetic structure

Table 1

	Fe				*Pd*		
	↑	↓	↑-↓		↑	↓	↑-↓
n	5.3700(5.5074)	2.6704(2.5247)	2.7096(2.9697)		5.1263(5.1523)	4.7863(4.8312)	0.3400(0.3211)
n_s	0.3268(0.3327)	0.3279(0.3327)	-0.0011(-.0043)		0.2834(0.2828)	0.2991(0.2949)	-0.0157(-0.0121)
n_p	0.3668(0.3676)	0.3756(0.3676)	-0.0088(-.0022)		0.3097(0.2995)	0.3291(0.2995)	-0.0194(-0.0091)
n_d	4.6864(4.8071)	1.9669(1.8309)	2.7195(2.9762)		4.5332(4.5700)	4.1581(1.5197)	0.3751(0.3423)
N(EF)	1.7855(1.2308)	12.6785(18.5056)			4.5332(4.9316)	1.9580(1.5197)	
N_T(EF)	14.4640(19.7364)				6.4913(6.4513)		
N_c(EF)				45.9223(27.3666)			
γ				7.9587(4.7428)			
EF				0.6789(0.6109)			
ΔQ	0.0453(0.0351)				-0.0892(-0.0161)		

Table 1. Calculated parameters of the 2Fe/Pd multilayers using spin polarization and self-consistent potentials. n is in electrons/spin; $N(E_F)$ is the density of states at the Fermi level in states/atom spin Ry, $N_T(E_F)$ in states/atom Ry and $N_C(E_F)$ in states/unit cell Ry. γ is the linear coefficient of specific heat in $mJ/molK^2$, E_F is the Fermi level in Ry and ΔQ is the charge transfer in electrons. The values calculated for 2Pd/Fe are shown in parentheses.

To study the magnetic properties of 2Fe/Pd and 2Pd/Fe multilayers we calculated the theoretical parameters listed in table 1 using spin-polarization at equilibrium volume. Clearly for both multilayers the calculated magnetic moments at the Fe sites are greater than the bulk values of FePd and Fe_3Pd, which are μ_{Fe}=2.25μ_B and μ_{Fe}=2.35μ_B , respectively [15]. This increase of the μ_{Fe} is in close agreement with the theoretical calculations of Kitcheter et al. [07], who showed that for Fe overlayers on Ag μ_{Fe} increses with decreasing coordination number. This also agrees with experimental results on Fe(001)/Pd(001)Fe(001) structures [08]. This increase in μ_{Fe} was also found for Fe/Pd (001) monolayers[16] where μ_{Fe}=2.87μ_B was obtained On the other hand at Pd sites, the calculated magnetic moments decrease from 0.79μ_B in the ordered FePd alloy[15], to 0.34μ_B and

$0.32\mu_B$ in the 2Fe/Pd and 2Pd/Fe multilayers as can be seen from table 1. These results are close to the value reported in the calculations of Blügel at al. [19], where they predict that Fe in Fe/Pd interfaces induces a magnetic moment on two adjacent Pd atomic layers with $\mu_{Pd} = 0.32\mu_b$ for the first Pd atomic layer. For the 2Fe/Pd and 2Pd/Fe multilayers the behavior of the magnetic moments at Fe sites are related to the poor degree of population of the spin-down d states. This can be seen comparing the occupation numbers give in table 1 with those given in table 3 and 4 of reference 15 for the ordered FePd and Fe_3Pd alloys. Also from table 1 we see that the main contribution for the DOS at the Fermi level $N(E_F)$ is due to the spin-down d electrons at Fe sites. There is an increase in the electronic specific-heat coefficient for the multilayers (table 1) reflecting a greater $N(E_F)$ than that of the ordered FePd alloy[15].

3. HYPERFINE PARAMETERS

The magnetic hyperfine field H that acts on the nucleus is given by:

$$H = H_{ext} + H_{FC} + H_{orb} + H_{dip}$$

where Hext is the external applied magnetic field, HFC is the Fermi contact field, which is given by

$$H_{FC} = -8/3\ \pi\ \gamma_n [\ |\Psi\uparrow(0)|^2 - |\Psi\downarrow(0)|^2]$$

where γ_n is the nuclear gyromagnetic ratio and $|\Psi(0)|^2$ is the density of s-electrons at the nuclear site. The other two contributions are the orbital and dipolar magnetic fields.

The isomer shift calculation is done using the following equation [20]:

$$\delta = 4/5\ \pi\ Ze^2\ \{\ |\Psi(0)|_A{}^2 - |\Psi(0)|_S{}^2\}\ (\Delta R/R)\ R^2,$$

where $|\Psi(0)|_A{}^2$ is the total s-electron density at the absorber nucleus, $|\Psi(0)|_S{}^2$ is the total s-electron density at the source nucleus, and $\Delta R/R$ is the nuclear radii fractional deviation. In our calculations we use the $|\Psi(0)|_S{}^2$ value given in [15]. Table 2 shows the H_{FC} and δ values for the multilayers 2Fe/Pd, and 2Pd/Fe. We can see that there is a large difference between the magnetic hyperfine parameters in going from 2Fe/Pd to 2Pd/Fe. As we go from one monolayer of Pd atoms to two monolayers of Pd atoms, there is an increase in both H_{FC} and δ. These results are in agreement with the magnetic moment values for the Fe sites which suggests a proportionality between H_{FC} and the magnetic moment due to the neighborhood of the atoms.

Table 2

		HF(KOe)	IS(mm/s)
2Fe/Pd	FeI	210.49	0.3169
2Pd/Fe	Fe	388.94	0.4565

Table 2. Calculated hyperfine fields and isomer shifts.

4. PRESSURE EFFECT ON THE MAGNETIC PROPERTIES

We now discuss the theoretical behavior of the magnetization under pressure for our multilayers by self consistently varying the lattice parameter in 2 % steps. The results can be seen in figure 1 for the Fe and Pd sites. We note that there is a distinct behavior for the magnetization of both systems. There is an abrupt drop of Fe magnetic moment in the 2Fe/Pd multilayers to at a near zero value. Such a transition from a FM state to a NM state is obtained for ordered alloys of Fe_3Ni [18], and also for Fe/Pd monolayers [05]. This kind of behavior characterizes an Invar type alloy [16-18]. In the same range of lattice parameter variation the magnetic moment of Fe atom in the 2Pd/Fe multilayers does not drop to zero, thus showing the same behavior as that of ordered alloys of FePd, which depends of the Pd concentration in the system [15]. Looking at the Pd site (Fig. 2) we see a behavior almost like that of the Fe site. However, in the 2Fe/Pd system an increase of pressure yields a small increase in the magnetization, and then a fast drop to zero; the same behavior was verified in Fe_3Pd [15]. On the other hand, in the 2Pd/Fe system this behavior is different and the magnetic moment does not drop to zero, there is still a sizable magnetization very like that in some FePd alloys [15].

The results in figures 3 and 4 show that there is a strong dependence of the hyperfine magnetic field and isomer shift upon the lattice spacing. Figure 3 shows that there is a decrease of the H_{FC} with increasing pressure in accordance with the decrease of the magnetic moment in both systems. The behavior of the isomer shift with the variation of the lattice parameter in both systems is shown in figure 4. As we expand the lattice the interaction between the layers decreases and the isomer shift becomes very similar in both systems.

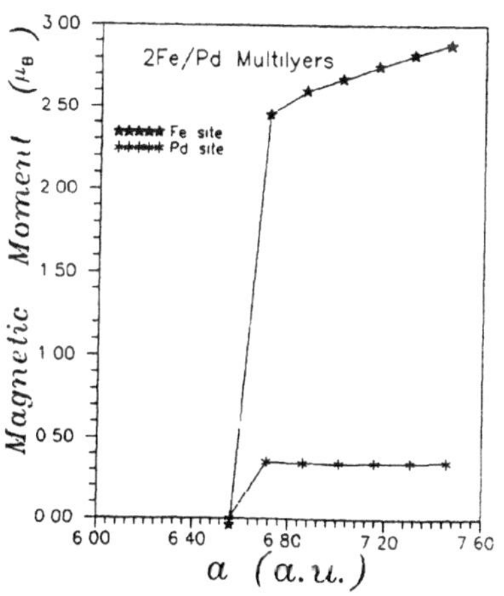

Figure 1. The magnetic moments (in Bohr magnetons) as a function of the lattice parameter for 2Fe/Pd multilayers.

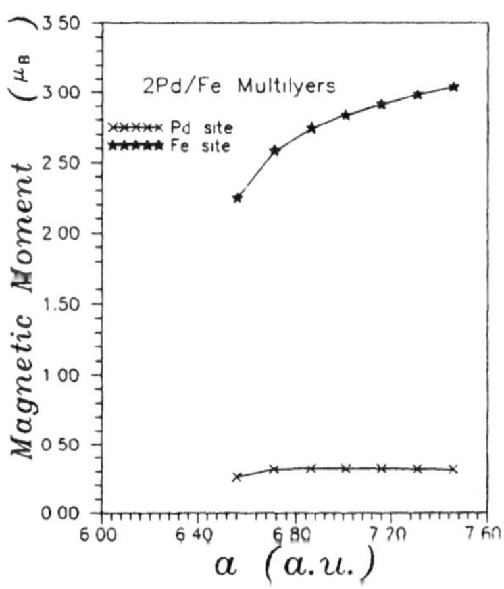

Figure 2. The magnetic moments (in Bohr magnetons) as a function of the lattice parameter for 2Pd/Fe multilayers.

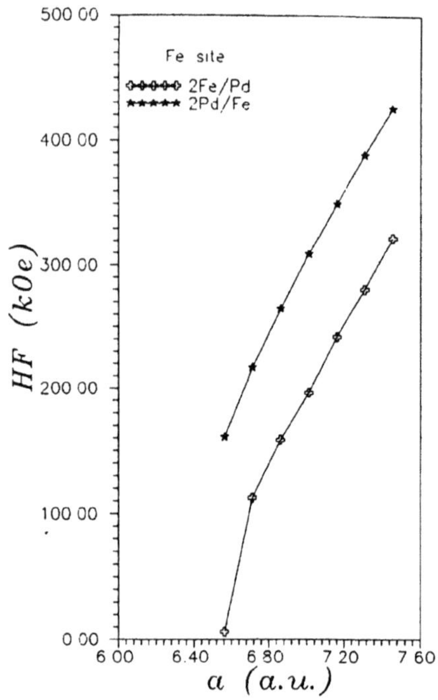

Figure 3. Calculated magnetic hyperfine field at Fe site as function of the lattice parameter.

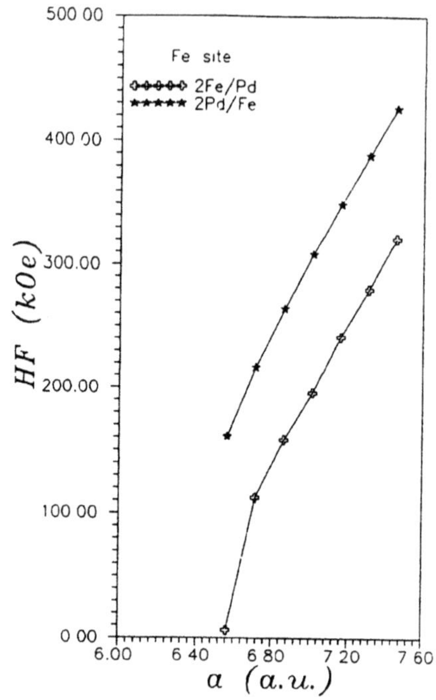

Figure 4. Isomer shift at Fe site as function of the lattice parameter.

5. CONCLUSIONS

We have obtained theoretically the band structures for 2Fe/Pd and 2Pd/Fe multilayers, and have shown that there are important changes in the electronic structure as we go from one to the other system. One of them is the increase in the magnetization at the Fe site, and a decrease in the magnetization at the Pd site with increasing lattice parameter. Also, the hyperfine structure is sensitive to the immediate neighborhood of the Fe site according to the results for the hyperfine magnetic field and isomer shift in both systems. These systems also display distinct magnetic behavior with pressure, as 2Fe/Pd multilayer being more sensitive to it.

Acknowledegments- We would like to thank to the Centro Nacional de Supercomputação (CESUP-UFRGS) for the of Cray Supercomputer. One of us (A. V. dos Santos) acknowledges the financial support of the Conselho Nacional de Desenvolvimento Científico e Tecnológico (CNPp)

References

[1] C.L. Fu, A.J. Freeman & T. Oguchi, Phys. Rev. Lett. 54 (1990) 2700.
[2] R. Richter, J.G. Gay & J.R. Smith, Phys. Rev. Lett. 54 (1990) 2704.
[3] H. Ebert, S. Rueg. G. Schutz, R. Wienk & W.B. Zepev, J. Mogn. Mogn. Mat. 93 (1991) 601.
[4] D. Knab & C. Koenig, J. Mogn. Mogn. Mat. 93 (1991) 398.
[5] A.V. dos Santos & C.A. Kuhnen, Solid State communications. 95 (1995) 537.
[6] T. Kraft, P.M. Marcus, & M. Scheffler, Phys. Rev B. 79 (1994) 11511.
[7] G. Bayreuther, Hyperfine Interaction. 47 (1989) 237.
[8] Z. Gelinski, B. Heinrich, J.F. Conhron, W.B. Muir, A.S. Arrott & J. Kirschner, Phys. Rev. Lett. 65 (1990) 1156.
[9] M.L. Neel, J. Phys. Rod. 15 (1954) 225.
[10] Q. Yong, S.F. Lee, P. Holady, R. Loloee, P.A. Schoeder, W.P. Pratt Jr. & J. Boss, Physic-B. 194 (1994) 327.
[11] Yau - Wei, Do, Ming - Xiong Xu, Jivan Mei - Ying - Bing Ghi & H. Woi - Xion Le, J. Appl. Phys. 70 10 (1991) 10.
[12] O. K. Andersen, Phys. Rev B, 12, 3060 (1975); H.L. Skriver, The LMTO Method: Muffin-Tin Orbitals and Electronic Structure Spriger, New York (1984).
[13] U. On Bath & L. Heeding, J. Pays. C5. (1972) 1629.
[14] L. Heeding & BI Linguist, J. Pays C4. (1971) 2064.
[15] C. A. Kuhnen and E. Z. da Silva, Phys Rev. B46 (1992) 8915.
[16] C.A. Kuhnen, and A.V. dos Santos, J. Magn. Mater. 130 (1994) 353.
[17] C.A. Kuhnen, and A.V. dos Santos, Solid State communication 85 (1993)273.
[18] J. Bonlond and C.A. Kuhnen, Brazilian J. Phys 23 (1993) 288
[19] S. Blügel B. Drittler, R. Zeller and P. H. Dederichs Appl. Phys 49 (1989) 547.
[20] T. C. Gibb, in *Principles of Mössbauer Spectroscopy*. Chapman and Hall, London (1976).

^{119}Sn MÖSSBAUER STUDIES OF GAS-SENSITIVE STANNOUS TUNGSTATE SEMICONDUCTORS

J.L. Solis [1,], V. Lantto [1], L. Häggström [2] and M. Wikner [2]*

[1] Microelectronics and Material Physics Laboratories, University of Oulu, P.O. Box 400, FIN-90571 Oulu, Finland

[2] Institute of Physics, Uppsala University, P.O. Box 530, 751 21 Uppsala, Sweden

ABSTRACT

α-$SnWO_4$ thin films were prepared by reactive dc and rf magnetron co-sputtering of Sn and W, respectively, and screen-printing technique was used to fabricate thick films both of the low-temperature (α) and high-temperature (β) phases of $SnWO_4$. CEMS spectra were measured from the sputtered thin films and transmission Mössbauer spectra from stannous tungstate powders. The results show that tin is present in the divalent form Sn^{+2} in both α and β phases. The responses of the sputtered thin films to CO and NO in synthetic air showed an unusual dual conductance behaviour. The α–$SnWO_4$ thick films were very sensitive to low concentrations of H_2S in synthetic air.

1. INTRODUCTION

Semiconductor gas sensors utilize the chemical sensitivity of semiconductor surfaces. Usually, when minor constituents of air are monitored, the sensors operate at elevated temperatures up to 500 °C. Tin dioxide, SnO_2, has very resistant surfaces against irreversible decomposition up to 500 °C and, therefore, it is an usual material in commercial sensors [1]. Tin is in the stannic form Sn^{+4} in the bulk SnO_2, even when a neighbour of ionized bulk oxygen-vacancy donors. The shallow donor levels, however, drop down in the band gap to deep traps states of the oxygen surface vacancies [2] and, therefore, the neighbouring tin atoms at the surface appear in the from Sn^{+2}. Reduction of SnO_2 surfaces by a reducing agent such as CO is now reflected in a change in the valence state of the surface tin atoms [3]. Here we report some results of ^{119}Sn Mössbauer studies of gas-sensitive stannous tungstate semiconductors together with some gas-response properties of the orthorhombic low-temperature (α) phase of $SnWO_4$. Above 670 °C the stannous tungstate appears in a cubic (β) phase. α-$SnWO_4$ is a wide gap (3.05 eV) n-type semiconductor where donors originate from nonstoichiometry (oxygen vacancies). Its conductivity is gas sensitive, e.g., to CO, NO [4] and H_2S [5]. Thin and thick films of stannous tungstate were prepared by sputtering and by the usual screen-printing technique, respectively. Conventional Mössbauer spectroscopy can be used to characterize bulk properties such as the valence state of Sn in stannous tungstates, while the conversion electron Mössbauer spectroscopy (CEMS) gives information which is surface sensitive.

* Permanent address: Facultad de Ciencias, Universidad Nacional de Ingenieria, P.O. Box 31-139, Lima, Peru

2. EXPERIMENTAL

The α-$SnWO_4$ thin films were grown by means of reactive co-sputtering, which has described in ref. [4]. The fabrication of stannous tungstate powders began with mixing equimolar quantities of SnO and WO_3 powders (reagent grade) and ball milling the mixture with acetone for 1 h. The α-$SnWO_4$ powder was prepared by heating the mixture at 600 °C for 15 h, sealed in an evacuated silica tube. The high-temperature β phase was obtained as metastable at room temperature by a rapid quenching of the mixture, sealed in the evacuated silica tube, after heating at 800 °C for 15 h.

The Mössbauer spectrometer used was working in the constant acceleration mode with a source on the each side of the vibrator and with velocities between ± 8 mm/s. The velocity calibration spectrum was measured with a ^{57}CoRh source and a α-Fe sample at room temperature. The source used for the ^{119}Sn spectrum was $Ca^{119}SnO_3$ (2 mCi). The measurements were done at room temperature. Mössbauer spectra in the transmission geometry were obtained from samples with the amount of Sn of around 30 mg/cm^2 in the radiation direction. To spread out the stannous tungstate powder over the whole measuring area the sample was mixed with BN into an homogeneous mass. This was, in turn, pressed into a 2 mm thick plate. CEMS experiments were made on thin-film samples mounted in a parallel plate avalanche counter.

For the electrical measurements, gold electrodes were evaporated over the sputtered thin films and the thick-film layers were printed on alumina substrates having preprinted gold electrodes and a Pt thick-film heating resistor in the other side of the substrate. A computer-controlled measuring system employing the flow-through principle was used to vary the gas concentration and the substrate temperature, and for data acquisition, handling and storage.

3. RESULTS AND DISCUSSION

Figure 1 shows the characteristic Mössbauer spectra measured of both the stannous tungstate thin films and the powders. Fitted values of the doublet splitting ΔE_Q, the centroid shift CS, the line width (FWHM) and the intensity are given in Table 1. Typical CEMS spectra from both as-grown and annealed (at 400 °C) α-$SnWO_4$ thin films are shown in Fig. 1(a). Table 1 gives for the as-grown film two (broad) doublets which can be closely assigned to SnO_2 and α-$SnWO_4$ phases. However, XRD (x-ray diffraction) pattern of the thin film showed only Si peaks from the substrate which means that the as-grown film was amorphous with a possible short-range order characteristic of the two above phases. The Mössbauer spectrum of the annealed film in Fig. 1(a) presents also two defined doublets which may be identified to originate in polycrystalline SnO_2 and α-$SnWO_4$ phases. The XRD pattern of the film [4] now revealed the presence of SnO_2 phase. However, the few SnO_2 peaks ((110), (101) and (211)) in the XRD pattern of the film were very weak as compared to the strong peaks of the α-$SnWO_4$ phase in the pattern. In addition, there was also some other weak peaks in the pattern which may relate to structures with stoichiometries between SnO and SnO_2, e.g., to Sn_2O_3 and Sn_3O_4 (ICDD 25-1259 and 20-1293). For instance, Raman scattering and x ray diffraction measurements revealed such structures when SnO films were oxidized in steps to SnO_2 [6]. In these structures there are both Sn^{+2} and Sn^{+4} ions. However, the peak intensities in the XRD pattern of the thin film indicated that α-$SnWO_4$ is the major phase in the film structure which is in strong disagreement with the Mössbauer spectrum of the film in Fig. 1(a). The high intensity of the doublet at the origin (assigned to SnO_2 phase) in Fig. 1(a) may show that the outermost surface of the film is converted into SnO_2 phase (or other phases between SnO and SnO_2) during annealing in air and, therefore, its gas-sensing properties may resemble those of sputtered SnO_2 films, to some extent.

Figure 1(b) shows transmission Mössbauer spectra from the α- and β-phase powders of $SnWO_4$ (Mössbauer spectra were not measured from the thick films). They also consist of two doublets where the small doublet is assigned to the SnO_2 phase and the larger doublet, with a little different intensities between two peaks, belongs to the doublet of both the α and β phases of $SnWO_4$. The obtained values for CS and ΔE_Q are in good agreement with the values reported by Ballard and Birchall [7].

Table 1. Results from the fit of the Mössbauer spectra in Fig. 1 for FWHM value Γ, the spectral intensity I, the centroid shift CS, and the electric quadrupole splitting ΔE_Q.

sample	G (mm/s)	CS (mm/s)	ΔE_Q (mm/s)	I (%)	phase
α−$SnWO_4$ thin film (as grown)	0.97 1.20	-0.05 3.05	0.50 1.28	27.0 73.0	SnO_2 α-$SnWO_4$
α-$SnWO_4$ thin film (annealed at 400°C)	0.78 0.90	-0.02 3.32	0.52 1.24	86.3 13.7	SnO_2 α-$SnWO_4$
α-$SnWO_4$ powder	0.58* 0.58*	0.02 3.42	0.55 1.18	19.7 80.3	SnO_2 α-$SnWO_4$
β-$SnWO_4$ powder	0.47* 0.47*	-0.01 3.50	0.49 1.30	14.0 86.0	SnO_2 β-$SnWO_4$

* absorber line width

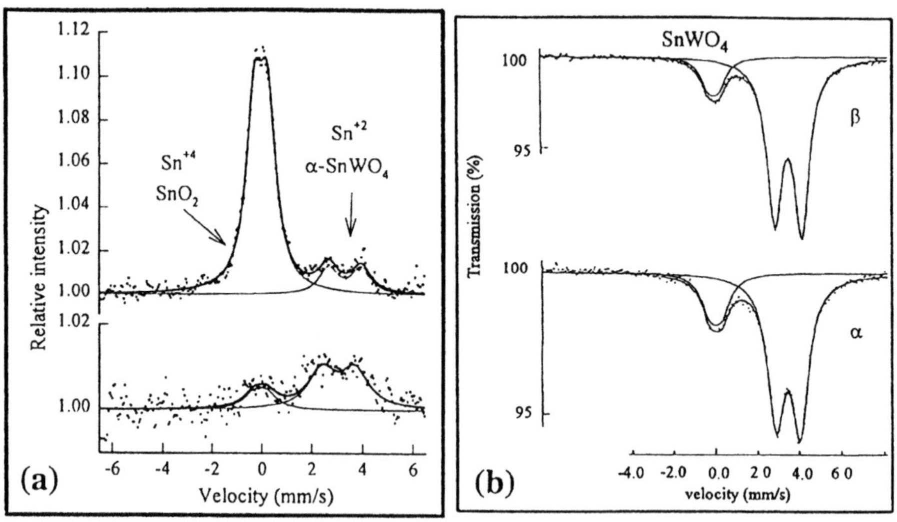

Figure 1. ^{119}Sn Mössbauer spectra measured from different $SnWO_4$ samples at room temperature. (a) CEMS spectra from as-grown (below) and annealed (top) (at 400 °C) α-$SnWO_4$ thin films. (b) Transmission Mössbauer spectra from α- and β-$SnWO_4$ powders.

The main difference in the Mössbauer spectra of Fig. 1 between the annealed thin film and powders is the intensity of the SnO_2 doublet from the SnO_2 component. The outermost surface of the thin film was probably converted into the SnO_2 phase (and other phases between SnO and SnO_2) during the annealing in air. This was seen also from RBS (Rutherford backscattering spectroscopy) results where a diffussion of W from the surface layer into the substrate was found and AFM (atomic force microscopy) pictures (but not SEM pictures) also revealed small growing grains on the surface of the annealed thin films. In the corresponding α-$SnWO_4$ powder the amount of the Sn^{+4} ions was small and it was not possible to reveal the SnO_2 phase from the XRD pattern [5]. During annealing of the powders in vacuum only a small amount of the starting powders was crystallized in the SnO_2 phase (the XRD pattern of the used SnO powder showed only SnO peaks [5]). There was no differences between the XRD patterns measured from the α-$SnWO_4$ powder and the α-$SnWO_4$ thick films. Therefore, we may expect large differences in the gas-response properties between our α-$SnWO_4$ thin and thick films.

The conductance of the α-$SnWO_4$ thin film, annealed in air, showed a dual behaviour in the case of exposure to 250 ppm of CO in synthetic air at different temperatures. The conductance of an n-type semiconductor usually increases upon exposure to a reducing gas such as CO. As is seen in Fig. 2(a) the transition temperature T* for the dual behaviour (above and below T* the conductance increases and decreases, respectively, at exposure to CO) is between 200 and 250 °C. A conductance decrease follows the exposure to 250 ppm of CO at 150 and 200 °C, while at 250 °C and higher temperatures a conductance increase follows the CO exposure. The amount of tin in the thin film seems to be an important factor for this unusual dual conductance behaviour [4].

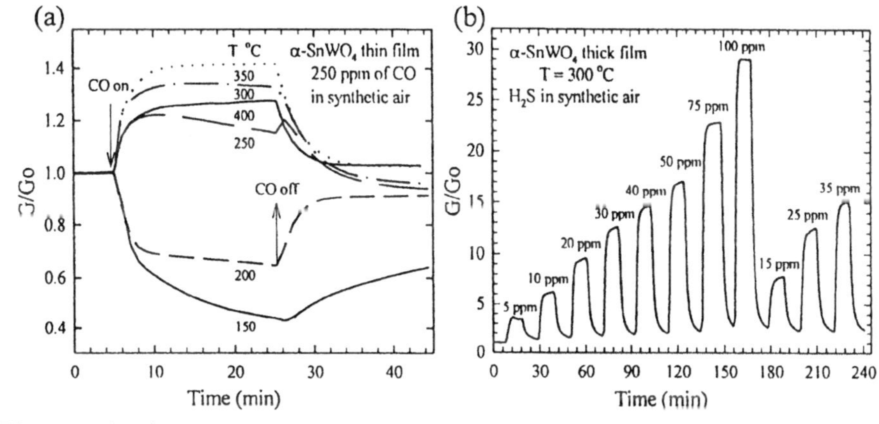

Figure 2. Conductance response G/Go of an α-$SnWO_4$ thin film (a) to 250 ppm of CO in dry synthetic air at temperatures of 150, 200, 250, 300, 350 and 400 °C and thick film (b) to different concentration pulses of H_2S in dry synthetic air at a film temperature of 300 °C. Go is the initial conductance at zero CO or H_2S concentrations.

The thick-film samples of α-$SnWO_4$ powder did not show the unusual dual conductance behaviour in the exposure to CO and their sensitivity to CO was low. These films were also very insensitive to CH_4 and SO_2 in the temperature range between 150 and 500 °C, but very sensitive to H_2S in the low ppm range between 150 and 300 °C (Fig. 2(b)). In the case of thick films the gas sensitivity originates mainly in the α-$SnWO_4$ phase, since only a small amount of SnO_2 was present in these samples. The thick-film samples from the metastable β phase had very low conductance (below 10^{-10} S up to 400 °C) and, therefore, they are out of practical interest for the gas sensing applications. The β phase also begins to decompose in air at temperatures above 450 °C which is also a limitation, since at low working temperatures the response of semiconductor gas sensors is slow.

4. CONCLUSIONS

The present study demonstrates that Mössbauer spectroscopy is a very valuable tool for studying the semiconductor gas sensors based on Sn oxides and can help to elucidate the gas-surface interactions. The presence of tin at the two valence states Sn^{+2} and Sn^{+4} at the surfaces of these oxides may have very important consequences to their gas-response properties. Large differences were found in the gas-response behaviour between α-$SnWO_4$ thin and thick films with similar XRD patterns but very different Mössbauer spectra (Fig. 1).

It was found that α-$SnWO_4$ thick films at around 300 °C are very applicable for the H_2S sensing in the low ppm range

ACKNOWLEDGEMENTS

J. Solis expresses his gratitude to International Science Programs at the Uppsala University, Sweden for the financial support.

REFERENCES

[1] V. Lantto, in: *Gas Sensor* , de G. Sberveglieri (Kluwer, Dordrecht, 1992) p. 117.
[2] T.S. Rantala, V. Lantto and T T. Rantala, Sensors and Actuators B 18-19 (1994) 716.
[3] E.W. Thornton and P.G. Harrison, J. Chem. Soc. 71 (1975) 1502.
[4] J.L. Solis and V. Lantto, Sensors and Actuators B 24-25 (1995) 591.
[5] J.L. Solis and V. Lantto, Physica Scripta (in press).
[6] J.Geurts, S. Rau, W. Richter and F.J. Schmitte, Thin Solid FIlms 121 (1984) 217.
[7] J.G. Ballard and T. Birchall, Can. J. Chem. 1153 (1975) 3371.

LIST OF PARTICIPANTS

Ana Esther Mijovilovich
Comisión Nacional de Energía Atómica
Av. del Libertador 8250
1429 Buenos Aires.
mijovilo@tandar.cnea.edu.ar
ARGENTINA

Bibiana Arcondo
Depto. de Física. Fac. de Ingeniería.
Univ. de Buenos Aires.
Paseo Colón 850
1063 Buenos Aires.
barcond@aleph.fi.uba.ar
ARGENTINA

Celia Saragovi
Comisión Nacional de Energía Atómica
Av. del Libertador 8250
1429 Buenos Aires.
saragovi@cnea.edu.ar
ARGENTINA

Claudia Elena Rodríguez Torres
Depto. de Física
Universidad Nacional de La Plata.
CC 67, 1900 La Plata.
torres@venus.fisica.unlp.edu.ar
ARGENTINA

Francisco H. Sánchez
Depto. de Física Fac. de Ciencias Exactas
Universidad Nacional de La Plata
CC 67, 1900 La Plata.
sanchez@venus.fisica.unlp.edu.ar
ARGENTINA

Marcela Beatriz Fernández Van Raap
Depto. de Física, Fac. de Ciencias Exactas
Univ. Nacional de La Plata
CC 67, 1900 La Plata.
raap@venus.fisica unlp edu.ar
ARGENTINA

Marcelo Fontan,
Depto. de Física. Fac. de Ingeniería.
Univ. de Buenos Aires.
Paseo Colón 850
1063 Buenos Aires.
mfontan@aleph.fi.uba.ar
ARGENTINA

Maria Andrea Ureña
Depto. de Física. Fac. de Ingeniería.
Univ. de Buenos Aires.
Paseo Colón 850
1063 Buenos Aires.
murena@aleph.fi.uba.ar
ARGENTINA

Roberto C. Mercader
Depto. de Física, Fac. de Ciencias Exactas
Univ. Nacional de La Plata
Calle 115 Y 49, CC 67
1900 La Plata.
rup@venus.fisica.unlp.edu.ar
ARGENTINA

Silvana Stewart
Depto. de Física. Fac. Ciencias Exactas
Univ. Nacional de La Plata.
CC. 67 - 1900. La Plata.
stewart@venus.fisica.unlp.edu.ar
ARGENTINA.

Stella Duhalde
Depto. de Física. Fac. de Ingeniería.
Univ. de Buenos Aires.
Paseo Colón 850.
1063 Buenos Aires.
sduhald@aleph.fi.uba.ar
ARGENTINA

Victoria Juliana Cremashi
Depto. de Física. Fac. de Ingeniería.
Univ. de Buenos Aires.
Paseo Colón 850
1063 Buenos Aires.
vcremas@aleph.fi.uba.ar
ARGENTINA

Anuar Abras
Depto. de Física
Univ. Federal de Minas Gerais.
ICEx/UFMG. C.P. 702
30161-970. Belo Horizonte. MG.
aabras@oraculo.lcc.ufmg.br
BRASIL.

Guido Langouche
Instituut voor Kern-en Stralingsfysika,
Katholieke Universiteit Leuven
Celestijnenlaan 200 D
B-3001 Leuven.
guido.langouche@fys.kuleuven.ac.be
BELGIUM.

Carlos Larica
Dpto. de Física, Centro de Ciencias Exatas
Universidade Federal de Espírito Santo.
Av. Fernando Ferrari s/n
CEP 29060-900. Vitoria. ES.
larica@cce.ufes.br
BRASIL

Paul De Bakker
SCK-CEN Reactor Materials Research.
Boeretang 200 B-2400 Mol.
pdbakker@sckcen.be
BELGIUM.

Carmen S. de Moya Partiti
Inst. de Física. Universidade de São Paulo.
Caixa Postal 66318
CEP 05389-970. São Paulo.
BRASIL.

Ada López Giménez
Centro Brasileiro de Pesquissas Físicas
Rua Dr. Xavier Sigaud 150, Urca.
CEP 22290-180. Rio de Janeiro. RJ.
alopez@novell.cat.cbpf.br
BRASIL

Dalber Sánchez Candela
Centro Brasileiro de Pesquisas Físicas
Rua Dr. Xavier Sigaud 150, Urca.
CEP 22290-180. Rio de Janeiro. RJ.
dalber@novell.cat.cbpf.br
BRASIL

Angel Bustamante Domínguez
Centro Brasileiro de Pesquisas Físicas
Rua Dr. Xavier Sigaud 150, Urca.
CEP 22290-180. Rio de Janeiro. RJ.
angel@novell.cat.cbpf.br
BRASIL

Edson Passamani Caetano
Centro Brasileiro de Pesquisas Físicas
Rua Dr. Xavier Sigaud 150, Urca.
CEP 22290-180. Rio de Janeiro. RJ.
edson@novell.cat.cbpf.br
BRASIL

Antônio Vanderlei Dos Santos
Instituto de Física
Univ. Federal do Rio Grande do Sul
Av. Bento Gonçalves 9500
Campus do Vale, Caixa Postal 15051
CEP 91501-970, Porto Alegre, RS.
BRASIL

Eduardo Ceretta Moreira
Inst. de Física
Universidade Federal de Rio Grande do Sul.
Av. Bento Gonçalves 9500
Campus do Vale, Caixa Postal 15051
CEP 91501-970, Porto Alegre, RS.
emoreira@if.ufrgs.br
BRASIL

Elisa M. Baggio-Saitovitch
Centro Brasileiro de Pesquisas Físicas
Rua Dr. Xavier Sigaud 150, Urca.
CEP 22290-180. Rio de Janeiro. RJ.
elisa@novell.cat.cbpf.br
BRASIL

F.A. Batista
Depto. de Física. ICEx - UFMG,
CP 702, 30161-970 Belo Horizonte. MG.
BRASIL

Genilson Pereira Santana
Univ. Federal de Minas Gerãis
R. Carlos Alves 105 /103. Bairro São José
Belo Horizonte MG CEP 31275-120
genilson @oraculo.lcc.ufmg.br
BRASIL

Geraldo Magela Da Costa,
Depto de Química
Univ. Federal de Ouro Preto. ICEB
Campus Universitario. Morro do Cruzeiro.
35400. Ouro Preto - MG.
magela@dedalus.lcc.ufmg.br
BRASIL.

Hans-Dieter Pfannes
Depto. de Física, ICEX
Univ. Federal de Minas Gerais, CP 702
30161-970 Belo Horizionte, MG.
hansdiet@bach.fisica.ufmg.br
BRASIL

Hercilio Rechenberg
Inst. de Física, Univ. de São Paulo
Caixa Postal 66318
05315-970. São Paulo
hercilio@if.usp.br
BRASIL

Izabel Azevedo de Souza
Centro Brasileiro de Pesquisas Fisicas
Rua Dr. Xavier Sigaud 150, Urca.
CEP 22290-180. Rio de Janeiro. RJ
izabel@novell.cat.cbpf.br
BRASIL

Javier A. Gómez
Centro Brasileiro de Pesquisas Físicas
Rua Dr. Xavier Sigaud 150, Urca.
CEP 22290-180. Rio de Janeiro. RJ.
gomez@novell.cat.cbpf.br
BRASIL

Jorge López Aguilar
Universidade Federal de Minas Gerais.
Belo Horizonte. BRASIL
Alameda dos Jacarandas Nro. 23,
Apto. 301. Bairro São Luiz
CEP 31275060. MG .
jorge@bach.fisica.ufmg.br
BRASIL

José Domingos Fabris
Emp. Brasileira de Pesquisa Agropecuária
- EMBRAPA
Centro Nacional de Pesquisa de Milho e
Sorgo - CNPMS
C. *Postal 151 - 35.701 970 -*
D. *Sete Lagoas - Minas Gerais. MG*
fabris@oraculo.lcc.ufmg.br
BRASIL

José Flavio M Borges,
Inst. de Física
Univ. Federal de Rio Grande do Sul.
Av. Bento Gonçalves, 9500
Campus do Vale. Caixa Postal 15051
CEP 91501-970. Porto Alegre. RS.
fborges@if.ufrgs.br
BRASIL

José Higino Dias Filho
Univ. Federal de Minas Gerais
Av. Antonio Carlos 6627.
Caixa Postal 702
Campus da Pampulha. Belo Horizonte.
MG.
higino@bach.fisica.ufmg.br
BRASIL

Klinger Marcos Barbosa Alves
Dpto. de Física, Centro de Ciencias Exatas
Universidade Federal de Espírito Santo
Campus Universitário
Av. Fernando Ferrari s/n
CEP 29060-900. Vitoria. ES.
klinger@cce.ufes.br
BRASIL

Magale Brückmann
Inst. de Física,
Univ. Federal de Rio Grande do Sul.
Av. Bento Gonçalves, 9500
Caixa Postal 15051 - Ag. Campus
CEP 91501.970 - Porto Alegre. RS.
mabruck@if.ufrgs.br
BRASIL

Magda Cristina Ferreira Pinto
Depto. de Química - ICEx
Univ. Federal de Minas Geráis
Av. Antonio Carlos 6627. Pampulha.
31270-90.1Belo Horizonte. MG.
magda@oraculo.lcc.ufmg.br
BRASIL

Marcos A. Morales
Centro Brasileiro de Pesquisas Físicas
Rua Dr. Xavier Sigaud 150, Urca.
CEP 22290-180. Rio de Janeiro. RJ.
morales@cbpfsu1.cat.cbpf.br
BRASIL

Miguel A. B. de Araujo
Dpto. de Física.
Universidade Federal de Ceará
Caixa Postal 6030.
60450-970 Fortaleza-Ceará.
araujo@fisica.ufc.br
BRASIL

Paulo Domingues
Inst. de Física. Univ. Federal do Rio de Janeiro. Centro de Tecnologia-Bloco A
Cidade Universitária, Ilha do Fundão.
21945-970 Rio de Janeiro, RJ
paulodom@if.ufrj.br
BRASIL

Rosa Scorzelli Bernstein
Centro Brasileiro de Pesquisas Físicas
Rua Dr. Xavier Sigaud 150, Urca.
CEP 22290-180. Rio de Janeiro. RJ.
scorza@novell.cat.cbpf.br
BRASIL

Sebastião Luiz Pereira,
Universidade Federal de Minas Gerais
Av. Antonio Carlos 6627. Caixa Postal 702
Campus da Pampulha. Belo Horizonte. MG.
spereira@bach.fisica.ufmg.br
BRASIL

Teresa Verônica Costa
Depto. de Física
Univ. Federal de Ceará. - UFCE
Caixa Postal 6030
60450-970. Fortaleza-Ceará.
veve@fisica.ufc.br
BRASIL

Walf Henning
Centro Brasileiro de Pesquisas Físicas.
Rua Dr. Xavier Sigaud 150, Urca.
CEP 22290-180. Rio de Janeiro. RJ.
BRASIL

Wagner da Nova Mussel
Depto de Química -Inst. de Ciências Exatas- ICEX
Universidade Federal de Minas Geráis
Av. Antonio Carlos 6627
Campus Pampulia - Belo Horizonte
CEP 31270-901 MG.
wdmussel@oraculo.lcc.ufmg.br
BRASIL

Alberto Bohorquez
Depto. de Física. Univ. Del Valle.
Sede Melendes, A.A. 25360. Cali .
bohorque@calima.univalle.edu.co
COLOMBIA

Alvaro Luis Morales
Depto de Física.
Univ. de Antioquia.
Apartado Aéreo 1226, Medellin.
amoral@fisica.udea.edu.co
COLOMBIA

Willian Alfonso Pacheco Serrano
Univ. Pedagógica y Tecnológica de Colombia
Avenida Central del Norte.
COLOMBIA

Germán Pérez Alcazar
Univ. del Valle. Ciudad Universitaria - Meléndez
A.A. 25360. Cali.
geperez@quantum.univalle.edu.co
COLOMBIA.

Alberto José Sánchez Asseff
Depto. de Física. Univ. del Valle. Sede Melendes
A.A. 25360. Cali.
ajsanche@quantum.univalle.edu.co
COLOMBIA

Jesús Anzelmo Tabares Giraldo,
Depto. de Física. Univ. del Valle. Sede Melendes
A.A. 25360. Cali.
jatabare@calima.univalle.edu.co
COLOMBIA.

Steen Morup
Deparment of Physics,
University of Denmark.
Building 307. DK-2800 Lyngby.
morup@fysik.dtu.dk
DENMARK

J.L. Solis
University of Oulu. Microelectronics and Material Physics Laboratories
P.O.Box. 400. FIN-90571 Oulu.
jose@ee.oulu.fi
FINLAND.

Agnes Traverse,
LURE / CNRS
Bâtiment 209ª - Centre Universitaire Paris-Sud, 91405 Orsay.
traverse@lure.u-psud.fr
FRANCE.

Jean-Marc Greneche
Laboratoire de Physique de lÉtat Condensé.
URA CNRS N° 807
Faculté des Sciences. Université du Maine
F72017 LE MANS cedex.
greneche@aviion.uni-lemans.fr
FRANCE

Hans Micklitz
II Phys. Inst. Universitaet du Koeln..
Zuelpicicher str. 77
D.50937 Koeln.
micklitz@colorix.ph2.uni-koeln.de
GERMANY.

Mohammad Ghafari
Materials Sciences Department. Thin Films Division
Technical University Darmstadt
Petersenstr. 23 a
D64287 Darmstadt.
dh5i@hrzpub.th-darmstadt.de
GERMANY.

Rafael Gómez Escoto
Materials Sciences Department. Thin Films Division
Technical University Darmstadt
Petersenstr. 23 a
D64287 Darmstadt.
dh5i@hrzpub.th darmstadt.de
GERMANY.

Ursel Wagner
Physik-Department E 15.
Technische Universität München.
D-85747 Garching,
uwagner@physik.tu-muenchen.de
GERMANY

Giuseppe Pedrazzi
Istituto di Scienze Fisiche, Univ. Parma.
Via M. Dázeglio, 85 - 43100 Parma.
mosslab@ipruniv.cce.unipr.it
ITALY

Juan Antonio Jaén,
Depto. de Química
Centro Inv. con Técnicas Nucleares (CITEN)
Fac. de Ciencias Naturales y Exactas.
Univ. de Panamá.
jjaen@ancon.up.ac.pa
PANAMÁ.

Alberto Bueno
Museo de Arqueología.
Univ. Nacional Mayor de San Marcos.
Casona del Parque Universitario. Lima 1.
PERÚ

Carlos Quiñonez Monteverde
Fac. de Ciencias Físicas y Matemáticas.
Univ. Nacional del Callao.
Ciudad Universitaria. Av. Venezuela s/n. Callao.
PERÚ

Cesar Chung Chung
Fac. de Ciencias Físicas
Univ. Nacional Mayor de San Marcos.
Apartado Postal 14 0149. Lima 14.
PERÚ

Eladio Casapía Almonte
Fac. de Ciencias Físicas y Matemáticas.
Univ. Nacional del Callao.
Ciudad Universitaria. Av. Venezuela s/n. Callao.
PERÚ

Emilio Medrano Atencio
Fac. de Ciencias Físicas
Univ. Nacional Mayor de San Marcos
Apartado Postal 14 0149. Lima 14
PERÚ

Francisco Vittor García
Fac. de Ciencias Físicas
Univ. Nacional Mayor de San Marcos.
Apartado Postal 14 0149. Lima 14.
PERÚ

Honorio Barrientos
Fac. de Ciencias.
Univ. San Antonio de Abad.
Ciudad Universidad. Cusco.
PERÚ

Ismael Saavedra Valdivieso
Lab. de Espectroscopía Mössbauer
Fac. de Ciencias Físicas
Univ. Nacional Mayor de San Marcos.
Apartado Postal 14 0149. Lima 14.
PERÚ

Jesús Flores Santibañez
Lab. de Espectroscopía Mössbauer
Fac. de Ciencias Físicas
Univ. Nacional Mayor de San Marcos.
Apartado Postal 14 0149. Lima 14.
PERÚ

Jorge Bravo Cabrejos
Fac. de Ciencias Físicas
Univ. Nacional Mayor de San Marcos.
Apartado Postal 14 0149. Lima 14.
decanofis@unmsm.edu.pe
PERÚ

Julio Larrea Jiménez
Fac. de Ciencias Físicas
Univ. Nacional Mayor de San Marcos.
Apartado Postal 14 0149. Lima 14.
PERÚ

Leovigildo Lastra
Fac. de Ciencias Físicas
Univ. Nacional Mayor de San Marcos.
Apartado Postal 14 0149. Lima 14.
PERÚ

Lorenzo Malpartida Contreras
Fac. de Ciencias Físicas
Univ. Nacional Mayor de San Marcos.
Apartado Postal 14 0149. Lima 14.
PERÚ

Malco Reyes Sifuentes
Lab. de Espectroscopía Mössbauer.
Fac. de Ciencias Físicas
Univ. Nacional Mayor de San Marcos.
Apartado Postal 14 0149. Lima 14.
PERÚ

Máximo Poma Torres
Fac. de Ciencias Físicas
Univ. Nacional Mayor de San Marcos.
Apartado Postal 14 0149. Lima 14.
PERÚ

Quirico Sibille San Miguel
Lab. de Espectroscopía Mössbauer.
Fac. de Ciencias Físicas.
Univ. Nacional Mayor de San Marcos.
Apartado Postal 14 0149. Lima 14.
PERÚ

Rubén Ponce Herrera,
Fac. de Ciencias Físicas
Univ. Nacional Mayor de San Marcos.
Apartado Postal 14 0149. Lima 14.
a3_user2@unmsm.edu.pe
PERÚ

Víctor Peña Rodríguez
Lab. de Espectroscopía Mössbauer.
Fac. de Ciencias Físicas.
Univ. Nacional Mayor de San Marcos.
Apartado Postal 14 0149. Lima 14.
a3_user1@unmsm.edu.pe
PERÚ

Yuri Zenitagoya
Fac. de Ciencias.
Universidad Nacional de Ingeniería
Casilla 31-139. Lima
PERÚ

Marek Przybylski
Faculty of Physics and Nuclear Technics.
University of Mining and Metallurgy.
al.Mickiewicza 30. 30-059, Krakow,
marprzyb@uci.agh.edu.pl
POLAND

Fernando González Jiménez
Depto. de Física, Fac. de Ciencias.
Univ. Central de Venezuela
Apartado Postal 47586. Los Guaramos.
Caracas 1041 A,
fgonzal@dino.conicit.ve
VENEZUELA

Josefa M. Borrego Moro
Depto. de Física de la Materia Condensada
Inst. de Ciencia de Materiales, C.S.I.C.
Univ. de Sevilla
Apartado 1065, 41080-Sevilla,
borrego@cica.es
ESPAÑA

José Ramón Gancedo
Inst. de Química Física "Rocasolano"
Consejo Superior de Investigaciones Científicas.
Calle Serrano 119. 28006 Madrid.
iqrjr12@cc.csic.es
ESPAÑA

Ercan Alp
Advanced Photon Source.
Argone National Laboratory
Argone IL 60439.
eea@aps.anl.gov
USA

James G. Mullen,
Department of Physics, Purdue University.,
Bldg1396, West Lafayette, IN 47907-1396.
jgm@physics.purdue.edu
USA

Author index

9 781402 008818

GPSR Compliance

The European Union's (EU) General Product Safety Regulation (GPSR) is a set of rules that requires consumer products to be safe and our obligations to ensure this.

If you have any concerns about our products, you can contact us on ProductSafety@springernature.com

In case Publisher is established outside the EU, the EU authorized representative is:

Springer Nature Customer Service Center GmbH
Europaplatz 3
69115 Heidelberg, Germany

Batch number: 10615117

Printed by Printforce, the Netherlands